GENETICS, DEVELOPMENT, AND EVOLUTION

STADLER GENETICS SYMPOSIA SERIES

GENE MANIPULATION IN PLANT IMPROVEMENT
Edited by J. Perry Gustafson

GENETICS, DEVELOPMENT, AND EVOLUTION
Edited by J. Perry Gustafson, G. Ledyard Stebbins, and Francisco J. Ayala

GENETICS, DEVELOPMENT, AND EVOLUTION

17th Stadler Genetics Symposium

Edited by

J. Perry Gustafson

University of Missouri
Columbia, Missouri

G. Ledyard Stebbins
and

Francisco J. Ayala

University of California, Davis
Davis, California

PLENUM PRESS • NEW YORK AND LONDON

ISBN 0-306-42268-9
Library of Congress Catalog Card Number 79-635774

A Division of Plenum Publishing Corporation
233 Spring Street, New York, N.Y. 10013

Printed in the United States of America

TO PROFESSOR G. LEDYARD STEBBINS
ON THE OCCASION OF HIS 80TH BIRTHDAY

G. Ledyard Stebbins

Professor G. Leydard Stebbins celebrated his 80th anniversary on January 6, 1986. More than 50 years of his long life have been dedicated to the study of evolution. It is difficult to think of plant evolution in particular, without evoking Stebbins' name and works. We collaborated with him on the conception and organization of the 1985 Stadler Symposium, but want to dedicate this volume to him in recognition of his unfailing commitment to science, of his outstanding scientific contributions, and of his energy and cheerfulness. May he for years to come continue to be a source of inspiration and a role model for other scientists. He will always be our cherished friend.

J. Perry Gustafson
Francisco J. Ayala

Columbia and Davis

FOREWORD

One outstanding question in biology is the problem of development: how the genetic instructions encoded in the DNA become expressed in the morphological, physiological, and behavioral features of multicellular organisms, through an ordered sequence of events that extend from the first cell division of the zygote to the adult stage and eventual death. The problem is how a one-dimensional array of instructions is transformed into a four-dimensional entity, the organism that exists in space and time. Understanding this transformation is, nevertheless, necessary for mastering the process of evolution.

One hundred and twenty-five years after The Origin of Species, we have gained some understanding of evolution at the genetic level. Genetic information is stored in the linear sequence of nucleotides in the DNA. Gene mutations, chromosomal reorganizations, and a host of related processes introduce variation in the sequence and the amount of DNA. The fate of these variations is determined by interactions within the genome and with the outside environment that are largely understood. We have recently gained a glimpse of how the genome of eukaryotes is organized and will learn much more about it in the future, now that we have the research tools for it.

The processes that modulate the evolution of whole organisms are also understood, in principle if not in their particular operation. Natural selection is the resulting vector of multiple component interactions that yield the net differential rate of reproduction of organisms. And we know about the structure of biological communities, about prediction and competition for resources, about mating and kin interactions, and about the multifarious relationships of organisms with their environments.

But we know little about the connection between the evolution of the DNA and the evolution of whole organisms. This ontogeny gap looms large in evolutionary theory, because the process of biological evolution cannot be satisfactorily comprehended until we know how form and function come about. The theory of evolution will take major forward strides, and will doubtless undergo major modifications, when ontogeny and its evolution are elucidated.

It seems likely that the field of biology is now poised to start filling the gap of ontogeny. The techniques of gene cloning and DNA sequencing, the restless advances of molecular biology, new advance in cell biology and in biochemistry, have all contributed to open up the problem of the development of organisms. It is this anticipation that moved the organizers of the 1985 Stadler Symposium to choose the topic "Genetics, Development, and Evolution". The impressive array of contributors approach the problem of ontogeny from diverse points of view and working with different organisms--animals, plants, and fungi. We can only hope that this gathering may become a significant landmark toward resolving the problem of ontogeny, the puzzle of how ontogenetic development came about in evolution and how, in turn, it modulates the evolutionary process.

Francisco J. Ayala

Davis, July 1985

ACKNOWLEDGEMENT

The editor gratefully acknowledges the generous support of the following contributors: College of Agriculture, University of Missouri; Division of Biological Sciences, University of Missouri; E. I. DuPont De Nemours & Company; Graduate School, University of Missouri, School of Medicine, University of Missouri; Monsanto Company; Northrup King Company; Pioneer Hi-Bred International, Inc. and Trans World Airlines, Inc. who made the 17th Stadler Genetics Symposium a success.

The speakers, who spent a tremendous amount of time preparing their manuscripts and lectures are gratefully acknowledged. Without their expertise and dedication the Symposium could not have taken place.

I wish to thank the local chairpersons for their effort to see that everyone in the respective sessions were well taken care of during the Symposium.

The behind-the-scene and on-site preparation was excellently handled by Joanne Fredmeyer and Joy Gasparovic from Conferences and Specialized Courses, University of Missouri, who tirelessly handled all of my peculiar requirements and made sure everything was extremely well organized.

Many thanks are due to Joyce Reinbott, University of Missouri, for her excellent secretarial help in handling all the correspondence and typing. A special thanks goes to A. J. Lukaszewski and K. Robertson.

J. P. Gustafson

January 15, 1986
Columbia, Missouri

CONTENTS

EVOLUTION AND MORPHOGENESIS: THE REGULATOR HYPOTHESIS

Gerald M. Edelman

The Rockefeller University
1230 York Avenue
New York, New York 10021

> "The nature of epigenetic sequences [in development] is one of the most important unsolved problems of biology and is of vital importance to students of evolution."
>
> Ledyard Stebbins

The quotation from a lucid and prescient paper by Ledyard Stebbins (1968) sets my theme: I am concerned here with confronting the problem of relating developmental genetics to evolution at the level of what he calls epigenetic sequences during development. These are programmed sequences occuring during ontogeny in which the action of a gene follows precisely upon that of an otherwise unrelated gene in temporal sequences that are prolonged. These sequences are precisely those confronted by the developmental biologist when he attempts to understand morphology--pattern formation in histogenesis and form. The fact that these sequences concern gene expression entails a molecular approach to this most fundamental problem of developmental biology. At the same time, interpretation of this approach must be consistent with the evolution of both animal macroscopic form and of microscopic complexity. The issue may be focused succinctly in two complementary questions:

(1) How can a one dimensional genetic code specify a three dimensional organism? (I shall call this the developmental genetic question)

(2) How can the mechanism proposed to answer this question be reconciled with the large changes in form that can occur in relatively short evolutionary times? (The evolutionary question)

Clearly it is the answer to the evolutionary question that would shed the most light on issues related to the evolution of complexity and on the part played by the embryonic milieu in providing strong selection pressure during evolution. But, as you will notice, it is a contingent question--the developmental genetic question must be answered first. Before considering the possible nature of the answer, it may be useful to summarize briefly some of the previous ideas and approaches that have linked developmental biology and evolutionary thinking.

Developmental Biology and Evolution

Looking back at the rather meagre role (Hamburger, 1980) that developmental biology has played in evolutionary thinking from the present vantage point, and in possession of cogent theories of evolution and genetics, it is tempting to surmise that one reason for this slight role is the lack of an equally cogent theory of development. The explanations for the absence of such a theory are various but certainly a major one is the molecular complexity of developmental processes, most of which remains to be explored; another is that such a developmental theory would itself have to depend upon the modern synthesis. Inasmuch as that synthesis is only four decades old and the tools for attacking morphogenesis at the molecular level are only recently available, the inadequacy of developmental theory is perhaps not surprising. Until we have a cogent theory at the molecular level, the relationship of development to evolution will remain incompletely described.

Nonetheless, Darwin himself recognized that -- "community in embryonic structure reveals community in descent." He was aware of von Baer's (1828) laws of development in establishing this criterion. On the developmental side, however, two important lines of interest nonetheless obscured the issue: Haeckel's now discredited theory of recapitulation (1879) and Roux's (1905) "entwicklungsmechanik" which, while it correctly emphasized the need for detailed mechanisms of development, avoided genetics and comparative issues by focusing on only a few animal examples. More recently, Waddington's (1966) interest in developmental genetics has been rekindled (Bonner, 1982; Raff and Kaufman, 1983), and we face the key questions in this field once again.

The current view is perhaps best represented by the idea that differences in morphology in related taxa can be explained by heterochrony (Haeckel's term): changes in the relative rate of appearance and development of characters already present in ancestors. Such changes are now interpreted as either local morphogenetic accelerations or retardations that lead to progenesis, neoteny or paedomorphosis (Gould, 1977). The putative explanation for these effects is that they result from mutations in regulatory genes that change the timing of biochemical events, the fates of embryonic cells, or the integration of structural gene expression.

This picture leaves open the molecular mechanisms by which such effects are established in morphogenetic terms and we are thus brought back to the developmental genetic question. To address that question effectively, let us first consider some fundamental aspects of embryonic development.

Primary Processes in Regulative Development

Regulative development is development in which cells of different history are brought together by morphogenetic movements resulting in the milieu-dependent (or trajectory-dependent) cell differentiation known as embryonic induction. The other main mode of altering differentiation according to position -- mosaic development -- achieves this by unequal partitions of cytoplasm prior to egg cleavage in successive cell divisions. The contribution of each differs in different species but in higher animals (e.g. the vertebrates) the main mode is regulative (Weiss, 1969).

One may distinguish a number of processes at the cellular level which combine in varying degrees as a function of time directly to influence pattern formation and histogenesis. Besides the two just mentioned (the movement of cells and tissue sheets, and cellular differentiation), there are cell division, cell adhesion, and cell death. Each of these primary processes is complex, consisting of many linked biochemical interactions.

What regulates the sequence and relative contribution of these processes? Some aspects of these primary processes are clearly under genetic control. Others, however, are epigenetic: certain sequences of events must occur in order for others to occur in certain regions of embryonic space at particular instants of time. But, inasmuch as the genes which _ultimately_ regulate the primary processes cannot directly specify space-time coordinates, means other than genes and additional mechanisms must also regulate these epigenetic events. In searching for regulatory principles, it is well to keep in mind that, to a certain extent, the primary processes are dissociable, as Needham (1933) first pointed out. This implies that to a certain extent they may be isolated experimentally from each other, facilitating tests of their relative contributions to the morphogenetic scheme.

Form arises as a result of the differential effects of the driving forces provided by the processes of cell division, cell movement, and cell death. The key issue is concerned with how these forces are controlled. Given the basic features of regulative development, it is clear that a central role is played by sequences of morphogenetic movements and by the places at which cells or collectives of cells interact as a result of these movements. The differentiation that occurs as a consequence of interaction can change the shape of a cell, the products it synthesizes, and the movements themselves. In some sense, such inductive sequences create the environment necessary for triggering ensuing sequences, all of which occur in a historical fashion over time periods that are long as compared to those of intracellular processes.

Inasmuch as genes cannot store the spatial or temporal coordinates of a local frame, and because cell motion cannot itself be genetically programmed to reach a specific target, the existence of a system of individual cellular addresses expressed at the molecular level (Sperry, 1963) in order to ensure appropriate cellular assembly seems highly unlikely (Edelman, 1984c). Moreover, even if information for such an addressing system could be stored in the genes, another problem would obtrude: the need to induce the addressing molecule precisely on a given cell at a given place and time would be almost impossible to meet, given observations on the plasticity of development (Weiss, 1969). For example, before certain differentiation events have occurred, embryonic cells can be mixed and exchanged with equivalent morphogenetic results. It is the local environment, not the particular cell that is critical. How could specific cellular addresses at the molecular level adapt to such maneuvers?

All of these considerations, which center upon the critical issue of positional specification of cells, make it highly unlikely that there is a prespecified, instructive, or precise informational mode by which cells know their place at the molecular level. The question of position and regulation must be answered in another way.

Position and Regulation

The view taken here is embodied in the so-called regulator hypothesis (Edelman, 1984b) which asserts that the regulation of movement and position is carried out by particular mechanisms of cellular adhesion. This hypothesis shall be considered in detail in a later section, but at this juncture some of the assumptions on which it is based are worth describing. The fundamental assumptions are:

1) The smallest unit by means of which place is recognized is the cell; this recognition occurs by interactions with other cells in collectives or additionally with various substrates such as basement membranes.

2) Recognition involves a mechanochemical event which occurs via special molecules and modulatory mechanisms at the cell surface.

3) The special molecules are cell adhesion molecules (CAMs) and the main primary process to be controlled is cell adhesion. Variations in such control alter the expression of CAMs at the cell surface which, in turn, alters the regulation of morphogenetic movements. CAMs exist in a relatively small number of specificities but are capable of a large number of alterations in their binding properties which are graded and non-linear.

4) Altered pressure, tension, or flow in the vicinity of a collective of cells held together by one kind of CAM in the neighborhood of another collective held together by a CAM of different specificity alters the expression or modulation of the CAMs at cell surfaces. A sequence of such expressions can alter morphogenetic movements and lead to specific form and pattern.

According to these assumptions, cell adhesion is the main primary process concerned with regulating morphogenesis. Inasmuch as cell adhesion molecules are likely to be proteins specified by particular genes, we have a candidate mechanism for how the one-dimensional genetic code might regulate three-dimensional form. In order to test such an idea experimentally, one must develop assays for CAMs, isolate them, demonstrate their function, identify their structural genes, and determine the sequence and mechanisms of expression of these genes. Of course, such a demonstration must relate this gene expression to time and place in the embryonic tissue or organ of interest. Some of this program has been carried out and the remainder is under way. Here, instead of describing the chemistry, molecular genetics, and structure of CAMs in detail, I shall focus on major properties salient to the developmental genetic and evolutionary questions. Reviews may be consulted for further information (Edelman, 1983a; 1984a; 1985).

CAMs and Cell Surface Modulation

Cell adhesion molecules were first isolated from chick embryonic tissues (Brackenbury et al., 1977; Thiery et al., 1977) by developing short-term adhesion assays for dissociated cells and then identifying their molecular binding function by blockade

Table 1. Some Characteristics of CAMs

CAM	Type	Molecular Weight	Carbohydrate Characteristics	Binding Mechanism	Ion Dependence
N-CAM (neural CAM)	**1°**	**180-250 Kd (E form) 180,140,120 Kd (A forms)**	**glycoprotein with unusual polysialic acid**	**Homophilic**	**none**
L-CAM (liver CAM)	**1°**	**124 Kd**	**glycoprotein**	**Homophilic**	**Ca^{++}**
Ng-CAM (neuron-glia CAM)	**2°**	**200 Kd 135 Kd 80 Kd**	**glycoprotein**	**Hetero-philic**	**none**

with antibodies to the cell surface that contained specificities against a given CAM. The specific inhibiting antibody was then used to immunoprecipitate the particular CAM. Since that time, CAMs have been identified in a large variety of vertebrate species. As shown in Table 1, three different CAMs of different specificities, molecular weights, binding dependence on ions, and time and place of first appearance have been chemically characterized. There is suggestive evidence that a fourth CAM exists on glial cells. CAMs are named according to the initials of the tissues from which they were first isolated (e.g. N-CAM, neural CAM) or additionally in terms of their heterotypic binding (e.g. Ng-CAM, neuron-glia CAM). This convention should not lead to the misconception that a named CAM is necessarily seen only in that tissue; nevertheless, the tissue is obviously representative of one of the germ layers upon which the CAM first appeared.

Prior to describing the orderly distribution of CAMs in embryonic and adult tissues, a brief recital of some of their biochemical properties may be useful. CAMs are large cell surface glycoproteins synthesized by the cells on which they appear; *in vitro* studies suggest that they turn over at the cell surface. Primary CAMs first appear in early embryogenesis on tissue derivatives of more than one germ layer. They persist concordantly on the descendents of such derivatives into adult life. Secondary CAMs appear in later histogenetic sequences on derivatives of a single germ layer.

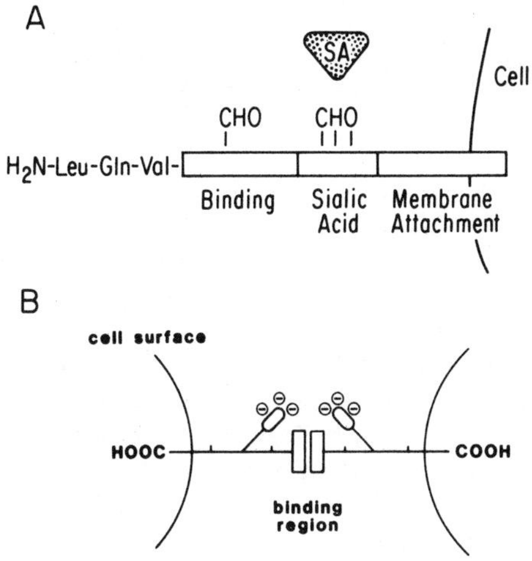

Fig. 1. N-CAM structure and binding. A. Linear structure of N-CAM. Three structural and functional regions of N-CAM (see Edelman, 1983a; 1984a) deduced from studies of the intact molecule and a series of fragments: the NH_2-terminal region includes a specific binding domain and carbohydrate (CHO) but little, if any, sialic acid; the neighboring region is very rich in sialic acid (SA), present mainly as polysialic acid; and the COOH-terminal region is associated with the plasma membrane. B. Binding is homophilic i.e. to the binding domain of N-CAM present on an apposing cell.

As illustrated by N-CAM (shown diagramatically in Figure 1) they bind in second order mode to link two cells (cell 1 and cell 2) in either a homophilic (A_1-A_2) or a heterophilic mode (A_1-B_2). So far, 3rd order interactions (A_1-B-A_2), in which a free molecule links molecules on two cells, have not been seen. CAMs undergo local cell surface modulation (Edelman, 1976; Edelman, 1983a), changing in amount, distribution or polarity at the cell surface or in their binding chemistry as a function of time and place in development (Fig. 2). The most striking example of the latter mechanism for N-CAM occurs in the nervous system. This molecule displays a large decrease in the charged polysialic acid attached at three locations in its middle domain (see Fig. 1), converting it from a polydisperse E form with lesser binding rates to three A forms with greater binding rates.

So far, only a small number of CAM's of different specificity have been discovered. However, very large changes in binding are seen with small changes in CAM surface density or chemical alteration such as E to A conversion and these changes occur in a graded manner. In other words, the regulation of CAM binding as a function of modulation appears to be continuous and highly nonlinear (Hoffman and Edelman, 1983).

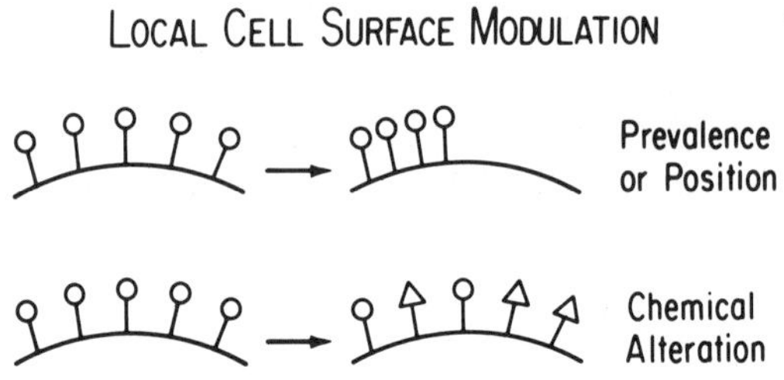

Fig. 2. Schematic representation of local cell-surface modulation. The various elements represent a specific glycoprotein (for example N-CAM) on the cell surface. The upper sequence shows modulation by alteration of both the prevalence of a particular molecule and its distribution on the cell surface. The lower sequence shows modulation by chemical modification resulting in the appearance of new or related forms (triangles) of the molecule with altered activities. Local modulation is distinct from global modulation, which refers to alterations in the whole membrane that affect a variety of different receptors independent of their specificity (see Edelman, 1976).

Table 2. Distribution of L-CAM and N-CAM in Three Epochs

0-3 day Embryo	5-13 day Embryo	Adult
L-CAM		
Ectoderm		
Upper layer	Epidermis	Skin: Stratum
Epiblast	Extraembryonic	germinativum
Presumptive	ectoderm	
epidermis		
Placodes		
Mesoderm		
Wolffian duct	Wolffian duct	Epithelium of:
	Ureter	Kidney
	Most meso- and	Oviduct
	metanephric	
	epithelium	
Endoderm		
Endophyll	Epithelium of:	Epithelium of:
Hypoblast	Oesophagus	Tongue
Gut primordium	Proventriculus	Oesophagus
and buddings	Gizzard	Proventriculus
	Intestine	Gizzard
	Liver	Intestine
	Pancreas	Liver
	Lung	Pancreas
	Thymus	Lung
	Bursa	Thymus
	Thyroid	Thyroid
	Parathyroid	Parathyroid
	Extraembryonic	Bursa
	endoderm	
N-CAM		
Ectoderm		
Upper layer	Nervous system	Nervous system
Epiblast		
Neural plate		
Placodes		
Mesoderm		
Notochord	Striated muscle	Striated muscle
Somites	Adrenal cortex	(end plate)
Dermomyotome	Gonad cortex	Testis
Somato- and	Some mesonephric and	
splanchno-	metanephric epithelia	
pleural	Somato- and splanchno-	
mesoderm	pleural elements	
Heart	Heart	
Mesonephric		
primordium		

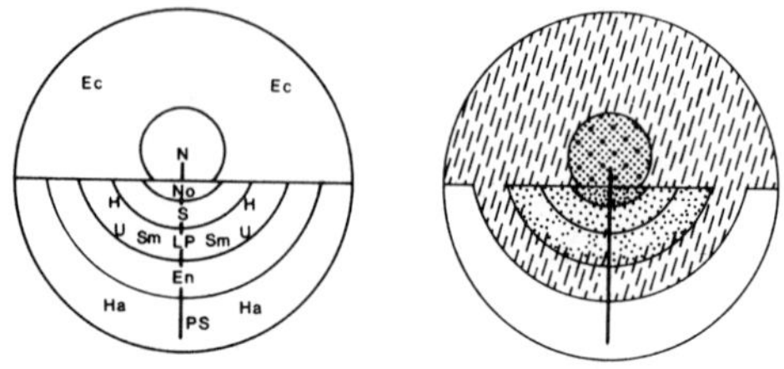

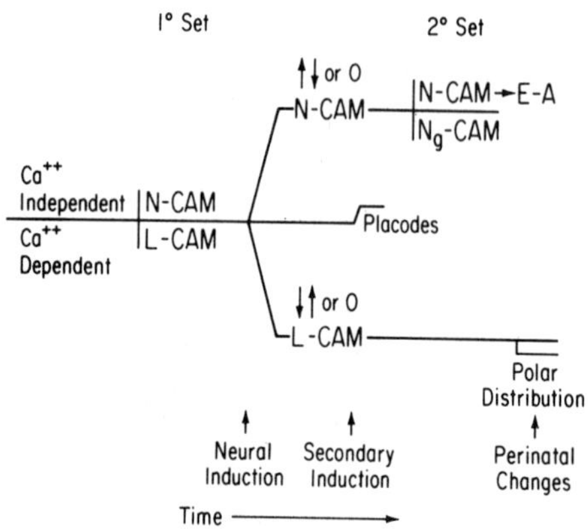

Both primary and secondary CAMs appear to have major roles in the regulation of movement (Edelman, 1984b; Grumet et al., 1985). These roles are various: i) permissive (down-regulation of cell surface CAM to allow cells linked in an epithelium to convert to a mobile, loosely linked mesenchyme), ii) formative (mediating interaction of two epithelia under tension and plastic flow), and iii) initially requisite (needed for movement of cell bodies on cell processes already laid down).

Fig. 3. Spatial and temporal distribution of CAMs. A and B. ← Composite CAM fate map. (A) Fate map of the blastodisc (Vakaet, 1984) showing areas of cells that will give rise to differentiated tissues (indicated by letters; see below for designations). (B) Map of cells that will express CAMs. The distribution of N-CAM (stippled) or L-CAM (slashed) in tissues at 5-14 days (stages 26-40) as determined by immunofluorescence staining is mapped back onto the blastodisc fate map. Cells that will give rise to the urinary tract (U) express both L-CAM and N-CAM. Smooth muscle (Sm) and hemangioblastic (Ha) tissues express neither N-CAM nor L-CAM; areas giving rise to these tissues are blank on this map. Ng-CAM is represented on neuroectoderm by (●) symbols; it is seen only after three days. The vertical bar represents the primitive streak (PS). Other abbreviations: Ec = intraembryonic and extraembryonic ectoderm; En = endoderm; H = heart; LP = lacteral plate (splanchnosomatopleural mesoderm); N = nervous system; No = prechordal and chordamesoderm; S = somite. C. Schematic diagram showing the temporal sequence of expression of CAM during development. After an initial differentiation event, N-CAM and L-CAM diverge in cellular distribution and are then modulated in prevalence (↑↓) within various regions of inductions or actually disappear (○) when mesenchyme appears or cell migration occurs. Note that placodes which have both CAMs, echo the events seen for neural induction. Just before appearance of glia, a secondary set CAM (NgCAM) emerges; unlike the other two CAM, this CAM would not be found in the map shown in Fig. 4 before 3.5 days. In the perinatal period, a series of epigenetic modulations occurs: E to A conversion for N-CAM and polar redistribution for L-CAM. The diagrammed events are based mainly on work on the chick.

Perhaps the most striking overall property of these molecules during development is that they are expressed and modulated in well-defined sequences in particular places in the embryo (Edelman, 1984b; Edelman et al., 1983b). The critical modulatory

Table 3. Sequences of Expression of N-Cam and Ng-CAM

N-CAM (1°)		AGE		NG-CAM (2°)
		DAYS	STAGE	
BLASTODERM		0	1	
NON-NEURAL DERIVATIVES	NEURAL DERIVATIVES			
NOTOCHORD, SOMITE; LATERAL PLATE MESODERM MYOCARDIUM; MYOBLAST; TRANSIENT APPEARANCE ON PLACODES AND MESONEPHRIC TUBULES; MYOTUBES	NEURAL PLATE; NEURAL TUBE; NEURAL CREST; DISAPPEARS FROM MIGRATING NEURAL CREST CELLS; REAPPEARS IN AGGREGATING NEURAL CREST CELLS;	2	13	
		4	23	VENTRAL NEURAL TUBE; CILIARY GANGLIA; DORSAL ROOT GANGLIA; SPINAL CORD: WHITE COLUMN; RETINA: GANGLION CELL FIBERS; TECTUM: FIBER LAYER UNDER PIA; SYMPATHETIC GANGLIA;
STRIATED MUSCLE				
		6	29	SPINAL CORD: GRAY MATTER; TECTUM: STRATUM OPTICUM AND ALBUM CENTRALE;
↓	↓	8	34	
		10	36	
				CEREBELLUM: FIBER TRACTS;
		12	38	CEREBELLUM: MOLECULAR LAYER;
		14	40	
		16	42	
	E ↓ A			CEREBELLUM: MIGRATING GRANULE CELLS;
		18	44	
		BIRTH	46	
N-CAM ON STRIATED MUSCLE CONCENTRATED AT THE END PLATE; CARDIAC MUSCLE; TESTIS; OVARY; OVIDUCT.	REMAINS ON ALL SUBSEQUENT NERVOUS TISSUES.	ADULT		DIMINISHES IN SPINAL CORD WHITE MATTER, CEREBELLAR FIBER TRACTS AND TECTAL STRATUM OPTICUM AND ALBUM CENTRALE.

changes are seen at all places of primary or secondary embryonic induction and taken together they constitute an expression sequence. Within this framework (schematized in Fig. 3), we may then place more detailed lists of sites of appearance (Table 2) and sequences (Edelman, 1985) of appearance and modulation (Table 3). It is hoped that these tables will convince the reader of both the generality of CAM distributions and of the specificity of the sequences of their expression.

The CAM fate map of the blastoderm of the chick (Fig. 3A,B) shows that throughout organogenesis, the two primary CAMs bear a definite topological relationship to each other. N-CAM which appears on neural ectoderm, somites, and lateral plate mesoderm forms a continuous singly connected region surrounded by a continuous ring of L-CAM expressed on non-neural ectoderm and endoderm. This topological relation in connection with the germ layers is neither trivial nor accidental. It reflects a series of events in which cells of different epithelial sheets linked by the separate CAMs are first delineated, moved, or folded together. The regulation of fundamental events of gastrulation leading to formation of the germ layers is accompanied by distinct changes in the surface expression of CAMs. For example, during gastrulation, when middle layer (mesoblast) cells move through the primitive streak between epiblast and hypoblast, the primary CAMs are markedly reduced or disappear from their surfaces. In most cases of primary and secondary induction, separate collectives of cells, each linked by a primary CAM of different specificity (L or N) are seen. Let us turn now to a more detailed discussion of the spatiotemporal distribution of the CAMs in the chick, for a consideration of their order of appearance is highly revealing.

CAM Expression in Space and Time

As has already been suggested, one way of relating the function of the CAMs to their specificity and to regulative development is to study their appearance in various germ layers and their derivatives, to map their appearance onto the classical fate maps used by embryologists, to summarize the fate of cells of a given origin through some defined time period in which topographic changes occur, and then to relate their temporal expression sequence to such a map. I attempt to do this here by describing in more detail the composite CAM fate map; this description must be considered in close connection with dynamic surface changes of the molecules including down-regulation or prevalence modulation accompanying the formation of loosely connected groups of cells called mesenchyme.

A striking event occurs at neural induction, as indicated in the expression sequence (Fig. 3C). Before this primary induction, all cells that are not moving in mesenchyme appear to have low levels of both N-CAM and L-CAM at their cell surfaces. At induction, there is a rapid establishment of the fundamental topology of the CAM fate map: a central region of the blastoderm corresponding to the neural plate shows increases in surface N-CAM while losing L-CAM, and the surrounding region of non-neural ectoderm and endoderm increases in surface L-CAM while losing N-CAM. This reciprocal change, which is maintained in the map throughout life (with some small local fluctuations and exceptions) must occur by alterations in gene expression because the cells of the final epithelia, each connected by separate CAMs, do not move prior to the topological change.

Additional changes in prevalence modulation occur subsequently as indicated in Fig. 3C, Table 2. Neural crest cells, which first have N-CAM on their surface, lose it when they become a mobile ectomesenchyme and migrate to various sites to form the peripheral nervous system and ganglia. When these cells reach their destination, N-CAM reappears on their surface and they aggregate and differentiate to form the ganglia. Somites express N-CAM in a striking fashion as they undergo segmentation. Placodes (neural and non-neural) recapitulate the (L-CAM) - (N-CAM) sequence. Moreover, L-CAM increases in each collection of cells destined to form an endodermal derivative at just the time of induction in particular sites of the gut (some examples are the liver, pancreas, and the bursa of Fabricius). In the kidney, a definite sequence of L- and N-CAM prevalence modulation is seen: L-CAM on Wolffian duct (the inductor), N-CAM on mesonephric mesenchyme as it is induced to form tubules, and reappearance of L-CAM as collecting tubules are extended.

It is important to recognize that these (L-CAM) - (N-CAM) sequences appear in many different regions that will become quite different tissues. These regions are as disparate as the blastoderm and each individual skin placode which will form a feather in a bird. In this latter case, L-CAM linked ectodermal cells come into apposition with N-CAM linked collectives of inducing mesodermal cells to produce distinct structures.

Secondary CAM's introduce a new richness into this picture, but the basic outlines remain the same. For example, when tissues of the nervous system first become postmitotic and neurons extend processes to form tracts and maps, moving both on other neurons and on glial cells which act as guides, a new secondary CAM appears on neurons (Fig. 3C, Table 3). This neuron-glia CAM is polarity-modulated (i.e. it is seen mainly on axons as they extend) but it also appears on neural cell bodies as they move on glia. It has definitely been shown to mediate neuron-glia binding (Grumet et al., 1985) and abrogation of its binding function by specific antibodies *in vitro* prevents cell movement of this type. Thus, in cell movement in one complex tissue (the brain), a CAM couple consisting of a primary CAM and a secondary CAM appears on neurons. This is presumably to solve the problem of sorting out neuron-neuron interactions from simultaneous neuron-glia interactions. In accord with this notion, N-CAM binding is homophilic whereas Ng-CAM binding is heterophilic to an as yet uncharacterized Gn-CAM (Table 1). This suggests that a combination of both modulation and specificity differences is required to generate rich patterns during differentiation.

The sequence of CAM expressions and modulation is further elaborated in the perinatal period: N-CAM undergoes E to A conversion at just the time that neural tracts are being fixated and myelination occurs (Chuong and Edelman, 1984). The entire macrosequence of primary CAM appearance followed by secondary CAM appearance as well as the extraordinary microsequences of the times of first appearances and of modulation for each of these molecules is illustrated in Table 3. Modulatory dynamism is particularly striking for Ng-CAM: each region of appearance of a fiber tract or of a cell movement is heralded by appearances of this CAM at neuronal cell surfaces. Subsequently, surface Ng-CAM diminishes or disappears and great reversals of its distribution in neural areas can be seen in comparing embryos prior to tract fixation and myelination with adult animals. Obviously, local signals in the collectives of cells corresponding to each of the designations in Table 3 must be responsible for the appearance and modulation of the CAMs.

It is important to notice that the CAM appearances in structural anlagen of organs occur prior to the cytodifferentiation that will occur in these regions. Accordingly, in their distribution, primary CAMs extend across germ layers (Fig. 3B) and the signals for their gene expression must therefore relate more to the trajectories, regions and other CAMs encountered than to genes which will later specify other molecules in specific organs. As shown by E to A conversion of N-CAM, however, genes other than those for CAMs can play important roles in the sequence. In this case, the genes coding for enzymes responsible for formation and attachment of polysialic acid enter the picture (Rothbard et al., 1982; Finne et al., 1983; Crossin et al., 1984).

This dynamic spatial and temporal pattern would not be convincingly interpretable in terms of an hypothesis connecting CAM gene expression, CAM function, and morphogenesis, were it not demonstrable _in vitro_ and _in vivo_ that disruption of CAM function leads to altered structure and morphogenesis. This has been done with tumor viruses that alter CAM expression and as well as with anti-CAM antibodies.

Perturbation and Altered Form

Form-producing movements in the embryo are various (Weiss, 1969; Raff and Kaufman, 1983; Slack, 1983; Trinkhaus, 1984). They include the movement of cells following epithelial to mesenchymal conversion (as mentioned above in mesoblast formation and neural crest cell movements), the movement of tissue sheets in tension upon each other (as seen in neurulation, the folding up of the neural tube driven by the extension of the notochord), and also the exquisite tract and map-forming movements of neurons on each other and glia following neural process extension. If the phenomenological descriptions of the expression sequences of CAMs are to be related confidently to molecular function, then one should be able to show that perturbation of this function alters movement and form.

Perturbation experiments have been carried out on cells and tissues *in vitro* and on tissues *in vivo*; the perturbants include transforming viruses and specific anti-CAM antibodies. Transformation of rat cerebellar cells with temperature sensitive mutants of Rous sarcoma virus allows one to ask whether the altered differentiation induced by the *src* gene product will alter the expression and function of N-CAM. When this experiment was performed (Brackenbury et al., 1984; Greenberg et al., 1984), it was found that at the non-permissive temperature, neuron-like cells had both normal morphology and the A forms of N-CAM at their cell surface. When placed at the permissive temperature the cells transformed, and within 12-24 hours after transformation, they lost the major portion of their N-CAM and were no longer competent to aggregate. Cells at the permissive temperature could be brought to the non-permissive temperature and would then express N-CAM and aggregate. Chick brain cells infected with the unmutated form of the Rous sarcoma virus also showed a drastic reduction in surface N-CAM, after which they rounded up and migrated extensively in culture (Brackenbury et al., 1984). Besides shedding light upon the permissive conditions that may occur prior to cancer cell metastasis, these experiments result in the creation of an *in vitro* homologue of epithelial to mesenchymal transformation (although undoubtedly the CAM control mechanisms differ in the two cases).

A strong effect of blockade by specific antibodies of N-CAM binding in tissue culture has been shown in the formation of fascicles of neurites by explanted dorsal root ganglion cells. Instead of orderly branching with side to side interactions, ganglion cells exposed to the antibodies extended neurites which did not interact to form fascicles but instead formed a tangled mesh (Rutishauser et al., 1978). Similar effects of anti-L-CAM antibodies have been observed *in vitro* on the histotypic aggregation of liver cells (Gallin et al., 1983). When anti-N-CAM was present for 6 days in organ cultures of chick retina explanted at 6 days of embryonic age, instead of the orderly layering of neural cells seen in the controls, a highly disordered mosaic structure was found in which cells invaded neurite bundles and layering was perturbed (Buskirk et al., 1980). Inasmuch as the establishment of neural laminae depends upon cell migration and process extension, it seems highly likely that N-CAM is centrally involved in these events.

The most refined expression of morphogenesis is seen in the formation of neural maps such as the retinotectal projection, which maps visual space (as projected on the retina) onto central neural laminae. This occurs by means of a complex series of neurite bundle movements, neurite extensions and competitions, and synapse formation followed by functional validation of refined connections and withdrawals of inappropriate ones (Purves and Lichtman, 1985). When anti-N-CAM antibodies in an agar plug were implanted *in vivo* into the tecta of frogs (*Xenopus laevis*), the order and the precision of the forming retinotectal map were both perturbed (Fraser et al., 1984). In time, after depletion of the antibody concentration in the tissue, the maps began to refine again, suggesting that the entire process is kinetically constrained and under dynamic control. Preliminary anatomic studies of the perturbed tecta showed a greater extension of the axonal arbors over the tectum than in normal controls, suggesting either that in the presence of antibody, the migration and extension of neurite branches was greater or that their retraction was impeded. As already mentioned, anti-Ng-CAM antibodies have been observed to block the migration of external granular layer cells down guide glia into lower layers of cerebellar slices *in vitro* (Grumet et al., 1985). Preliminary observations on agar plugs containing anti-Ng-CAM suggest that similar effects also occur in frog tecta *in vivo*.

All of these *in vitro* and *in vivo* experiments support the notion that control of CAM expression and binding functions plays a key role in the morphogenetic movements leading to form and histogenesis. Clearly, the functions of CAMs differ in different contexts of movement, cell type, and tissue production. Nonetheless, there can be little doubt that, as a molecular family, CAMs are major regulators of movement. This brings one back to the main issue: how can such regulation be related to the expression of CAM genes on the one hand and to epigenetic sequences on the other?

The Regulator Hypothesis

The preliminary considerations underlying this hypothesis have already been stated in the section on position and regulation. Briefly put, the hypothesis states (Edelman, 1984b) that CAMs are the key regulators of morphogenetic movements leading to induction and histogenesis. CAMs are expressed at the cell surface under the influence of various forms of local modulation (Fig. 2); their binding mechanisms are non-linear and their distribution and binding strengths can be continuously graded by these mechanisms. By altering movements, these effects regulate the sequences of induction.

Although these propositions address the question concerning epigenetic sequences, we must specifically consider the connection of the hypothesis to the developmental genetic question. To this end, the hypothesis states that the control of regulatory genes for CAMs is independent from that connected with the expression of genes specifying particular tissue-specific products. This is equivalent to saying that, for the most part, CAM gene expression precedes and is independent of that of other structural genes in tissue and organ anlagen. Later, however, as other genes are expressed, they affect epigenetic sequences and pattern formation by altering cell shape, cell movement types, and the expression of enzymes connected to CAM modulation (e.g. the sialyl transferases involved in E to A conversion).

The entire set of assumptions of the regulator hypothesis can be summarized in the CAM cycle shown in Fig. 4. The outer loop of the cycle expresses the precedence of CAM gene control; the inner loop indicates the parallel and dependent expression of tissue-specific and other genes. Although a direct experimental demonstration that CAMs affect embryonic induction remains to be made, the evidence reviewed above certainly supports the idea that CAMs are regulators of morphogenetic movement and that they are controlled by surface modulation.

This is not the place to consider all of the tests and predictions of the hypothesis, but it is useful to consider some major supporting evidence and some critical predictions. It is particularly important to consider the milieu-dependent character of induction and to take up the critical issue of the local signals which feed back to regulate the CAM regulators themselves. This unresolved matter of ultimate control in the CAM cycle is signaled by the two large horizontal black arrows in Fig. 4. The one on the left, pointing to induction, indicates the possibility that small molecules acting as morphogens might be released by collectives of cells bound by CAMs. A more parsimonious idea is indicated by the right hand arrow which points to surface modulation: cell surface modulation itself may initiate induction directly or after one traverse of the cycle. By cross linking CAMs, this form of signal, known as global cell surface modulation (Edelman, 1976), would affect the cytoskeletal regulation of cell growth and movement. While not yet shown for CAMs, such

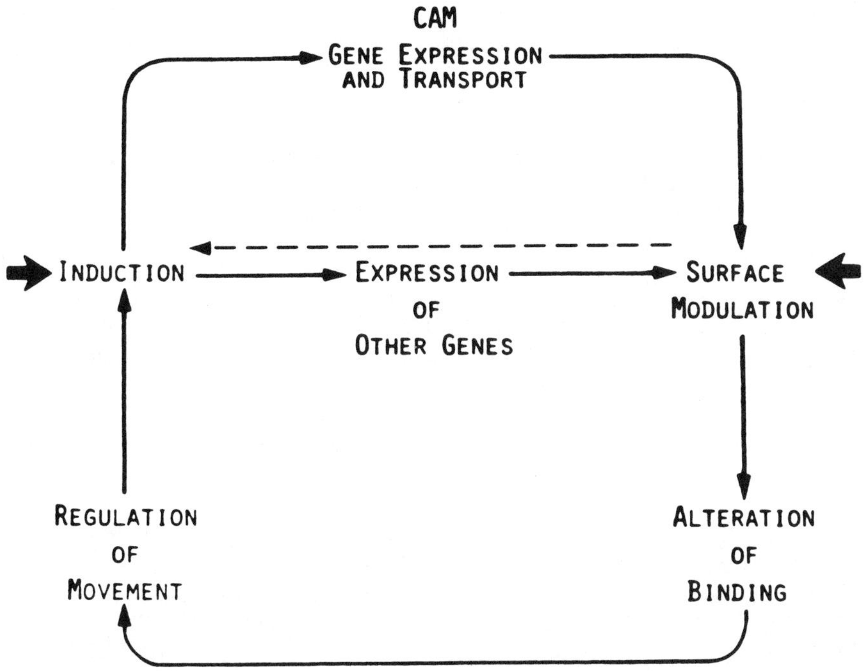

Fig. 4. CAM regulatory cycle. This schematic diagram illustrates the regulator hypothesis (Edelman, 1984b). After CAM gene expression and transport to the cell surface, a variety of modulation mechanisms (see Fig. 2) act to alter the binding of CAMs on apposing cell surfaces. These non-linear mechanisms regulate morphogenetic cell movements which affect epigenetic sequences of embryonic induction. Induction can lead to further alterations of CAM gene expression as well as the expression of other genes. A key assumption of the regulator hypothesis is that the regulatory genes for CAMs are under separate control from those for other proteins. (Note the parallelism of the branches subsequent upon induction.) The large black arrows on the left and right indicate sites for signals (so far unidentified) that may initiate the inductive process. Small molecules (or possibly peptides) may act as morphogens (large arrow left) affecting induction. Alternatively, as shown by the large arrow on the right, global cell surface modulation (Edelman, 1976) arising by fluctuation and alteration of local modulation may affect the passage through the cycle. Such global modulation could act directly on induction (dotted arrow) affecting the expression of induction by altering cytoskeletal states or it could act through the cycle itself by first altering cell binding.

transmembrane control has been extensively documented in a series of *in vitro* experiments in which lectins were used to cross link cell surface glycoproteins (Edelman, 1976). Cell division, receptor motion, cell motion, and transformation have all been shown to be affected by this cross-linkage. Although we cannot exclude the simultaneous action of both morphogens and global modulation, in any case, global modulation has the additional virtue of taking into account the possible role of substrate adhesion molecules (SAMs) such as laminin and fibronectin during morphogenetic sequences. In view of their permissive role in certain migrations and their binding to cells, SAMs may well be involved in global modulation. Nonetheless, according to the hypothsis, SAMs do not play the same specific role as CAMs in organizing cellular collectives.

A picture of global modulation induced by the specific local modulation occurring in collectives of cells linked by different CAMs and influencing each other inductively is consistent with the existence of (L-CAM) - (N-CAM) couples seen in primary induction and at all sites of secondary induction. It is also consistent with the view that the same CAM couples are used in different modulatory modes to make feathers and kidneys. What differs in these locales is the dynamic signals, the epigenetic sequences influencing other primary processes, and above all the mechanochemical boundary conditions relating macroscopic tensions, pressures, and cell movements to global cell modulations. Indeed, it is possible to imagine certain tensions and pressures at the cell surface turning CAMs on or off, with feedback leading to spatial periodicity. Recent experiments on the appearance during feather formation of cell collectives of ectoderm linked by L-CAM in conjunction with inducing collectives of mesodermal cells linked by N-CAM are entirely in accord with this view.

How does this view of patterning in epigenetic sequences differ from previous ideas? The most radical difference from conventional views is that cell recognition does not depend upon a very large number of pre-assigned gene products but rather upon kinetically constrained dynamic and graded modulations (Edelman, 1976; 1983a; 1984b,c). While not dismissing the specificity of CAMs, the main emphasis of the regulator hypothesis is upon selection during epigenesis. "Place" is not recognized by a microscopically assigned marker; it is established as a consequence of modulation and control. CAMs as gene products, through their role as regulators of cellular mechanochemical events, link the one dimensional genetic code to the higher order assemblages of cells and tissues.

The control of CAM binding makes the necessary rapprochement between developmental genetics and mechanochemistry at cell and tissue levels. CAMs are the bridge between "entwicklungs-mechanik" (Roux, 1905) and genetics (Waddington, 1966; Bonner, 1982; Raff and Kaufman, 1983). Nevertheless, it is clear that while the CAMs are necessary as key regulators of pattern, other molecules such as SAMs and other tissue specific gene products must also be present to ensure cellular migration and specialization in histogenesis. In this dynamic model, CAMs themselves are clearly insufficient to generate a rich diversity of form; the regulatory cycle (Fig. 4) is the key. Nonetheless, these molecules provide a potential link between answers to the developmental genetic question and the evolutionary question. We turn finally to a brief discussion of this link.

Heterochrony Reconsidered

The evidence (King and Wilson, 1975) that mutations in regulatory genes are likely to be the major locus of morphological differences within classes and that heterochrony provides a major means by which morphological change occurs during evolution leaves open the question of the kinds of regulatory genes that are involved. The regulator hypothesis asserts that CAM regulatory genes are the _minimal_ set likely to be instrumental. The key idea here is the occurrence during evolution of independent but _covariant_ alterations in morphogenetic cell movements and in timing of CAM regulatory gene expression (Edelman, 1984b). Any mutational variation of either that fails to lead to successful embryonic induction will be selected against. On the other hand, any evolutionary variation in morphogenetic motion that is consistent with an epigenetic sequence or with a variation in the timing of gene expression that results from a mutation in CAM regulatory genes may be selected. Whether that selection takes place or not depends upon the phenotypic expression of the altered form and upon its adaptive advantage. A strong prediction of the regulator hypothesis related to this idea of covariance is that the CAM fate maps of the chick and the frog will be highly similar despite the large differences in early morphogenetic movements seen in the two species of embryo.

It is clear that as a potential answer to the developmental genetic question, the regulator hypothesis is consistent with the demands of the evolutionary question. Small changes in the timing or influence of CAM regulatory genes could have large non-linear effects upon form and pattern in short periods of evolutionary time while still obeying Stebbins' requirement (1968) of selective inertia. The parallel nature of the loops of the CAM cycle (Fig. 4) allow florid variations within the same general morphological scheme. It is not difficult to incorporate mechanisms leading to progenesis, neoteny and paedomorphosis into this scheme. In this discussion, most of the emphasis is upon animals and indeed mainly upon vertebrates, for most of the present evidence has been obtained from these species. While one expects to see CAMs in complex metazoans such as insects, not all animal forms (e.g. vertebrates, Dictyostelium or sponges) need be related by the same kinds of adhesion molecules or modulation mechanisms. Nonetheless, even in cases of convergent evolution, I would expect that general considerations similar to those expressed in the regulator hypothesis would hold.

The evolution of the regulators themselves deserves brief mention within the framework set by this discussion. If CAMs act dynamically in epigenetic sequences, one would expect them to be found in all vertebrates in similar distributions. Moreover, although different molecules will have different binding specificities, there will not be huge numbers of different CAMs. Inasmuch as the emphasis is on the regulation of these molecules, one would expect that the binding specificities of a given evolved CAM would not be changed in widely divergent species. This is exactly the case (Hoffman et al., 1984): frog N-CAM binds to mouse N-CAM as well as each CAM binds to itself in the homophilic binding that is seen within each species. Rather than emphasizing the elaboration of molecular specificities, the regulator hypothesis directs that the bulk of the effort to discern the molecular basis of form should be placed on the sequence of CAM gene expressions and their controls. The recent identification of cDNA probes (Murray et al., 1984; Gallin et al., 1985) for CAM genes makes this a feasible long range experimental goal.

Coda: On Complexity

Some remarks appended to this review of the regulator hypothesis may correct any impression that CAM regulation is unaccompanied by other means of control. It is clear that there are a number of other means to alter the complex interaction of the primary processes of development, particularly cell division and cell death. In some organisms, these mechanisms can play dominant roles at certain periods of development, and overshadow CAM regulation. In the large, however, these means of control of cell division and death would be adjunct to the CAM regulatory cycle. Mosaic development (in which these other means of control may be more dominant) and regulative development itself form a spectrum of related means for controlling form (Weiss, 1969). Evolution will opportunistically vary the emphasis on these means; this implies that a whole spectrum of genes in addition to CAM genes will be selected for in a covariant fashion to yield the complexities of adaptive animal form. How do such complexities emerge and is there a trend towards increasing complexity?

In his analysis of development and evolution, Stebbins (1968) considered this question: how do we account for the apparent trend of evolution toward greater complexity? I paraphrase his argument here. He pointed out that, overall, one sees no progressive trends but "rather, successive cycles of adaptive radiation followed by wholesale extinction, or by widespread evolutionary stability." While increases in complexity have been rare, attainment of greater simplicity has been rarer so that the net consequence is an increase in complexity. He suggests that mutations that can be established in populations for some significant period are not random relative to the genetic and developmental integration of any organisms in which they are found. Any mutations adversely affecting this integration would, however, be quickly eliminated and "may never reach the point where they affect the interaction between the adult and its environment." Mutations which do not affect this integration or which lead to even more complex integrations are much more likely to become established. In long stretches of time, complexities which survive are those which render the organism increasingly free of the short term fluctuations of their environment. They make possible evolution in directions determined increasingly by the kind of organized structures which they themselves have helped to evolve.

While the regulator hypothesis is entirely in accord with these ideas, and alterations in the relative timing of CAM regulatory gene expression could lead to increases in complexity, these evolutionary principles apply not only to the covariant changes between these genes and those regulating morphogenetic motion, but also to genes affecting the other primary processes of development.

This provides a basis for understanding how complexity arises epigenetically during development. The most striking impression left by developmental processes aside from their plasticity is of their parallel nature. Independent parallel events can interact in such systems in a non-linear and complex fashion. In the scheme reflected by the CAM cycle (Fig. 4), this parallelism is subserved by the shunt pathway with the heading "other genes". But some of these genes, of course, specify and regulate key primary processes such as cell division and indirectly cell death, both of which are major driving forces in morphogenesis. As I have already mentioned, under certain circumstances, and at certain times, such processes can temporarily wrest control of morphogenesis away from the regulator loop.

More specific molecular examples are provided by the functions of substrate adhesion molecules (SAMs) which are important for cell migration and by cell junctional molecules (found in gap junctions, tight junctions etc.) which specify cell linkages and communications. The action of these molecules is required for orderly morphological development. The example of intermodulation of N-CAM and fibronectin in the movement of neural crest cells makes the point: fibronectin is required for crest cell migration but is not produced by crest cells. N-CAM down-regulation at the cell surface is a permissive condition for motion; concomitant expression of fibronectin is a necessary condition.

These few examples suggest that we may expect to see parallel processes in epigenesis acting to give concerted patterns in a complex fashion. But this does not relieve us of the necessity to account for a major control of form at the molecular level. The evidence so far suggests that the bulk of that control rests with the CAMs and not with the SAMs or CJMs. Thus a main prediction of the regulator hypothesis is that CAMs will appear and be modulated at every site of induction in complex tissue structures, however microscopic. In contrast, SAMs will not necessarily take part in all of these processes but will be parallel adjuncts to certain stages of these morphogenetic events. The basic idea remains that induction will be mediated by cell collectives linked by L-CAM interacting with other collectives linked by N-CAM.

The complexities that emerge will be a result of sequential repeated courses through the CAM cycle, each traverse representing a new context resulting from epigenetic conditions which ensue from expression of new structural genes or from the effects of new locales established by the primary processes. It will be a challenging task for the future to determine how many traverses of the CAM regulator cycle and how many regulatory genes are engaged in the morphological determination of a complex appendage such as a feather or of a complex central apparatus such as a brain.

ACKNOWLEDGEMENT

The work of the author cited here was supported by U.S. Public Health Service Grants HD-09635 and AM-04256.

REFERENCES

Baer, K.E. von, 1828, Entwicklungsgeschichte der Thiere: Beobachtung und Reflexion (Borntrager, Konigsterg).

Bonnner, J.F., 1982, Evolution and development, Dahlem Konferenzen, Life Sciences Research Report 22, Springer Verlag, Berlin, Heidelberg, New York.

Brackenbury, R., Thiery, J.-P., Rutishauser, U., and Edelman, G.M., 1977, Adhesion among neural cells of the chick embryo. I. An immunological assay for molecules involved in cell-cell binding, J. Biol. Chem., 252:6835.

Brackenbury, R., Greenberg, M.E., and Edelman, G.M., 1984, Phenotypic changes and loss of N-CAM mediated adhesion in transformed embryonic chicken retinal cells, J. Cell Biol., 99:1944.

Buskirk, D., Thiery, J.-P., Rutishauser, U., and Edelman, G.M., 1980, Antibodies to a neural cell adhesion molecule disrupt histogenesis in cultured chick retinae, Nature, 285:488.

Chuong, C.-M. and Edelman, G.M., 1984, Alterations in neural cell adhesion molecules during development of different regions of the nervous systems, J. Neurosci., 4:2354.

Crossin, K.L., Edelman, G.M., and Cunningham, B.A., 1984, Mapping of three carbohydrate attachment sites in embryonic and adult forms of the neural cell adhesion molecules (N-CAM), J. Cell Biol., 99:1848.

Edelman, G.M., 1976, Surface modulation in cell recognition and cell growth, Science, 192:218.

Edelman, G.M., 1983a, Cell adhesion molecules, Science, 219:450.

Edelman, G.M., Gallin, W.J., Delouvee, A., Cunningham, B.A., and Thiery, J.-P., 1983b, Early epochal maps of two different cell adhesion molecules, Proc. Natl. Acad. Sci. USA, 80:4384.

Edelman, G.M., 1984a, Modulation of cell adhesion during induction, histogenesis, and perinatal development of the nervous system, Ann. Rev. Neurosci., 7:339.

Edelman, G.M., 1984b, Cell adhesion and morphogenesis: The regulator hypothesis, Proc. Natl. Acad. Sci. USA, 81:1460.

Edelman, G.M., 1984c, Cell surface modulation and marker multiplicity in neural patterning, Trends Neurosci, 7:78.

Edelman, G.M., 1985, Cell adhesion and the molecular processes of morphogenesis, Ann. Rev. Biochem., in press.

Finne, J., Finne, U., Deagostini-Bazin, H., and Goridis, C., 1983, Occurrence of 2-8 linked polysialosyl units in a neural cell adhesion molecule, Biochem. Biophys. Res. Commun., 112:482.

Fraser, S.E., Murray, B.A., Chuong, C.-M., and Edelman, G.M., 1984, Alteration of the retinotectal map in _Xenopus_ by antibodies to neural cell adhesion molecules, _Proc. Natl. Acad. Sci. USA_, 81:4222.

Gallin, W.J., Edelman, G.M., and Cunningham, B.A., 1983, Characterization of L-CAM, a major cell adhesion molecule from embryonic liver cells, _Proc. Natl. Acad. Sci. USA_, 80:1038.

Gallin, W.J., Prediger, E.A., Edelman, G.M., and Cunningham, B.A., 1985, Isolation of a cDNA clone for the liver cell adhesion molecule (L-CAM), _Proc. Natl. Acad. Sci. USA_, in press.

Gould, S.J., 1977, "Ontogeny and Phylogeny", The Belknap Press of Harvard University Press, Cambridge, Mass., London, England.

Greenberg. M.E., Brackenbury, R., and Edelman, G.M., 1984, Alteration of neural cell adhesion molecule (N-CAM) expression after neuronal cell transformation by Rous sarcoma virus, _Proc. Natl. Acad. Sci. USA_, 81:969.

Grumet, M., Hoffman, S., Chuong, C.-M., and Edelman, G.M., 1985, Polypeptide components and binding functions of neuron-glia adhesion molecules, _Proc. Natl. Acad. Sci. USA_, 82:7989.

Haeckel, E., 1879, "The Evolution of Man: A Popular Exposition of the Principal Points of Human Phylogeny and Ontogeny," D.A. Appleton and Co., New York.

Hamburger, V., 1980, Embryology and the modern synthesis in evolutionary theory, _in_: "The Evolutionary Synthesis," E. Mayr and W.B. Provine, eds., Harvard University Press, Cambridge, Mass., pp. 96-112.

Hoffman, S., and Edelman, G.M., 1983, Kinetics of homophilic binding by E and A forms of the neural cell adhesion molecule, _Proc. Natl. Acad. Sci. USA_, 80:5762.

Hoffman, S., Chuong, C.-M., and Edelman, G.M., 1984, Evolutionary conversation of key structures and binding functions of neural cell adhesion molecules, _Proc. Natl. Acad. Sci. USA_, 81:6881.

King, M.C., and Wilson, A.C., 1975, Evolution at two levels in human and chimpanzees, _Science_, 188:107.

Murray, B.A., Hemperly, J.J., Gallin, W.J., MacGregor, J.S., Edelman, G.M., and Cunningham, B.A., 1984, Isolation of cDNA clones for the chicken neural cell adhesion molecule (N-CAM), _Proc. Natl. Acad. Sci. USA_, 81:5584.

Needham, J., 1933, On the dissociability of the fundamental process in ontogenesis, _Biol. Rev._, 8:180.

Purves, D., and Lichtman, J.W., 1985, "Principles of Neural Development," Sinauer Association, Sinderland, Mass.

Raff, R.A., and Kaufman, T.C., 1983, "Embryos Genes and Evolution, The Developmental Genetic Basis of Evolutionary Change," Macmillan Publishing Co., New York.

Rothbard, J.B., Brackenbury, R., Cunningham, B.A., and Edelman, G.M., 1982, Differences in the carbohydrate structures of neural cell-adhesion molecules from adult and embryonic chicken brains, J. Biol. Chem., 257:11067.

Roux, W., 1905, Die Entwickelungsmechanik der organismen, W. Engelmann, Lupzig, pp. 283

Rutishauser, U., Gall, W.E., and Edelman, 1978, Adhesion among neural cells of the chick embryo. IV. Role of the cell surface molecule CAM in the formation of neurite bundles in cultures of spinal ganglia, J. Cell Biol., 79:382.

Slack, J.M.W., 1983, "From Egg to Embryo," Cambridge University Press, New York.

Sperry, R.W., 1963, Chemoaffinity in the orderly growth of nerve fiber patterns and connections, Proc. Natl. Acad. Sci. USA, 50:703.

Stebbins, G.L., 1968, "Integration of Development and Evolutionary Progress," in: "Population Biology and Evolution," R.C. Lewontin, ed., Syracuse University Press, New York., pp. 17-36.

Thiery, J.-P., Brackenbury, R., Rutishauser, U., and Edelman, G.M., 1977, Adhesion among neural cells of the chick embryo. II. Purification and characterization of a cell adhesion molecule from neural retina, J. Biol. Chem., 252:6841.

Trinkhaus, J.P., 1984, "Cells Into Organs. The Forces that Shape the Embryo," 2nd Edition, Prentice Hall, Englewood Cliffs, New Jersey.

Vakaet, L., 1984, Early Development of Birds, in: "Chimeras in Developmental Biology," N.M. Le Douarin and A. McLaren, eds., Academic Press, London, pp. 71-87.

Waddington, C.H., 1966, "Principles of Development and Differentiation," Macmillan Publishing Co., New York.

Weiss, P., 1969, "Principles of Development, a Text in Experimental Embryology," (Facsimile 1939 edition), Hafner Publishing Co., New York.

GENE ACTION AND MORPHOGENESIS IN PLANTS

G. Ledyard Stebbins

Department of Genetics
University of California
Davis, CA 95616

One of the most important unanswered questions facing evolutionists is: "How do those genes act that are responsible for morphogenetic or developmental patterns in multicellular eukaryotes?" The importance of this question lies in the fact that answers to it will provide the most significant links possible between those morphological changes recorded by paleontologists and differences that geneticists can analyze. At present, the gap between the historical documentation of the fossil record and the analysis of evolutionary processes at the level of DNA-coded information is wide and almost completely unbridged, as is clearly shown by the content of the most recent books dealing with development and evolution (Gould, 1977; Raff and Kaufman, 1983).

Recent comparisons between biochemical, morphological and chromosome evolution (Wilson et al., 1974; Wilson, 1975; Valentine and Campbell, 1975; Cherry et al., 1978; Maxson and Wilson, 1979; Turner et al., 1979) have shown that this question will never be answered by more intensive investigations of those genes to which developmental geneticists have paid the greatest attention, i.e. those that code for enzymes concerned with general processes of cellular metabolism, or for proteins such as globins and immunoglobulins, that function in transport or bodily defense systems. Neither will evolutionists arrive at a solution by depending chiefly upon the oft repeated dichotomy between "structural" and "regulatory" genes. In eukaryotes, the latter category is poorly defined and probably contains a great variety of different elements. Moreover, many regulatory elements control the action of structural genes that are concerned with cellular metabolism, as is revealed by the complex web of interactions that affect the activity of alcohol dehydrogenase in *Drosophila* (McDonald, 1983). How, then, shall we proceed?

In my opinion, the most fruitful hypothesis with which to start is to assume that genes concerned with morphogenesis exert their effects via the action of proteins that are their primary products. A second question to ask is, therefore, what proteins are likely to exert the greatest effects upon morphogenesis? Many such proteins can be postulated to have morphogenetic effects. In Metazoa, one can think of those concerned with mitotic cell division, polarity of growth and stratification of cytoplasmic elements, with movements of cells and supramolecular aggregates through the developing embryo, with cell to cell adhesions, and with inductive interactions between tissues. In plants, the roster is considerably smaller, since movements, adhesions and inductive interactions between tissues do not occur. This is one reason why plants are more favorable objects for investigating this problem than are animals.

A number of other features make plants more favorable than animals. As Gottlieb (1984) has pointed out, higher plant species contain many conspicuous morphological variants that are controlled by single genes, while such situations are far less common in animals. Furthermore, as is evident from experiments reviewed below, some of these differences can be mimicked by applying known chemical agents to developing plants under controlled conditions. Also, the developing embryonic or meristematic tissue of plants can be modified by microsurgery at various stages of development (Cutter, 1971). To these favorable circumstances can be added the fact that many different plant species, that differ with respect to several morphological characteristics, can be crossed to produce partly fertile F_1 hybrids followed by widely segregating F_2 progenies (Clausen, 1951; Grant, 1971). Finally, a review by this author (Stebbins 1974) shows that all of the morphological differences that taxonomists use to separate orders and families of flowering plants occasionally exist as differences at the species or intraspecific level. In the past, a drawback to using plants for this purpose has been the difficulty of maintaining cell and tissue cultures, particularly of those floral meristems in which many of the most conspicuous morphological differences arise. Nowadays, however, techniques for culturing plant tissues and organs have been improved to the extent that even this type of investigation may become possible via further research on appropriate techniques.

A METHOD OF APPROACH: MORPHOLOGICAL MUTANTS

As is suggested by the above discussion, the search for significant primary gene products must begin at the cellular level. The most appropriate material with which to start consists of single gene differences having conspicuous effects upon adult morphology. While for evolutionary purposes the most useful genes would be those for which different alleles have been established in natural populations of different races or species, these are difficult to locate in genetically known material, so that mutations that are known only in cultivated plants must, at present, serve as our principal source of information. I shall illustrate this approach by reviewing three examples that were analyzed in my laboratory during the 1960's.

The first is hooded barley (Hordeum vulgare L.), a complex syndrome of morphological effects that is produced by the incompletely dominant allele K, located on the short arm of the 4th chromosome (Tsuchiya et al., 1960).

Using the method of allometry, Ezra Yagil showed that the difference between the genotypes KK and kk is initiated in primordia of the fertile scale or lemma that are between 150 and 200 micra long. At this stage, all of the cells are in the meristematic condition (Stebbins and Yagil, 1966; Yagil and Stebbins, 1969). Since the adult lemma with its awn (in kk phenotypes, reaches a length of 12-16 cm., the stage at which differentiation begins corresponds to early embryonic differentiation in an animal. Its first manifestation is an increase in the frequency of cells that are replicating DNA, followed immediately by mitosis. This increase is localized to a particular area on the adaxial or upper surface of the embryonic appendage. It is accompanied by shifts in the orientation of mitotic spindles. By reference to relationships between the frequency of mitosis and the orientation of cell division, elaborated on other material, as well as to the hypothesis that the histological structure of embryonic meristems creates cellular environments that promote the differentiation of organ primordia, a plausible sequence of events was hypothesized, leading to the conclusion that the localized and stage specific increase in frequency of nuclei that have entered the S phase of DNA replication could account for all of the diverse morphological characteristics of the hooded (KK) phenotype. This sequence of events is discussed below, after two other examples have been presented.

The second example is the curled mutation in tomato (Lycopersicon esculentum Mill), of which the development was described by Smith and Stebbins (1971). Plants homozygous for this recessive mutation have leaves much smaller than normal, appressed to the stem and tightly curled. As in hooded barley, a difference between curled and normal phenotypes is evident in young, fully meristematic leaf primordia, and consists of an increase in frequency of cells that are replicating DNA and entering mitosis, in this instance near the embryonic midrib of the leaf.

These two mutations, therefore, agree in that the first detectable difference between mutant and non-mutant phenotypes is a change in frequency of nuclei that are replicating DNA and undergoing mitosis in a particular region at a particular stage of development. If these differences at the nuclear and cell level are, in fact, the first manifestations of differential gene action, how can one explain their divergent effects on the adult phenotype?

This divergence can be explained by the very different histological makeup of the organs in which the mutation has produced more rapid cell proliferation. In the tomato, the adult leaf is strongly dorsiventral, having the densely packed cells of palisade tissue on the upper or adaxial surface and the loosely organized spongy mesophyll on the lower or abaxial surface. Given this histological structure, more rapid proliferation of the adaxial cells would, during the stage of cell enlargement, produce stresses that would cause the upper surface to become curled over the lower surface. No other conditions are necessary.

To explain the effects of localized cellular proliferation in altering subsequent histological development and hooded barley, an additional principle is necessary. This principle was developed by P. B. Green (1963, and 1980). On the basis of his experiments with Nitella and other plants, he supported the hypothesis of Kny (1902) and others (French and Paolillo, 1975; Gunning, 1982; Miller, 1960) that mitotic spindles become oriented parallel to tensions created by directional growth of cells and tissues, and at right angles to pressures exerted by the firm cell walls of plant cells, or by the growth of neighboring cells. In plant meristems, this effect causes spindles to be oriented parallel to and new cell walls at right angles to the long axis of the cell. The effect is exerted, prior to spindle formation, by the close connection between incipient spindle fibers and elements of the cytoskeleton, particularly microtubules, that exist before the onset of spindle formation or of any other manifestation of mitosis.

This principle can be applied directly to the hooded gene in barley since at developmental stages later than that of increase in mitotic frequency, direct observation shows that in the region of high mitotic frequency, spindles are oriented in many planes, while at the corresponding stage in the awned genotype, they are nearly all parallel with the long axis of the primordium.

In addition, the hooded gene produces effects that, superficially, appear quite different from the conversion of the narrow, straight awn into the broader, curved hood. This is the appearance of a dome or cushion of meristematic tissue that is structured in the same way as the reproductive meristem that at an earlier stage gives rise to a floret. From this cushion are differentiated an inner floral scale or palea, and rudimentary floral structures. Moreover, as the hood grows, a central portion develops that, in its pattern of cell structure, has a reversed polarity compared to the lower part of the hood, as well as to the entire awn produced by the kk genotype. This reversal is a consequence of the prior formation of secondary reproductive meristem, since in barley and other plants, polarized gradients are formed between meristematic and adult or maturing tissues. This reversed polarity might be explained by reference to biophysical research by Jaffe and his associates (Jaffe, 1981; Nuccitelli, 1983). These workers, using a highly sensitive vibrating probe, have shown that transcellular currents are established in organs of many different plants and animals by differences in membrane permeability such that near the inner surface of a "leaky" membrane, that is highly permeable to cations, chiefly Ca^{2+} and H^+, a positive pole is set up, while at the opposite end of the cell or organ a negative pole, associated with an energy-driven ionic pump, aids in establishing weak currents that traverse cells or tissues of organs.

In plant roots, currents are initiated either in the elongating epidermal cells proximal to the terminal meristem, or at the tips of growing root hairs (Weisenseel et al., 1979). These regions are similar in that cell walls are growing and elastic (Rayle and Cleland, 1977). The apical cells of a growing awn primordium in kk, as well as the cells immediately proximal to the newly formed reproductive meristem in the lemma primordium in KK, are similar with respect to wall condition to the root cells just mentioned. A reasonable hypothesis is, therefore, that similar currents are temporarily set up, that in primordia of awns and grass leaves are basipetal, and in the primordia of the hoods of KK genotype are alternately basipetal and acropetal, and responsible for the reversed polarity.

Another gene in barley, calcaroides produces an adult phenotype superficially similar to that of hooded, but its development is totally different (Stebbins and Price, 1971).

In the young lemma primordia of this mutant, no significant differences in mitotic frequency were observed. Instead, the earliest difference between mutant and normal genotypes was a retardation of cell elongation during the transition from actively mitotic meristem to the stage of cell elongation. Nevertheless, the result with respect to cell shape was similar. In calcaroides as in hooded, a shift from unidirectional to multidirectional orientation of mitotic spindles is associated with a shift from elongate to more nearly isodiametric cell shape.

ISOLATING SINGLE GENE DIFFERENCES FROM HYBRIDS

The examples presented above present one difficulty as vehicles for understanding differences between natural populations. The mutant phenotype is in each example less fit than that of its normal allele, so that it could not survive in nature. Hence geneticists need to isolate from the array of genotypes found in segregating F_2 or backcross generations from interspecific or intervarietal hybrids particular phenotypic differences that are controlled by one or at most two gene pairs. So far as I am aware, no such genetic stocks have been isolated from hybrid progeny. Nevertheless, by use of the backcross method that has long been employed for transferring genes for disease resistance from one cultivated variety to another, this could be done. The nearest approach known to me is the research of Hiorth (1942) on the genus *Clarkia* ("Godetia") on which he transferred the allele for basal position of the dark colored spot on the petals from *C. rubicunda* ("Godetia amoena") to the genetic background of *C. amoena* ("Godetia Whitneyi"), in which the dark spot is characteristically in the middle of the petal. Such transfers are, therefore, feasible, even between species which, like these *Clarkias*, form highly sterile F_1 hybrids. Several of the differences mentioned by Gottlieb (1984) could be manipulated in this way.

What is the likelihood that single gene differences isolated from hybrids between existing species, varieties or races will prove to agree with the mutations described above with respect to site specific and stage specific alterations in the frequency of mitoses? Based upon the numerous examples cited by Gottlieb (1974) of single genes that affect adult morphology, I believe that such species-differentiating mutations could be found with little difficulty. I would like to call particular attention to differences that in some groups are diagnostic of different species, but in other groups homologous differences are alternate states within a single species. The six examples listed in Table 1, taken from Table 1 of Gottlieb's paper, appear to me to be the most likely prospects, since in them, development would be relatively easy to study at the cell and tissue level. Detailed investigations of these differences at the cell level, emphasizing the presence or absence of asymmetrical mitoses and particularly sequences of mitosis that differ with respect to the two genotypes should be most rewarding.

Table 1. Morphological and architectural differences governed by one or two genes in natural populations of flowering plants. (Extracted from Gottlieb 1984.)

Difference	Species
Petioles and floral organs elongate or not	*Nicotiana tabacum*
Fruit wingless or not	*Plectritis congesta*
Achene wings entire vs. fimbriate	*Coreopsis grandiflora* v. *saxicola*
Fruit triangular vs. ellipsoid (4x) (2 loci)	*Capsella bursa-pastoris*
Pappus floccose vs. nonfloccose (2 loci)	*Layia platyglossa*

PHYLETIC PHENOCOPIES

Another way of tackling this problem is to make use of phyletic phenocopies. This term was recently invented (Stebbins and Basile, in press), for phenotypes that have been elicited from the normal genotype of a particular species by controlled experimental manipulation and that mimic character states of related species. Two sets of examples exist, both of them the result of experiments in the laboratory of D. V. Basile. The first series showed that leaf arrangement in liverworts can be altered from the derived distichous or bilaterally symmetrical pattern to the less specialized condition of triradial symmetry by applying reagents that affect hydroxylation of proline residues belonging to proteins that are associated with the carbohydrates or lipids of the cell wall (Basile, 1969; Basile and Basile, 1984).

The second series of experiments was performed on a highly specialized angiosperm, Streptocarpus prolixus (gracilis) (Rosenblum and Basile, 1984). Normal phenotypes of this species completely lack vegetative stems and leaves. From the embryonic and seedling stage to the time of flowering they consist only of roots, hypocotyl and two cotyledons, one of which aborts, while the other grows to a gigantic size. Pedunculate inflorescences arise directly from the base of the large cotyledon. The experimenters were able, by the use of specific growth substances, to produce phenotypes which more nearly resembled the probable phyletic ancestors of S. prolixus. Applying to germinating seeds exogenous gibberellin, they produced plants having stems and leaves similar to those found in other, less specialized species of the family Gesneriaceae. Applications of benzyl adenine elicited phenotypes having two normal sized isodiametric cotyledons.

Based upon these experiments, one can conclude that two kinds of phenotypic changes were induced by mutations that became established during the evolution of Streptocarpus prolixus from its less specialized ancestors. One was the reduction in endogenous content of gibberellin, a growth substances that usually induces the formation of stems and leaves. The other was the reduction in the production of cytokinins, that in less specialized species play a major role in normal development of embryos and seedlings.

Can one or both of these changes be ascribed to mutations like those reviewed in a previous section, that alter frequencies of mitoses in a site specific and stage specific manner?

The following hypotheses suggest that this is likely for the example of Streptocarpus. With respect to the absence of endogenous gibberellin, Rosenblum and Basile point out that in S. prolixus vegetative meristems are much smaller and more highly localized than they are in species that have more normal development. Since gibberellins are synthesized in meristematic and submeristematic tissues (Jones and Phillips, 1966), paucity of these tissues could be directly responsible for low gibberellin content. Differences in the amount of meristematic tissue are, by definition, localized differences in the frequency of mitotically dividing cells. With respect to the influence of benzyl adenine, its effect is to stimulate directly mitotic divisions, so that its effect on the phenotype is most probably correction of a localized scarcity of cell divisions, that was induced by genetic mutations during the evolution of anisocotyly in the genus Streptocarpus and other Gesneriaceae.

FROM HISTOLOGICAL STRUCTURE TO CELL DYNAMICS AND ALTERED PROTEIN STRUCTURE

The lesson drawn from the examples presented above is that the primary products of genes that control morphogenesis cannot be identified directly. The search for them must be guided by inferences from histological structure, cell dynamics, an the nature of the processes and individual proteins that are involved. Guidelines for this search have been clearly stated by Hardham (1982): "The characteristic features of different tissues and organs are a reflection of the properties of constituent cells: their sizes and shapes and their functional capabilities. These properties arise through control of the timing and direction of cell division and cell expansion; but for the development of polarity, some of these properties are more important than others."

For understanding the mechanisms that control these properties and their interrelationships, both biochemical and biophysical principles must be invoked. Biochemical principles are based principally upon the synthesis and localized action of growth substances, in plants chiefly auxins, gibberellins and cytokinins. The localization of their action rendering them both stage-specific and tissue-specific, must be based chiefly upon biophysical principles that govern their transport and/or diffusion through the plant body. Their action probably determines the intensity of growth, and the relative participation in the growth process of cell proliferation as compared to cellular expansion. Additional biophysical principles are needed to explain the direction of growth, and the ways in which these directions can shift in order to produce complex patterns. As stated above, these are chiefly the effect of factors that orient microtubules and mitotic spindles, and the establishment of electrostatic gradients based upon differential permeability of membranes, followed by asymmetrical distribution within the cell of both nuclei and cytoplasmic elements.

Applying these principles to the examples presented above, the following hypothesis can be proposed. Hooded barley and the curl mutant of tomato both point to the primary importance of differential permeability of intercellular membranes in determining pattern. The calcaroides mutant of barley and the effect of proline hydroxylation in liverworts both suggest that altered histological patterns can be generated by retardation or acceleration of cell wall growth. The extraordinary pattern of organ development in the advanced species of Streptocarpus has been traced to unusual patterns of meristematic and adult tissues, that is related to hormone action. Recognizing that native auxins, gibberellins and cytokinins are most probably synthesized in different tissues, and hence are initially separated from each other, their effect on pattern formation depends greatly upon the rate at which they move from one tissue to another. The situation in Streptocarpus might be explained by the hypothesis that the more specialized species have evolved stronger barriers to movement of hormones, so that neighboring tissues can exhibit greater contrasts with respect to the relationship between cell proliferation and cell elongation; i.e. sharper boundaries between meristematic and maturing tissues, as was observed in the developing tissues of these plants.

All of these examples point to the primary importance for the evolution of histological patterns of the degree to which cell to cell communication is promoted or inhibited by the properties of intercellular membranes and walls. Proteins that are included in the structure of these membranes are, therefore, the most logical candidates for the primary products of genes that control histological patterns, including the direction and polarization of cell growth, and therefore many features of the morphological pattern as a whole.

Although histological patterns of cellular organization are central to the problem of morphogenesis in plants, they are not the entire problem. First, we must consider the relationship of histological pattern to the pattern formed by relationships between tissues and organs, and to the plant as a whole. Most important are overall size, the relative sizes and shapes of different organs, and the numbers of organs of a particular kind that the plant produces.

Overall size can be a function of the amount of hormone or growth substance that is available, or the duration of the developmental stage of its activity. Proportions have a twofold basis, (1) The establishment of differential growth rates at an early stage in ontogeny and (2) The duration of the stage or stages during which differential growth persists (Sinnott, 1960; Gould, 1977; Raff and Kaufman, 1983). While the second stage may be determined, as is overall size, by either the amount of growth substance that is available, or by a balance between different growth substances during a particular succession of developmental stages, differential growth rates of different organs may well be initially determined by the effects of polarized electrostatic gradients, and therefore by differential permeability of cell membranes. Detailed patterns at the level of tissues and individual cells are unlikely to be determined either by the amount of a particular growth substance, the length of its period of activity, or by any overall balance between growth substances. Localized centers of mitotic activity are most probably determined by equally localized concentrations of growth substances. These, in turn, depend principally upon differential rates of movement through the organism.

As pointed out elsewhere (Stebbins, 1974), numbers of parts that are produced simultaneously, such as calyx lobes, corolla lobes, or locules of the gynaecium, probably depend upon the relationship between the amount of meristematic tissue at the time of their differentiation and the amount of meristem needed to produce a single part. Hence their number can easily be altered by stage-specific alterations in the amount of cytokinin present in the meristem, and so can be controlled by genes that affect the amount and distribution of these growth substances. The number of serially produced organs, such as leaves on a stem, as well as stamens and separately produced carpels in an apocarpus flower, is probably the result of temporal shifts in the histological structure of the shoot apex, that may be the result of shifting balance between different growth substances. Proteins, probably enzymes, that affect the synthesis, transport and degradation of growth substances are the most likely candidates for the primary products of genes responsible for morphogenesis at the level of organs and the plant as a whole.

Finally, the cytoplasmic structure within individual cells plays an important role in determining patterns of histogenesis (Lloyd and Barlow, 1982; Marchant, 1982; Robinson and Quader, 1982). Of particular significance are proteins that form microtubules (Hyams, 1982), proteins and RNA that are associated with microtubule organizing centers (MTOC's, Brown, Stearns and Macrae 1982), and the components of the terminal complexes that are associated with the plasma membrane and the formation of cellulose microfibrils along its surface (Montezinos, 1982).

All of the proteins mentioned in the last few paragraphs possess in common a property that distinguishes them sharply from the enzyme proteins that are the primary products of the genes about which geneticists have the most knowledge. These enzymes often consist of several different polypeptide chains of two or more different kinds. Nevertheless, the whole enzyme is a separate structure, that can be isolated and perform its function in vitro in much the same manner as in vivo. By contrast, the proteins that are included in cellular membranes and cytoskeletons of individual cells, form complex lattices or networks, and usually function only in vivo, when complexed with specific carbohydrates, lipids or both. After having been synthesized by transcription and translation, they must pass through a complex process of self assembly, that is often site specific. This fact is largely responsible for technical difficulties in isolating them, and establishing connections between particular polypeptide chains and the genes that code for them. In my opinion, the most important problem that must be solved before the genetic basis of morphogenesis can be fully understood is to overcome this technical difficulty.

The hypothesis proposed in this section is, therefore, that the primary products of genes that in plants play dominant roles in determining patterns of morphogenesis and form are most often proteins associated with intercellular membranes and the intracellular cytoskeleton. Other proteins, particularly those that become complexed with growth substances and guide them to targets in the nucleus, may often be involved. Harmonious, adaptive differences in form are often determined by complex interactions between products of mutant genes that code for these particular proteins.

DOES THIS HYPOTHESIS APPLY ALSO TO ANIMALS AND MICROORGANISMS?

Reviewing some of the voluminous literature on morphogenesis and embryology in Metazoa, as well as research at the cellular and molecular level in Protozoa, bacteria and viruses, I believe that the generalized answer just suggested for plants is generally valid. An additional but functionally similar category, that is particularly important in Protozoa and other unicellular Eukaryotes, consists of those proteins that, via the processes of self assembly, form the cortical features of the outer cell membrane, such as flagella and cilia.

My belief is based upon the following comparisons. First, differences in rate or frequency of cell proliferation via mitoses are just as distinctive of different locations within developing embryos, and just as characteristic of different stages of development in Metazoa as they are in Metaphyta (Raff and Kaufman, 1983). Second, recent investigations reviewed by Jaffe (1981) and Nuccitelli (1983) have shown that transcellular ion currents, generated by membrane differences that set up the ionic leak-ionic pump mechanism, are as common and widespread in animals as in plants. Therefore, polarized gradients as well as shifting balances between independent processes of gene activation, protein synthesis, cell enlargement and cell proliferation affect patterns of growth and differentiation in animals much as they do in plants.

The situation in viruses, bacteria and unicellular eukaryotes appears to be particularly clear. In these forms, characteristics of cell membranes that affect cellular adhesion in Metazoa and intercellular communication in all multicellular organisms are, of course, irrelevant. On the other hand, evolution of outward form, that may be of minor importance, as in most bacteria; or highly elaborate and functionally significant, as in ciliate Protozoa, depends exclusively upon mutation-based alterations in the structure of those proteins that produce the cortical features of their outer cell walls. The only function of genes in the morphological evolution of these organisms is to produce proteins having specific structural characteristics that permit them to carry out, without further genic influence, elaborate sequences of self assembly. The topic is well reviewed by Fulton and Klein (1976). In my opinion, morphogenesis in all organisms will break out from the constraints that have surrounded it for decades, and make progress toward final answers, to the extent that experiments can be designed that will reveal the structure, genetic basis, mutational capacity and cellular functions of these particular proteins.

IMPACT OF THE HYPOTHESES EXPRESSED UPON RATES AND DIRECTIONS OF EVOLUTION

If the hypotheses proposed here are correct, they will require drastic revisions of current concepts about structural genes and regulatory genes or elements. An often repeated suggestion in current literature on development and evolution is that "molecular evolution" progresses by establishment of mutations of structural genes that code for cellular proteins, while the evolution of form depends upon mutations in "regulatory genes" or "regulatory elements." To a geneticist trained in the molecular discipline as exemplified by bacteria, viruses and yeasts, the term regulatory genes connotes those genes that, in response to either environmental or internal signals, determine the timing and location of the action of structural genes. I have suggested that the genes that regulate form and pattern, like those that regulate cellular metabolism, code also for particular proteins. With respect to both kinds of genes, regulatory genes (or "modifier genes," to use an older terminology) can determine timing and location of gene action. The difference between the two categories of genes lies not in the manner in which they act, but in the kinds of proteins for which they code. Structural genes, as they are usually recognized, code for enzymes and other globular proteins that control metabolic processes, chiefly in cells that are fully differentiated. Those genes that affect form and patterning code for proteins that form the structure of the cells themselves, particularly elements of the cytoskeleton membranes and out surfaces. Some genes affect form by coding for enzymes that catalyze the synthesis of non-proteinaceous macromolecules that are complexed with proteins in cell walls and cell membranes, such as lipids and the polymerized carbohydrates of which plant cell walls largely consist. Still other genes could have indirect effects on form and pattern via an enzymatic role in the synthesis of hormones and other growth substances, as well as of the proteins that are often complexed with hormones and direct their transport through the organism.

The two principal categories of proteins just mentioned, metabolic and morphogenetic, are by no means mutually exclusive. They represent opposite ends of a more or less continuous spectrum, as Goodwin (1984) has stated. For instance, genes that code for enzymes necessary for the synthesis of growth-regulating substances, such as hormones, clearly have morphogenetic effects. In plants and fungi, that synthesize most of the amino acids incorporated into their proteins, mutations of genes that code for enzymes catalyzing steps in amino acid synthesis will affect the presence and abundance of all kinds of proteins, and so will alter both cellular metabolism and morphogenesis. Nevertheless, the distinction made above serves to point out profound differences in pathways from gene to character with respect to genes that code for proteins having very different functions.

One of the major barriers to testing the present hypothesis is that most of the proteins just mentioned are extremely difficult to isolate and analyze. Nevertheless, some progress has been made in this direction (Edelman, 1983; Hoffman et al., 1984) and further progress may be made in the near future. Meanwhile, developmental geneticists should keep their minds open with respect to the older vs. the present concept of structural and regulatory elements in morphogenesis.

The evolutionary implications of the two hypotheses are radically different. Wilson and his associates (Wilson, 1975; Cherry et al., 1978; Maxson and Wilson, 1979) have shown that two different classes of vertebrates, mammals and amphibia, differ greatly from each other with respect to the relative rates of "molecular" vs. "anatomical" evolution. Kimura (1983) maintains that "molecular" evolution consists largely of the chance establishment of mutations that are either neutral or nearly so with respect to adaptation. He nevertheless recognizes that many differences in phenotype are differentially adaptive and have become established via Darwinian natural selection. In his view, the neutrality theory of "molecular evolution" is not a substitute for but a supplement to the traditional hypothesis of Darwin, as modified by the modern synthetic theory.

If the present hypothesis is correct, the dichotomy between "molecular" and "organismal" or "phenotypic" evolution is a false one. All evolution includes a molecular element, a cellular element, and often an element of selection or sorting out that, along with chance events particularly in small populations, affects directions and rates of evolution. In addition the evolution of organismic form and pattern includes distinctive principles and patterns. As I have pointed out (Stebbins, 1983), effects at different hierarchical levels are often independent of each other. Overall evolutionary rates, and to a lesser extent even directions, are a mosaic of different individual rates and trends, only a few of which have been followed by any particular observer.

In my opinion, the dichotomy that Wilson and his associates have emphasized exists because the majority of mutations of genes that code for enzymes and other globular proteins that participate in cellular metabolism have only indirect effects on the adaptations of organisms, and many of them are essentially neutral. They are, therefore, buffered against the diversity of environments and environmental changes that organisms face during evolutionary time spans, so that successive mutations can become established with greater regularity. On the other hand, membrane and cell wall proteins form the shield between the organism and its environment, and participate directly in organism-environment interactions. Hence mutations of genes that code for these proteins are much more likely to alter organism-environment interactions, and hence to have relatively strong positive or negative adaptive values. The rates of evolution of these proteins and the structures for which they are responsible are more likely to be diverse, and to affect strongly such organism-environment interactions as animal behavior.

We do not know what proportion of the total number of different kinds of proteins that make up the organism in many-celled animals and plants belong to the cell metabolism category as compared to membranes, walls and growth regulators. An educated guess can be based upon the encyclopedic lists of gene loci and their mutants by Lindsley and Grell (1968) and by Treat-Clemons and Doane (1984). Of the 887 gene loci recorded, 600 include mutants that affect size, form and/or pattern and structure, while at 287 loci the alternate alleles affect enzymes or pigments that are synthesized by enzyme action. These figures suggest that if it were technically possible to analyze the action of morphogenetic genes, allelic differences that affect non-enzymatic proteins would prove to be more common than those that code for differences in enzymes and other proteins that are primarily concerned with cellular metabolism. At present, we have no basis for deciding whether neutral or nearly neutral allelic differences would prove to be as common at loci that affect morphogenesis as at those that affect cellular metabolism. Nevertheless, since about one third of the morphogenetic mutants recorded by Lindsley and Grell (183 out of 600) produce side effects of partial sterility or lethality, a reasonable probability exists that the proportion of neutral allelic differences would prove to be lower among loci that concern morphogenesis than among those that concern cellular metabolism. Moreover, since correct, adaptive self assembly requires close structural correspondence between several different kinds of proteins, we might expect to find a greater amount of epistatic interaction among genes that affect morphogenesis, and so the shifting balance theory of evolutionary change would apply more strongly to alterations of morphogenetic pattern than to alterations of cellular metabolism.

Given the hundreds of gene loci that affect morphogenesis, and the hundreds, thousands or tens of thousands of generations that, according to the fossil record, are usually required for changes even at the level of species, even if only a small percentage of mutations at morphogenetic loci could increase fitness or aid populations to become adapted to changing environments, the genetic resources would be present to enable natural selection to bring about the changes that the fossil record and the diversity of modern organisms demand. The synthetic theory that was current during the middle of this century will undoubtedly require drastic modification. Its nature will become evident when the present newly found welter of variability in populations at genic and chromosome fine structure levels becomes integrated with developmental information that should be the major task of innovative evolutionists during the next few decades. Nevertheless, the basic tenets of this theory are firmly established. Evolution progresses to a large extent via modification of adaptive syndromes that are controlled by harmonious interactions between many genes. These shifts require shifting balances of gene action that are produced by natural selection of new mutants and chromosomal changes of certain particular kinds. Evolutionary theory will be accommodated to modern facts not by wholesale revolutionary rejections and substitutions, but by carefully evaluated changes of basic tenets that already exist.

I am much indebted to my colleagues, L. D. Gottlieb and R. Nuccitelli for helpful comments. Nevertheless, all ideas and hypotheses expressed are strictly my own.

REFERENCES

Basile, D. V., 1969, An experimental approach to the systematics and phylogeny of leafy liverworts, _in_: "Current Topics in Botany: Symp. Torr. Bot. Club Centennial," J. E. Gunckel, ed., Academic Press, New York, pp. 120-133.

Basile, D. V., and Basile, M. R., 1984, Probing the evolutionary history of Bryophytes experimentally, _J. Hattori. Bot. Lab._, 38:91-98.

Brown, D. L., Stearns, M. E., and Macrae, T. H., 1982, Microtubule organizing centres, _in_: "The Cytoskeleton in Plant Growth and Development," C. W. Lloyd, ed., Academic Press, London, pp. 55-83.

Cherry, L. M., Case, S. M., and Wilson, A. C., 1978, Frog perspective on the morphological difference between humans and chimpanzees, _Science_, 200:209-211.

Clausen, J., 1951, "Stages in the Evolution of Plant Species," Cornell Univ. Press, Ithaca.

Cutter, E. G., 1971, "Plant Anatomy: Experiment and Interpretation. Part 2. Organs," Addison-Wesley, Reading.

Edelman, G. M., 1983, Cell-adhesion molecules, _Science_, 219:450-457.

French, J. C., and Paolillo, D. J., Jr., 1975, The effect of the calyptra on the plane of guard cell mother cell division in *Funaria* and *Physcomitrium* capsules, *Ann. Bot.*, 39:233-236.
Fulton, C., and Klein, A. O., 1976, "Explorations in Developmental Biology," Harvard, Cambridge, MA.
Goodwin, B. C., 1984, A relational or field theory of reproduction and its evolutionary implications, *in*: "Beyond Neo-Darwinism," M.-W. Ho and P. T. Saunders, eds., Academic Press, London, pp. 219-241.
Gottlieb, L. D., 1984, Genetics and morphological evolution in plants, *Amer. Nat.*, 123:681-709.
Gould, S. J., 1977, "Ontogeny and Phylogeny," Harvard Univ. Press, Cambridge.
Grant, V., 1971, "Plant Speciation," Columbia Press, New York.
Green, P. B., 1963, On mechanisms of elongation, *in*: "Cytodifferentiation and Macromolecular Synthesis: 21st Symp. Soc. Devel. and Growth," M. Locke, ed., Academic Press, New York, pp. 203-231.
Green, P. B., 1980, Organogenesis. A biophysical view, *Ann. Rev. Plant Physiol.*, 31:51-82.
Gunning, E. S., 1982, The cytokinetic apparatus: its development and spatial regulation, *in*: "The Cytoskeleton in Plant Growth and Development," C. W. Lloyd, ed., Academic Press, London, pp. 229-292.
Hardham, A. R., 1982, Regulation of polarity in tissues and organs, *in*: "The Cytoskeleton in Plant Growth and Development," C. W. Lloyd, ed., Academic Press, London, pp. 377-403.
Hiorth, G., 1942, Zur Genetik und Systematik der *amoena*-Gruppe der Gattung *Godetia*, *Zeitschr. Ind. Abst.-u. Vererbungsl.*, 80:289-349.
Hoffman, S., Chuong, C.-M., and Edelman, G. M., 1984, Evolutionary conservation of key structures and binding functions of neural cell adhesion molecules, *Proc. Nat. Acad. Sci. U.S.A.*, 81: 6881-6885.
Hyams, J. S., 1982, Microtubules, *in*: "The Cytoskeleton in Plant Growth and Development," C. W. Lloyd, ed., Academic Press, London, pp. 32-53.
Jaffe, L. F., 1981, The role of ionic currents in establishing developmental pattern, *Phil. Trans. Roy. Soc. London B*, 295: 553-566.
Jones, R. L., and Phillips, I. D. J., 1966, Organs of gibberellin synthesis in light grown sunflower plants, *Plant Physiol.*, 41: 1381-1386.
Kimura, M., 1983, "The Neutral Theory of Molecular Evolution," Cambridge Univ. Press, Cambridge.
Kny, L., 1902, Uber den Einfluss von Zug und Druck auf die Richtung der Scheidewande in sich theilenden Pflanzenzellen, (Zweite Mittheilung), *Jahrb. Wiss. Bot.*, 37:55-98.
Lindsley, D. L., and Grell, E. H., 1968, "Genetic Variations of *Drosophila melanogaster*," Carnegie Inst. Wash. Publ. no. 627.

Lloyd, C. W., and Barlow, P. W., 1982, The co-ordination of cell division and elongation: The role of the cytoskeleton, *in*: The Cytoskeleton in Plant Growth and Development," C. W. Lloyd, ed., Academic Press, London, pp. 203-228.

Marchant, H. J., 1982, The establishment and maintenance of plant cell shape by microtubules, *in*: "The Cytoskeleton in Plant Growth and Development," C. W. Lloyd, ed., Academic Press, London, pp. 295-319.

Maxson, L. E. R. and Wilson, A. C., 1979, Rates of molecular and chromosome evolution in salamanders, *Evolution*, 33:734-740.

McDonald, J. F., 1983, The molecular basis of adaptation: a critical review of relevant ideas and observations, *Ann. Rev. Ecol. Syst.*, 14:77-103.

Miller, J. H., 1980, Orientation of the plane of cell division in fern gametophytes: the roles of cell shape and stress, *Amer. J. Bot.*, 67:534-542.

Montezinos, D., 1982, The role of the plasma membrane in cellulose microfibril assembly, *in*: "The Cytoskeleton in Plant Growth and Development," C. W. Lloyd, ed., Academic Press, London, New York, pp. 147-162.

Nuccitelli, R., 1983, Transcellular ion currents: signals and effectors of cell polarity, *Modern Cell Biol.*, 2:451-481.

Raff, R. A., and Kaufman, T. C., 1983, "Embryos, Genes and Evolution," McMillan, New York.

Rayle, D. L., and Cleland, R., 1977, Control of plant cell development by hydrogen ions, *in*: "Current Topics in Developmental Biology. Vol. II. Pattern Development," A. Moscona and A. Monroy, eds., Academic Press, New York, pp. 187-214.

Robinson, D. G., and Quader, H., 1982, The microtubule-microfibril syndrome, *in*: "The Cytoskeleton in Plant Growth and Development," C. W. Lloyd, ed., Academic Press, London, pp. 109-126.

Rosenblum, I. M. and Basile, D. V., 1984, Hormonal regulation of morphogenesis in *Streptocarpus* and its relevance to evolutionary history of Gesneriaceae, *Amer. J. Bot.*, 71:52-64.

Sinnott, E. W., 1960, "Plant Morphogenesis," McGraw-Hill, New York.

Smith, A. T., and Stebbins, G. L., 1971, A morphological study of the tomato mutant "curl," *Amer. J. Bot.*, 58:517-524.

Stebbins, G. L., 1974, "Flowering Plants: Evolution Above the Species Level," Harvard Univ. Press, Cambridge.

Stebbins, G. L., 1983, Mosaic evolution: an integrating principle for the modern synthesis, *Experientia*, 39:823-834.

Stebbins, G. L., and Price, H. J., 1971, The developmental genetics of the calcaroides gene in barley. I. Divergent expression at the morphological and histological level, *Genetics*, 68:527-538.

Stebbins, G. L., and Yagil, E., 1966, The morphogenetic effects of the hooded gene in barley. I. The course of development in hooded and awned genotypes, *Genetics*, 54:727-741.

Treat-Clemons, L. G., and Doane, W. W., 1984, Biochemical loci of the "fruitfly" *Drosophila melanogaster*, *in*: "Genetic Maps, Vol. 3,"

S. H. O'Brien, ed., Cold Spring Harbor Laboratory, New York, pp. 309-323.

Tsuchiya, T., Hayashi, J., and Takahashi, R., 1960, Genetic studies in trisomic barley, II. Further studies on the relationships between trisomics and the genetic linkage groups, Jap. J. Genet., 35:153-160.

Turner, J., Johnson, M., and Eanes, W., 1979, Contrasted modes of evolution in the same genome: allozymes and adaptive change in Heliconius, Proc, Nat. Acad. Sci. U.S.A., 76:1924-1928.

Valentine, J. W., and Campbell, C. A., 1975, Genetic regulation and the fossil record, Amer. Sci., 63:673-680.

Weisenseel, M. H., Dorn, A., and Jaffe, L. F., 1979, Natural H^+ currents traverse growing roots and root hairs of barley (Hordeum vulgare L.), Plant Physiol., 64:512-518.

Wilson, A. C., 1975, Evolutionary importance of gene regulation, in: "Stadler Genetic Symp," Univ. Missouri, 7:117-133.

Wilson, A. C., Sarich, V. M., Maxson, L. R., 1974, The importance of gene rearrangement in evolution: evidence from studies on rates of chromosomal, protein and anatomical evolution, Proc. Nat. Acad. Sci. U.S.A., 71:3028-3030.

Yagil, E., and Stebbins, G. L., 1969, The morphogenetic effects of the hooded gene in barley. II. Cytological and environmental factors affecting gene expression, Genetics, 62:307-319.

MOBILE ELEMENTS IN MAIZE: A FORCE IN EVOLUTIONARY AND PLANT BREEDING PROCESSES

Peter A. Peterson

Agronomy Department
Iowa State University
Ames, IA 50011

SUMMARY

Mobile elements in maize _Zea mays_ L. are DNA segments of various sizes that insert into DNA sequences. When a mobile element is inserted into a functional gene, the gene becomes incapacitated by the element interrupting its activity. This interruption in the genes' functional capacity is caused by precocious termination of the transcript or by an early stop signal because it is possible to form a chimeric transcript.

Mobile elements are varied in size and functional capacity and are distinguished as different systems composed of a regulatory and receptor element. A receptor element originates as a defective regulatory element that has lost the capacity to induce excision events. When an element (regulatory or receptor) is inserted, the target-site DNA duplicates a DNA sequence of three or more DNA bases and the size of the element is excised, several events may occur, including the retention of the duplication, a return to wild type, or the alteration of the sequences including the duplication. This alteration in DNA sequences leads to codon changes, or frameshifts in the transcript that have consequence on the protein that is formed from that gene. This would contribute to genetic diversity.

In a canvass for active mobile elements of a limited sample of an important corn breeding population, the BSSS series (Iowa Stiff Stalk Synthetic), two (_Uq_ and _Mrh_) of seven active mobile elements tested for were found. These were present in four of the six cycles of BSSS sampled, though only the C3 cycle had both elements present. The others did not expose the elements, probably due to a limited sample.

It was anticipated that the excessive alterations that were induced by mobile elements in maize will generate the type of diversity leading to genotypes with selective attributes. None of the inbreds tested for had elements indicating that their effects, coupled with breeding protocol, would tend to eliminate mobile element presence in inbred lines.

EVOLUTION AND PLANT BREEDING

Evolutionary processes and plant breeding technology share many similarities. Both are dependent on genetic diversity and selective amplification of favorable variation. Since the pronouncement of Darwin's revolutionary thesis in 1859, there have been ample and continued demonstrations of a plentiful amount of genetic diversity in both domesticated and wild organisms (Darwin, 1868; review by McDonald, 1983). The main basis for a plant breeders existence is dependent on an ever providing and continual input of genetic diversity that is capable of being translated into heritable and selectable phenotypic variation. A plant breeders' main focus has been to provide an effective selection screen that will concentrate and thereby amplify favorable genotypes.

Though the basis and meaning for the origin and persistence of this genetic diversity is still debated by opposing camps (Dover, 1982; Gould, 1982; McDonald, 1983; Sober, 1984) there has been an unequivocal demonstration of a never-emptying cornucopia of variation at the genetic, protein (Shaw, 1965; Hubby and Lewontin, 1966; Lewontin and Hubby, 1966; Perutz, 1983; Stuber and Goodman, 1984; Stuber et al., 1980, 1982) and, currently, at the nucleotide level with the abundance of restriction site polymorphisms (Lucotte et al., 1982; Hudson, 1982; Robinson et al., 1983; Johns et al., 1983; Hudson, 1982; Burr et al., 1983; Files et al., 1983; Robinson et al., 1983), inverted sequences, and inserts that are revealed with almost every scrutiny of DNA fragments (Gupta et al., 1983, 1984) or nucleotide sequence (Schwarz-Sommer et al., 1984, 1985).

Ever since Darwin's observation of the abundance of phenotypic variation in his 1868 book on *The Variation of Plants and Animals Under Domestication*, it has been abundantly clear that genetic diversity is not a limitation in evolutionary change. Nor in plant breeding as is evident from the unlimited selection options available in elite populations (Russell and Vega, 1973; Hallauer et al., 1983). Very likely, the extensive demonstrations of populations subjected to long-term selection in corn (Dudley, 1977) and the almost limitless flexibility in response among both plants and animals of such a unique phenomenon as genetically adapted mimicry (Wickler, 1968) are ample support for the available heterozygosity or the facility of DNA response to selection that is so pervasive.

There are, therefore, several points on this evolutionary debate. How this genetic diversity becomes selectively amplified is not a subject of this presentation. Natural or human selection is an adequate mechanism. Nor will this appraisal be concerned with the molecular processes that drive evolution (Dover, 1982; Ohta, 1984a; Ohta and Dover, 1984) nor of adaptation that represents the kinds of changes that accommodate the genotypes that survive (McDonald, 1983). What will be considered is one way in which this genetic diversity can be generated and its possible role as a major contributor to the pool of genetic diversity. Put another way, if we can simplify this discussion of evolutionary and plant breeding processes into two parts, [namely, (1) the origin of phenotypic variation (genetic diversity) and (2) the selective forces that amplify this variation in a population], then the aim of this paper is confined to part 1. The following discussion and relevant data will be concentrated on this single parameter (part 1)--the pervasiveness of genetic diversity and how one component (namely mobile elements) could contribute to genetic diversity. What fraction of genetic diversity is contributed by mobile elements and is this amount adequate for the occurrence of adaptive evolution (Syvanen, 1984).

RATIONALE: PLANT MOBILE ELEMENTS AND THEIR CHARACTERISTICS

Plant mobile elements, though recognized for their mobility for the last 30 years (McClintock, 1947, 1951; Brink and Nilan, 1952) have been observed as variegated flowers, leaves, and other plant parts (Fig. 1A) by breeders, horticulturalists and naturalists for more than the last 300 years (see bibliography in Nevers, et al. 1985; Peterson, 1985). The role of mobile elements in the genome has been variously assignable to the assorted aspects of a genomes activity. Could it be more than the generalized reactions to the early demonstrations of gene mobility, that of unique aberrant behavior in a geneticist's field nursery? Or could it be playing a vital role in genome processes (Hunkapiller et al., 1983, Honjo, 1983) such as development (McClintock, 1961), or was the mobility of genetic material important in unique but important aspects of multiple

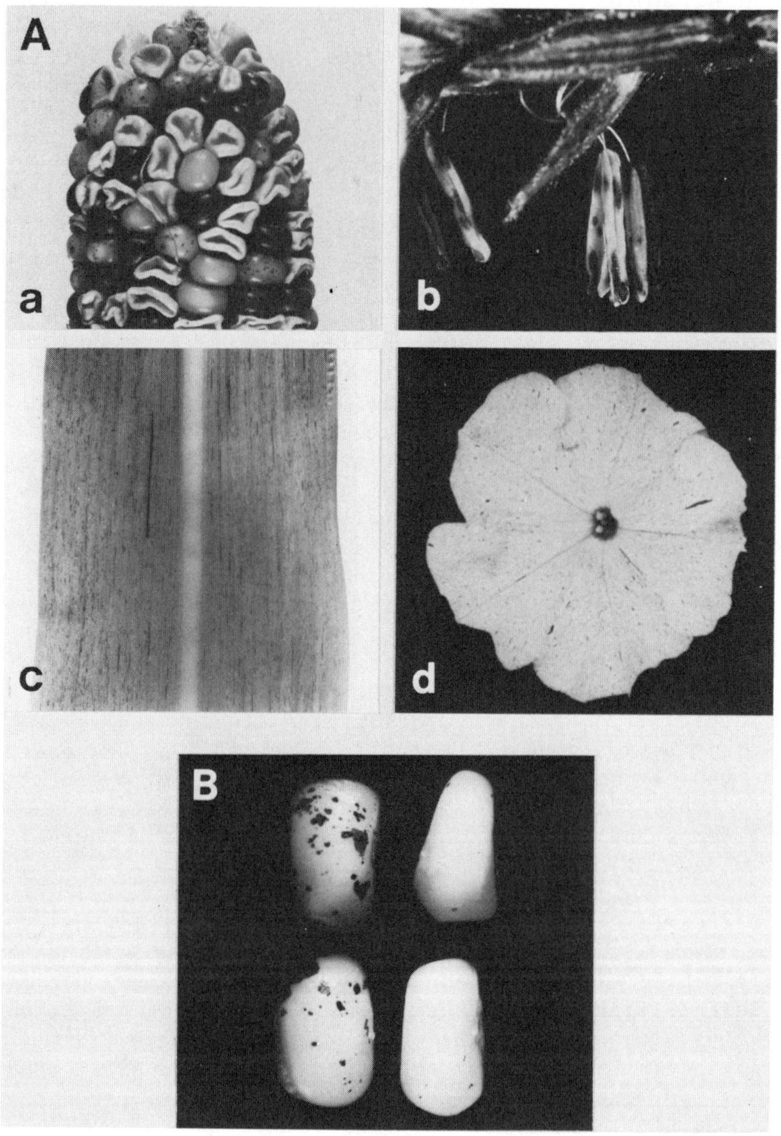

Figure 1. A. An assortment of variegating plant types illustrating variegation among different species. a. Maize ear illustrating instability of the *a-m* alleles (Peterson, 1961); b. *a-m* showing stripping in anther walls (Peterson, 1961); c. *pg*^{m} in maize (Peterson, 1960); d. *an1*$^{2/+}$ allele in *Petunia hybrida* (From A. Gerats) b. Two states of variegated phenotypes in the *En* system in maize.

gene roles such as the immune system or could they be inserts as foreign elements in a gene apparatus (Taylor, 1963; Peterson, 1967, 1970; Jordan et al., 1968). According to a commonly accepted rule, the survival of appendicular structures or genetic components is clearly a function of utility, but Ohta (1984b) has provided evidence and persuasive arguments that mobile elements have a survival role irrespective of positive selection.

Gene segments are mobile. The crossover process (Sturtevant, 1913) is no longer the only means to separate contiguous genes. The transfer of genetic material by the mobility of gene segments has been amply demonstrated in normal processes such as the immune system that reacts to multiple, unknown, and unanticipated invading systems. This has been amply documented.

But what is the role of mobile elements in the hierarchy of programmed development (McClintock, 1951, 1965). The distinct timing of events as depicted by states (Fig. 1B) was illustrative of expressions associated with mobile elements and developmental signals. Distinct turn-on and turn-off of gene expression of mobile element associated mutants (Peterson, 1966) were further in support of signals associated with development. Yet, more telling and convincing support came from enzymatic studies of loci with an assayable enzyme. Such an opportunity came with the studies with the _bz_ locus by Dooner (1981) for the tissue differential enzymatic expression UFGT (UDP-glucose:flavonoid 3-O-glucosyltransferase) in the endosperm by a mobile element. There was a mistiming and misplacement of UFGT expression. Yet, this might be a consequence of unusual structural changes in the vicinity of the gene in question (_Bz_) because, with the removal of this mobile element (_Ds_), the deviant gene action expression persists (Gerats et al., 1983). Yet, molecular support for and against this has not been adequately considered.

It has been well documented that mobile elements are pervasive in diverse and indigeneous genomes. They are present as active forms based on their triggering of receptor elements as realized in genetic tests (Gonella and Peterson, 1975; Salamini, 1981; Friedemann and Peterson, 1982). More significantly, they are extensively distributed as discrete related units as realized from southern blot hybridization using recently acquired probes (Fedoroff et al., 1983; Peacock et al., 1983; Sachs et al., 1983; Schwarz-Sommer et al., 1984; Bonas et al., 1984; Dellaporta et al., 1984; Johns et al., 1985; Hehl et al., 1985). In some cases, quasi-unique DNA components of elements such as _Ac_ will light up 5 to 10 bands in a genomic southern (Fedoroff et al., 1983). In other cases, the same procedures using other probes light up genomic DNA that appears as a variable middle repetitive distributed element. This includes the Cin elements (Shepherd et al., 1982; Gupta et al., 1984; Hehl et al., 1985), Ds (Courage-Tebbe et al., 1983), I (Schwarz-Sommer et al., 1984), Mu-1 (Bennetzen et al., 1984) and En (Pereira et al., 1985). Some of the elements that have not been genetically described have also been found in varied frequencies: Tz with several copies and Bsl with 2-5 copies (Johns et al., 1985). The point of emphasis is that these elements or sequences associated with these elements are widely distributed in corn varieties and have further appeared in corn relatives (Blumberg, et al, 1985) and to other members of the Gramineae (Dellaporta, Cold Spring Harbor, N.Y. pers. comm.) With such a wide distribution and with a dubious application to a vital developmental role, their pervasiveness and general distribution demand some consideration of their survival in plant genomes.

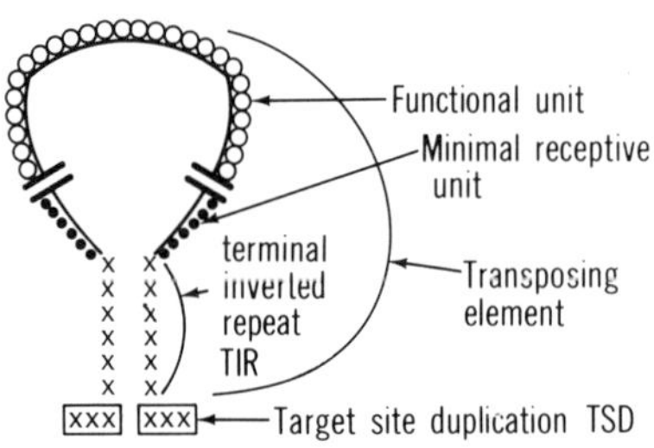

Figure 2. General composite figure of a plant transposable element illustrating several features. TIR = Terminal Inverted Repeat; TSD = Target Site Duplication; TE = Terminal End; Regulatory Element; Minimal Receptive Unit.

The recent revelations of cloned highly mobile elements have revealed several general features that are of significance in any consideration of mobile element induced genetic diversity. The different elements, at least determined to be different by genetic means have several features in common, though they differ in dimensions. A general characterization is

Table 1. General characteristics of some cloned elements in maize and Antirrhinum

	TSD (bp)	TIR (bp)	References
Ac	8	11	Fedoroff et al., 1983
Ds	8	11	Pohlman et al., 1984 Weck et al., 1984
Spm-I8	3	13	Schwarz-Sommer et al., 1984
En-1	3	13	Pereira et al., 1985
Mul	9	215	Bennetzen et al., 1984
Tz86	10	none	Dellaporta et al., 1984
Taml	3	13	Bonas et al., 1984
Cinl-001	5	6	Shepherd et al., 1982, 1984

illustrated in Fig. 2 and Table 1. Each element has the important terminal inverted repeat (**TIR**) and is associated with a target site duplication (**TSD**) that is generated (the duplicate segment) coincident with the insertion event. In essence, mobile elements add nucleotides to the maize genome.

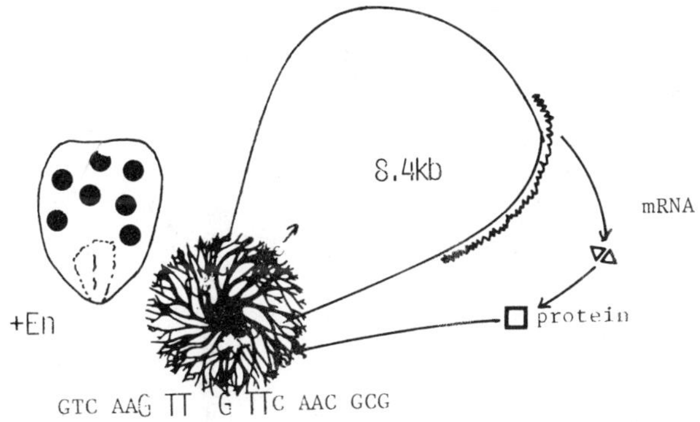

Figure 3. Visualization of excision of the Rg element yielding the spots. The element is shown excising with the TIR intact.

The excision event illustrated in Fig. 3 is visualized genetically as, for example, spots on a colorless background (Fig. 1A, 1B). This is coincident with an excision of the resident mobile element from the gene and is often accompanied

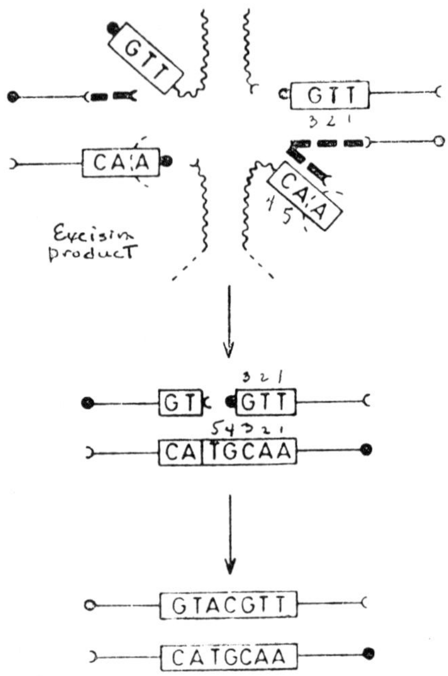

Figure 4. Adaptation of the Saedler-Nevers model of transposon excision leading to alterations in the DNA sequence of the TSD. This illustrates the DNA sequences surrounding the target site immediately following excision of the mobile element. The wavy lines represent the termini of Spm-I8 integrated within the wx gene. Following transposase induced nicking and consequent single stranded ends, some exonuclease degradation occurs (thin dashed half circled). Boxed-in sequences are target site sequences (TSD). Polymerase action (thick dashed line). Numbers refer to the template information of new sequences. 5' end and 3' end of DNA.

by alterations in the host target site duplication because of imprecise excision. This results in a deviation in the DNA sequence that has consequences if this is in a codigenic region that can alter the coding sequence with subsequent alteration in the final protein and, of course, a phenotypic change. The importance of this series of events (excision, alterations in **TSD**) in an evolutionary sense has been considered (Schwarz-Sommer et al., 1985).

In their recent proposal for the role of mobile elements as an evolutionary force, Schwarz-Sommer et al. (1985) have presented evidence based on "footprints" in introns of the wx gene that certain sequences are a result of visitation by transposons. This is based on a model proposed by Saedler and Nevers (1985) that provides an explanation for the origin of changes in the amino acids of proteins from mobile element excisions. This Saedler-Nevers (1985) model describes a process of transposon excision and the ensuing involvement of the target site duplication (**TSD**) leading to alterations in the nucleotides of the **TSD**. In this particular process, as the junction point is left with free ends (Fig. 4), exonucleotide degradation takes place. In the course of this event, a foldback occurs and these nucleotides are copied, leading to altered sequences (Fig. 4).

Another revelation of DNA sequence alteration came from the examination of the first 800 bases of the wx gene. Here, a striking series of direct duplications were found. These included four, five, or six or more nucleotides directly duplicated. All were found in introns. Such duplication sets appeared in a consecutive series in the course of sequencing the waxy gene in a comparison between the wild type and a mutant form. Strikingly, the duplications only appear in introns, and this is realistic because they would not be subjected to the selective process of changes that would occur in exons. The authors do concede that these particular duplications may be a consequence of polymerase slippage.

It would seem therefore axiomatic that, for mobile element induced effects to be consummated, it is necessary to have the following points fulfilled.

(a) that the model is a realistic appraisal for the mutation events considered;

(b) that the elements are present and mobile to impose these changes on the genome; and

(c) that their frequency is such that the events take place in a predictable manner.

PREDICTION ON MOBILE ELEMENT INVOLVEMENT IN GENOME CHANGES

The Number of Elements

Points (b) and (c) on elements and their frequency will be considered.

Table 2. The transposable element systems in maize listed as Regulatory-receptor combinations. Homologous elements are listed in parentheses (Peterson, in press). The receptor alleles (MERL) used in the BSSS series of crosses are identifed with an asterisk*;--indicates no identifiable named receptor

	Regulatory	Receptor element	Receptor allele	References
1	Dt	Dt	a(rDt)*	Rhoades (1936)
2	Ac (Mp)	Ds	CDs*	McClintock (1947); Brink and Nilan (1952)
3	En (Spm)	I	a(mr)*	Peterson (1953); McClintock (1954)
4	Fcu (Spf)	rcu	rcu*	Gonella and Peterson (1977)
5	Uq	ruq	a-ruq* c-ruq*	Friedemann and Peterson (1982) Oberthur and Peterson (1984)
6	Bg	rBg	o2m(r)*	Salamini (1981)
7	Mrh	rMrh	a(rMrh)*	Rhoades and Dempsey (1982)
8	Mut	rMut		Rhoades and Dempsey (1982)
9	c2-m3	--		Peterson (1984)
10	Cy	rcy		Schnable and Peterson (1984)
11	Mul	--		Robertson (1978)
12	M-st	I-R		Kermicle (1984)
13	BS1	--		Johns et al. (1985)
14	Tz	--		Dellaporta et al. (1984)

There are currently 14 mobile element systems under study (Table 2). The complete diallel has not been done, but the distinctiveness of most of the systems (Table 2) has been established (Fig. 5). For purposes of further discussion, it would be assumed that 15 systems exist. And it will be assumed that the 15 systems could all be mobile in a given population.

Approximation of Mobile Elements

In general, the frequency in most tests made to rescue a gene target with an element as an insert is approximately 1 in 250,000 (Table 3). This is a rescue as a zygote developing from a mutant gamete with the insert from a population with one to three active elements. With 10 elements, a particular target

Rcp (columns) × **Rg** (rows)

Rg		a-dt	Ds	Ac-Ds	a-m(r)	a-m1	r-cu	R#10	rug	o2-m(r)	a-mrh	bz-mut	bz-2m	bz-rcy
1	Dt	+	-	-	-	-	-	-	-		-	-	-	-
2	Ac	-	+	+	-	-	-	-	-	-	-	-	+	-
	Mp	-	+	+	-	-	-	-	-	-	-	-	-	-
3	En	-	-		+	+	-	-	-	-	-	-	-	-
	Spm	-	-		+	+	-	-	-					
4	Fcu	-	-		-	-	+	+	-	-	-	-	-	
	Spf	-	-		-	-	-	+	-					
5	Uq	-	-		-	-	-	-	+	-				-
6	Bg		-	-	-	-			-	+	-			
7	Mrh	-	-		-		-			-	+	-	-	
	Ac2	-	-		-						-	-	+	
8	Mut	-	-		-						-	+	-	
9	c2-m3	-	-		-	-	-		-	-	-			
10	Cy	-	-		-	-			-		-			+

Figure 5. A generalized scheme identifying the homology or non-homology of a mobile element system in maize; + = mutability as illustrated in Figure 1; - = no mutability.

Table 3. The frequency of incidence of instability of specific target loci induced in strains of maize carrying transposons in tested populations

mutants	System established	Plot	#	Rates x 10^6	References
a2	En	En	25	2.4-11.9	Peterson (1978)
c	En	En	47	1.5-9.4	Peterson (1978)
c	Uq	Uq	4	2.2	Oberthur and Peterson (1984)
c2	not known	En	5	4.0	Peterson (1983)
c2	not known	En	4	2.5	Peterson (1984)
bz	En	En	1	1.7	Peterson (1983)
wx	not known	En	5	1.2	Peterson and Cormack (1984)
wx	En wx84-4	En	1	.53	Peterson (1985b)
wx	En wx84-11	En	1	2.6	Peterson (1985b)
wx	not known	Mul	7	4.0	Robertson and Stinard (1984)
yl	not known	Mul	412	170.0	Robertson and Stinard (1984)

hit would be 1 in 25,000. In most tests, that gene target would be exposed when any element inserts so that the effective rate is still 1/250,000. If there are 100,000 genes, and there were 10 elements present, then four genes would gain an insert in a population of gametes. Though it would be difficult to recognize these new sites for inserts genetically unless they were all dominant mutations, such changes would be recognized as shifts in southern hybridization blots. Generalized shifts were seen in the recovery of nivea revertants in Antirrhinum majus by Saedler et al. (1983) that could indicate the massive movement of elements by a generalized derepression.

Of course, the assumption of one in three genes being visited by an insert in every population of gametes is very approximate. Any one of the parameters could be an over- or under-estimate. These include the number of elements, the activity of the specific element, target hit frequency, the size of the target site duplication, and number of genes.

THE MOBILITY PROCESS--EXCISION--TRANSPOSITION-INSERTION--EXCISION, ETC.

When the DNA sequence of the target-site duplication (**TSD**) of the excised element has been scrutinized after excision such as of a revertant (ex a-m to A^{rev}), a residue of the **TSD**, changed or unchanged, is left behind as the insertion transposes to a new site (Fig. 4). The alteration often leads to a diminution of the **TSD** or to an inversion or other type of alteration. It is these changes that evolved from an original sequence that duplicated (the **TSD**) following the insertion and, in turn, that became a "medium" for a new sequence organization following the excision event. It is this alteration in DNA sequences that could generate into a new codon and that can be translated to an altered protein. This can then become the substance on which evolutionary pressures could be selective. Certainly the abundance of multiple alleles at so many loci in maize could be an outward expression of such changes (Stuber and Goodman, 1984).

MATERIALS AND METHODS

1. LINE DEVELOPMENT--BSSS SERIES

1.1 Development of BSSSCO

The Iowa Stiff Stalk Synthetic population was developed by intercrossing 16 lines, each of which contained several desirable attributes (see, Hallauer et al., 1983). It was the

SC	1 x 2 3 x 4 5 x 6 7 x 8
	9 x 10 11 x 12 13 x 14 15 x 16
DC	[(1 x 2) x (3 x 4)][a] [(5 x 6) (7 x 8)][b]
	[(9 x 10) x (11 x 12)][c] [(13 x 14) x (15 x 16)][d]
4-way cross	(a x b) x (c x d)

Figure 6. Development of the original BSSSCO breeding population in the Iowa State University corn breeding program. Generalized crossing program of SC Single Cross, DC Double Cross, 4-way Cross.

purpose of the originators of this crossing scheme to develop a population that would combine the desirable attributes of all the lines that could readily be "siphoned off" in a breeding protocol (Sprague, 1946). The lines included in the development of this composite were IaI159, IaI224, Ia05420, IaWD456, Ind.461-3, Ill.12E, CI.617, CI540, Ill.Hy, OH3167B, Ind.AH83, Ind.Tr9-1-1-6, F1B1-7-1, A3G-3-1-3, CI187-2, LE23 (Hallauer et al., 1983) and will be designated 1, 2, 3, 4, etc. as a general categorical identification in Fig. 6. The intercrossing was manipulated by making 8 single crosses such as 1 x 2, 3 x 4, 5 x 6, etc. (Fig. 6). This was followed by four double crosses such as (1 x 2) x (3 x 4), (5 x 6) x (7 x 8), etc. In the final crossing two different 4-way crosses [(1 x 2) x (3 x 4)] x [(5 x 6) x (7 x 8)] and [(9 x 10) x (11 x 12)] x [(13 x 14) x (15 x 16)] were made. The bulked seed of these final 4-way crosses yielded the BSSS population and was identified as BSSSCO.

1.2 BSSSC1

To develop the BSSSC1 population, 169 BSSSCO plants were crossed to the unrelated line, Ial3, in a test of their performance capability and they were also selfed (Fig. 7). On the basis of the Ial3 test performance, the 10 best S1 lines were saved and combined in a diallel that yielded 45 crosses (Fig. 7 and 8). An equal amount of bulked seed of each cross was added into a population that made up the BSSSC1 population. (Note that the original 169 plants were reduced to 10.)

```
                       16 lines

                        ↓    Combined via 8 SC, 4 DC and 2 4-way cross

BSSS                   B'SSCO                    BSSSCO → B14, B37
                        ↓
                       167 plants testcrossed (TC) on Ia13
Cycle                   ↓                        ↓
Co to C1               167 So                  Populations evaluated at locations
                        ↓
                       10 S1 selected as best performers
                        ↓
                   10 x 10 diallel  gave 45 single crosses → equal quantities of each
                        ↓                                    cross contributed to bulk
                                                             seed = BSSSC1

               Bulk seed random mating (RM) in rotation ⟶ = BSSSC1
C1 to C2
               The BSSSS1 → So → 10 S1 → 10 x 10 diallel  - bulk = BSSSC2
                          ↘ TC on IA13

               After C2 - RM eliminated                           RM →
etc.                  →   etc. to BSSSC7                          BSSSC5  → B73
                                                                  BSSSC7  → B84

               BSSSC7  → So → 10 S1 → 10 x 10 diallel  - bulk = BS13(S)C1 = BS13CO

               BSSCO → So → 10 S1 → 10 x 10 diallel  - bulk = BSBC1

               etc.
```

Figure 7. General diagram of the origin of the various derivatives of the Iowa Stiff Stalk Synthetic population (BSSS-) (Hallauer et al., 1983; Penny and Eberhart, 1971; Eberhart, Debela and Hallauer, 197?). 139, 108, 84, 91, 86, and 90 plants evaluated between C1 and C6, respectively. SC, DC-single cross, double cross; TC-Test Cross; So (self) yields S1; RM - random mating; C1, C2 etc. represent the various cycles.

1.3 BSSSC2, C3, C4, C5, C6, and C7

So (beginning of selfing that yielded the next generation, S1) plants of BSSSC1 were similarly tested against Ia13. Selfing yielded S1 plants, and again, the 10 best performers in the Ia13 test were chosen to make up a diallel, again yielding 45 crosses (Fig. 7 and 8) that when bulked gave rise to the BSSSC2 population. This protocol was repeated for the BSSSC3 to BSSSC7 populations (Fig. 7).

	1	2	3	4	5	6	7	8	9	10
1		x	x	x	x	x	x	x	x	x
2			x	x	x	x	x	x	x	x
3				x	x	x	x	x	x	x
4					x	x	x	x	x	x
5						x	x	x	x	x
6							x	x	x	x
7								x	x	x
8									x	x
9										x
10										

Figure 8. The diallel used in the BSSS series to combine genotypes in order to progress to the next cycle. Equal amounts of seed from each cross were combined in bulk to yield the next population cycle C1, C2, C3, etc.

1.4 BS13

The next step in the progression of the BSSS series is the origin of the BS13 population series. Because of the development of these lines via the testing with Ia13, they were then termed BS13(S)C0, which is equal to BSSSC7 (Fig. 7). The nomenclature changed from BSSSC7 to BS13(S)C0. This was the next step following BSSS(Ia13)C7 in the cycling protocol (see Fig. 7).

1.5 Genome shrinkage

Note that, in the series BSSSC0 to BS13(S)C2, the sampling and selection of individual plants followed the seriation-169-10-(10x10)-Bulk-139-10-(10x10)-Bulk-108-10(10x10), etc. The reduction of the sample size to 10 following the Ia13 selection in each cycle would reduce the sample to 6 to 10% in each cycle and in five cycles, this shrinkage would be approximately 10 percent to the fifth power. With no input of genomic change,

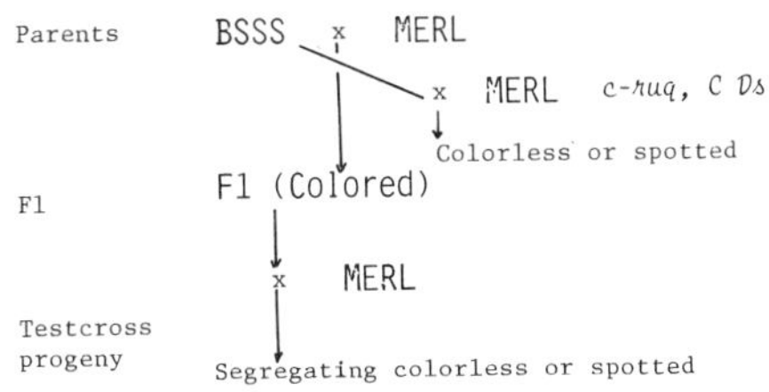

Figure 9. Crossing scheme used to uncover regulatory elements in BSSS lines in crosses to Mobile Element Receptor lines (MERL) (Tables 2 and 3).

this concentrates the genome approximately $.10^{-5}$ of the original starting genomic content.

1.6 Inbreds

The three inbreds used included B14, B37, and B84. The inbreds are promising lines selected out of the various cycles (Fig. 7), and these are made uniform by selfing for approximately six generations.

2. The BSSS lines used for the presence of active regulatory elements

Seed obtained from W. A. Russell (Iowa State University, Ames) included the following populations and inbreds: BSSSC0, BSSSC1, BSSSC3, BSSSC5, BSSSC7, BS13(S)C2, B14, B37, B73, B84. Twelve seed of each of these lines were planted in rows. Plants in these rows were then crossed by Mobile Element Receptor Lines (MERL) (Peterson, 1981). Attempts were made to cross two plants from each row with each of the MERL. Thus, these crosses tested two, four or six gametes from each of the BSSS lines with each MERL (Table 4).

3. Mobile Element Receptor Lines (MERL) and the Assay for Regulatory Elements

3.1 MERL x BSSS Series

To test for the presence of regulatory elements in the BSSS series, crosses were made with the MERL lines (Table 2), and are listed in Table 4.

Table 4. The distribution of active Regulatory Mobile Elements (Dt, Ac, En(Spm), Fcu, Uq, Mrh, Bg) as deduced from tests on receptor elements in source populations and derivative inbreds. Under F_2 progenies is shown the number of progenies with segregating spotted kernels over the total number of F_2 progenies. ++ = heavy mutability. Numbers refer to number of plants tested

No. of plants tested and the distribution of F_2 progenies with elements is given for Uq and Mrh.

	Dt	Ac	En (Spm)	Fcu	Uq	#	F_2 prog.	Mrh	#	F_2 prog.	Bg
A. 2 to 4 gametes of BSSS series tested.											
BSSSC0	0	0	0	0	0	2		0	2		0
BSSSC1	0	0	0	0	0	2		++	1	4/8	0
BSSSC3	0	0	0	0	++	2	7/8,7/11	++	1	9/11	0
BSSSC5	0	0	0	0	++	1	4/10	0	2		0
BSSSC7	0	0	0	0	0	1		0	2		0
BS13(S)C2	0	0	0	0	++	2	2/12,2/11	0	2		0
B14	0	0	0	0	0	2		0	2		0
B37	0	0	0	0	0	2		0	2		0
B84	0	0	0	0	0	2		0	2		0
B. 4-6 gametes, tested directly.											
BSSSC0		0			0	6					
BSSSC1		0			0	3					
BSSSC3		0			++	3	1/3				
BSSSC5		0			0	3					
BSSSC7		0			0	3					
BS13(S)C2		0			++	3	3/3				
B14		0			0	3					
B37		0			0	3					
B84		0			0	3					

C. Summary of no. of plants with elements among totals tested.

	Uq		Mrh	
	Total	With	Total	With
BSSSC1	5	0	1	1
BSSSC3	5	3	1	1
BSSSC5	4	1	2	0
BS13(S)C2	5	5	2	0

	Ac	En
Ds	+	-
I	-	+

Figure 10. The specificity of interaction of receptor and regulatory elements. *Ac* is specific for *Ds* and *En* is specific for *I*. + = mutability and - = no mutability.

The colored F1s were handled by selfing or backcrossing to the same MERL or a recessive line to uncover the specific recessive used with MERL. The MERL contained receptor elements at the loci indicated *dt* (*a-dt*); *Ds* (*CDs*); *I* (*a-m-1* and *a-m(r)*); *cu* (*rcu*); *ruq* (*a-ruq*, *c-ruq*); *bg* (*o2m[r]*); *mrh*, (*a-mrh*); *rcy* (*bz-rcy*).

3.2 *Assay for Active Regulatory Elements in the BSSS Series*

The exposure of any MERL to an active Regulatory element will trigger a spotting pattern in the resulting progeny (Fig. 9). The cross MERL line x BSSS line gave a colored F1 (Fig. 9). Because BSSS is *cc rr* and therefore lacks anthocyanin color, a direct reading of active elements could be obtained for the *C Ds c-m(r)* and the *c-ruq* lines in some of the crosses (Fig. 9). These are noted in the results. A typical example of the specificity of interaction is illustrated in Fig. 10.

THE DISTRIBUTION OF ACTIVE REGULATORY ELEMENTS IN BSSS AND BSSS DERIVATIVE POPULATIONS

In a canvass of 10 BSSS and BSSS derivative populations for the presence of active mobile elements, only four of the BSSS series showed active mobile elements. These were BSSSC1, BSSSC3, BSSSC5, and BS13S(C2) (Table 4). Only two of the seven elements tested were present in these lines. These were *Uq* (Friedemann and Peterson, 1982) and *Mrh* (Rhoades and Dempsey,

Table 5. The distribution of active Regulatory Mobile elements as deduced from tests on receptor elements in tribal and commercial varieties or lines of corn

	Active Regulatory Elements				
	Dt	Ac	En(Spm)	Fcu	Uq
A. Varieties-Tribal					
Quaw Paw (PI 213757)	-	-	++	-	-
KIOWA (PI 213753)	-	-	++	-	-
Cuna	+	+	++	++	-
B. Varieties (lines)** Commercial***					
A619-Plot 98	-	0	0	-	0
A619-Plot 100	-	0	0	-	0
A619-Plot 103	-	-	0	-	0
B68-Plot 85	-	0	0	-	0
B68-Plot 87	-	0	0	-	0
B68-Plot 90	-	0	0	-	0
Mo17-Plot 437	-	0	0	-	0
Mo17-Plot 106	-	0	0	-	0
Mo17-Plot 109	-	0	0	-	0
OH43-Plot 110	-	0	0	-	0
OH43-Plot 112	-	0	0	-	0
OH43-Plot 115	-	0	0	-	0
WF9-Plot 116	-	0	0	-	0
WF9-Plot 118	-	0	0	-	0
WF9-Plot 121	-	0	0	-	0
M14-Plot 122	-	0	0	-	0
M14-Plot 124	-	0	0	-	0
M14-Plot 127	-	0	0	-	0

- = not tested
0 = not present
++ = present-very active
+ = present-weakly active
** Bogenschutz, 1984
*** Cy, Mrh are currently being evaluated

1982). The other elements were not present in an active form in these lines.

There are several interesting features to be considered. Note that there was only a small number of gametes tested in the A set (Table 4). Despite this, the plants sampled for the BSSSC3 group appeared to be either homozygous for Uq or there were more than two nonallelic copies of the element with nearly all the progeny tested showing the element. However, in the B set (B set - same population source as the A set) where the presence of the element could be tested directly, only one of three plants had the element (Table 8), attesting to the heterogeneity of this group of plants. The BSSSC5 group was also heterogeneous.

The particular BSSSC5 plant that was tested was heterozygous in the A set, and none showed Uq in the B set. Table 4).

The BS13(S)C2 population was surprising. This was especially noteworthy because the BS13(S)C2 population was a selective group out of six previous cycles of selection (Fig. 7). (Note the genome shrinkage discussed in the M and M, section 1.5.) Here, in BS13(S)C2, all the plants had Uq though it was heterozygous in each instance (Table 4, B and C).

The Mrh element was present in both BSSSC1 and BSSSC3. In the plant of the BSSSC1 series it was heterozygous, and in the BSSSC3, it was obviously homozygous and, again, possibly contained more than two copies. This element was not present in the latter series.

Viewed the other way, BSSSC3 had both Uq and Mrh. (The manner of crossing with the use of the female as the BSSS plant did not allow for the determination of the presence of the different elements in each plant, though these questions can be answered with the studies in progress.)

Given the very limited sampling, it was surprising that the Uq and Mrh elements were present in the lines tested, which attests to the pervasiveness of the elements in the lines. This, despite the absence of the element in equal samples of lines that appeared in the selected population earlier in the cycling process.

The other interesting feature was that the inbred lines that developed out of these population cycles did not themselves contain any elements of those tested. This includes B14, B37, and B84. In these lines Uq was canvassed in the B set (Table

4), with more than a thousand kernels of the population analyzed which consisted of two or three plants but represented a direct test of that same population that composed the A set.

The original BSSS set (CO) was not extensively tested; however, neither of the two regulatory elements (Uq and Mrh) were present. Extensive testing is necessary to assure a more adequate sampling for the distribution of active elements.

THE DISTRIBUTION OF ACTIVE REGULATORY ELEMENTS IN ASSORTED VARIETIES AND LINES

Two tribal varieties, Quaw Paw and KIOWA (Table 5A) from the continental United States were tested for En by using the receptor allele a-m(r) (Peterson, 1961). Both lines contained En and the segregation pattern indicates that these varieties either had a large number of copies or it was homozygous (En/+, En-2/+, En-3/+ could appear like En-1/En-1).

The Cuna tribal corn (N.W. Colombia) was the original source of the Fcu system (Gonella and Peterson, 1977). In addition to the Fcu-rcu elements, this tribal corn also included a weak Dt and a weak En element (Gonella and Peterson, 1977) (weak, in that there was a low frequency of spots). In further studies with the weak En (few spots), it was possible to isolate a stronger-acting En (higher frequency of spots, very likely from a change in state).

In a survey of 23 maize varieties from Mexico, Colombia, Bolivia and Venezuela, eight were confirmed to have En. Other elements were not tested (Gonella and Peterson, 1975).

Another series of commercial lines was assayed for Ac, En, and Uq content (Table 5B). These particular lines were chosen because of their high and low levels of heterogeneity determined by the extent of instability of specific characteristics (Bogenschutz, 1984). It was considered possible that those lines showing high levels of heterogeneity might possess active mobile elements leading to their "volatility" in attempts to stabilize these lines. None of the lines listed in Table 5B had En, Ac, or Uq, elements that could be readily detected in these lines on the basis their appearance in the BSSS series. Other systems are being tested here.

OTHER LINES AND OTHER ELEMENTS

An active Uq was found in the Rhoades-Dempsey "high-loss" line (Rhoades and Dempsey, 1982). Further, the C allele in the "high-loss" line was receptive to Uq and could be considered a

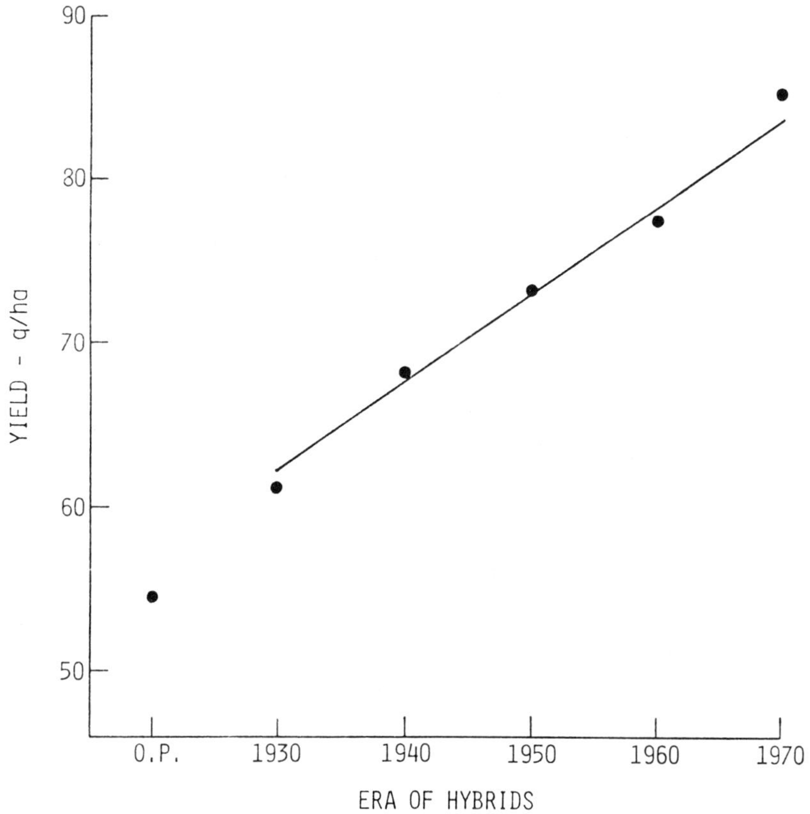

Figure 11. The predicted yields for hybrids at 10-year intervals, 1930 to 1970, based on the actual yields in Russell's study of maize hybrids representing different eras of maize breeding (after Russell, 1974).

C-m(r), a dominant allele responsive to Uq by being partly suppressed in pigmentation (Peterson, 1985a). An A allele, acting similarly, was found in the Cuna tribal corn and was responsive to En (Gonella, 1976).

Cy (Schnable and Peterson, 1984) has been found in several lines tested (Schnable and Peterson, 1985). Though this allele is not as ubiquitous as Uq, it nevertheless is present in two widely distinctly separated maize populations.

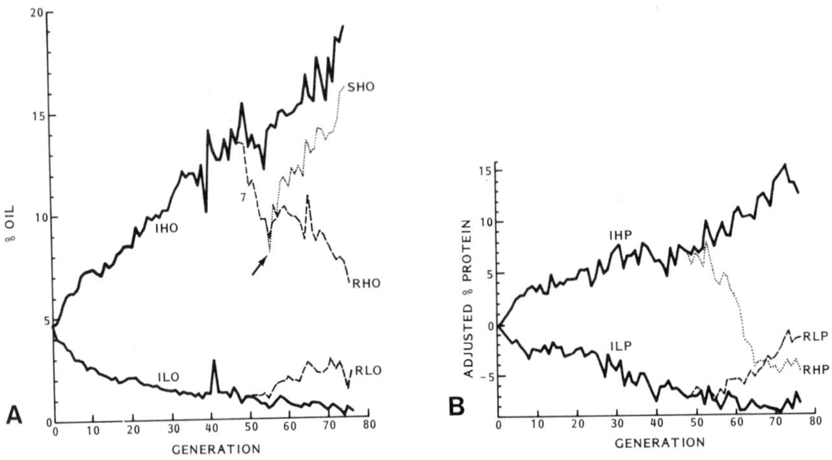

Figure 12. The manipulation of Illinois oil and protein inheritance after 76 generations of selection (Dudley, 1977). A High (IHO) and Low (ILO) Oil selection. RHO and RLO represent reverse high and low oil respectively. SHO is switchback high oil after 7 generations of RHO.

DISCUSSION

Mobile elements, both active and inactive are pervasive in the maize genome. All features, the distribution, frequency, and their individual entities, vary among lines and varieties.

In the survey of selected units of the BSSS series, an opportunity was made available to examine what would happen to such elements, active in this case, that were subjected to directed selection in a plant breeding protocol. The manner of handling the progression of the BSSS series (Smith, 1983) and the appearance of these elements in specific poulations is significant.

Despite the application of very stringent selection (approximately 10% of each cycle in a BSSS population series is saved), two active mobile elements have survived and in a significant frequency. Though they were present in a heterozygous condition, or even if two elements were present, they continue to appear in later cycles when selection contracted the gene pool from BSSSC3 to BS13 (5 cycles) by $.10^{-5}$.

It is inescapable that, for such elements to survive such a contraction of the gene pool, their role in the population under the test conditions imposed could not be random. Very likely, the favorable effect on the genome as realized in the test with the Ia13 population favors their continuation as the survivable components in the cycles of selection.

There is little doubt that corn breeders have been able to tap a continual supply of variation in their selection programs. Russell (1974, 1984, 1985) has surveyed the improvement of breeding lines over "eras" (decades) of breeding. The amount of improvement generated is substantial and amounts to 3/4 of a bushel per year (Fig. 11). The flexibility of the Illinois high oil and protein lines to respond to selection programs (Dudley, 1977, 1984) requires an input of variation that is flexible to selection demands (Figs. 12A and B). Whether the variation is generated by mobile element activity is a question that can be satisfied by future molecular probes.

ACKNOWLEDGMENT

The author thanks his colleagues at the Max Planck Institute, Egelspfad, Cologne, West Germany, for the generous use of unpublished data. These colleagues include Professor Heinz Saedler, Drs. Zs. Schwarz-Sommer, Alfons Gierl, and Francisco Salamini.

Journal Paper No. J-11828 of the Iowa Agriculture and Home Economics Experiment Station, Ames, Iowa, Project No. 2381.

REFERENCES

Bennetzen, J. L., Swanson, J., Taylor, C., and Freeling, M., 1984, DNA insertion in the first intron of maize Adhl affects message levels: cloning of progenitor and mutant Adhl alleles, Proc. Natl. Acad. Sci. USA,13:4125-4129.

Blumberg vel Spalve, J., Shepherd, N. S., and Saedler, H., 1985, The Teol insert from Zea mexicana teosinte Guerrero is a composite of the Cin2 and Cin3 elements, Maize Genet. Coop. Newsl., 59:35.

Bogenschutz, T. G., 1984, A comparison between reproduction by full-sib mating and by self-pollination to maintain genetic stability of maize inbred lines, Unpublished M.S. thesis, Library, Iowa State University, Ames, Iowa.

Bonas, U., Sommer, H., Harrison, B., and Saedler, H., 1984, The transposable element Taml of Antirrhinum majus is 17 kb long, Mol. Gen. Genet., 194:138-143.

Brink, R. A., and Nilan, R. A., 1952, The relation between light variegated and medium variegated pericarp in maize, Genetics, 37:519-544.

Burr, B., S., Evola, S. V., Burr, F. A., and Beckmann, J. S., 1983, The application of restriction fragment length polymorphism to plant breeding. in: "Genetic Engineering: Principles and Methods 5," J. K. Setlow and A. Hollaender, eds., pp. 45-59.

Courage-Tebbe, U., Döring, H. P., Fedoroff, N., and Starlinger, P., 1983, The controlling element Ds as the Shrunken locus in Zea mays; structure of the unstable sh-m5933 allele and several revertants, Cell, 34:383-393.

Darwin, C., 1868, "The variation of animals and plants under domestication." Orange Judd & Co. N.Y. 494 pp. London 1:411 pp.

Dellaporta, S. L., Chomet, P. S., Mottinger, J. P., Wood, J. A., Yu, S.-M., and Hicks, J. B., 1984, Endogenous transposable elements associated with virus infection in maize, Cold Spring Harbor Symp. Quant. Biol., 49:321-328.

Dooner, H. K., 1981, Regulation of the enzyme UFGT by the controlling element Ds in bz-m4, an unstable mutant in maize, Cold Spring Harbor Symp. Quant. Biol., 45:457-462.

Dover, G., 1982, Molecular drive: a cohesive mode of species evolution, Nature, 229:111-117.

Dudley, J. W., 1977, 76 Generations of selection for oil and protein percentage in maize, in:"Proc. Int. Conf. Quant. Genet.," Iowa State Univ. Press, Ames, Iowa, pp. 459-473.

Dudley, J. W., 1984, Theory for identification and use of exotic germplasm in maize breeding programs, Maydica, 29: 391-407.

Eberhart, S. A., Debela, S., and Hallauer, A. R., 1973, Reciprocal recurrent selections in the BSSS and BSCB1 maize populations and half-sib selection in BSSS, Crop Science, 13: 451-456.

Fedoroff, N., Wessler, S., and Shure, M., 1983, Isolation of the transposable maize controlling elements Ac and Ds, Cell, 35: 235-242.

Files, J. G., Carr, S., Hirsh, D., 1983, Actin gene family of Caenorhabditis elegans, Jour. of Mol. Biology, 164:355-375.

Friedemann, P., and Peterson, P. A., 1982, The Uq controlling-element system in maize, Mol. Gen. Genet., 187: 19-29.

Gerats, A. G. M., Groot, S. P. C., Peterson, P. A., and Schram, A. W., 1983, Regulation of UFGT activity in the bz-m4 allele of Zea mays: a possible case of gene fusion, Mol. Gen. Genet., 190:1-4.

Gonella, J. A., 1976, Controlling elements in a tribal maize from Colombia Fcu, a two-unit system, Unpublished Ph.D. Thesis, Iowa State University, Ames, Iowa.

Gonella, J., and Peterson, P.A., 1975, The presence of En among some maize lines from Mexico, Columbia, Bolivia and Venezuela, Maize Genet. Coop. Newsl., 49:73.

Gonella, J. A., and Peterson, P. A., 1977, Controlling elements in a tribal maize from Columbia: Fcu, a two unit system, Genetics, 85:629-645.

Gould, S. J., 1982, Darwinism and the expansion of evolutionary theory, Science, 216:380-387.

Gould, S. J., 1984, Darwin's untimely burial, in:"Conceptual Issues in Evolutionary Biology," E. Sober, ed., The MIT Press, Cambridge, London, 725 pp.

Gupta, M., Bertram, I., Shepherd, N.S. and Saedler, H., 1983, Cinl. a family of dispersed repetitive elements in Zea mays, Mol. Gen. Genet., 192:373-377.

Gupta, M., Shepherd, N. S. Bertram, I., and Saedler, H., 1984, Repetitive sequences and their organization on genomic clones of Zea mays, Eur. Mol. Biol. Organ. J., 3:133-139.

Hallauer, A. .R., Russell, W. A., and Smith O. S., 1983, Quantitative analysis of Iowa stiff stalk synthetic, Stadler Genetic Symp., J. P. Gustafson, ed., 15:83-104.

Hehl, R., Shepherd, N.S., and Saedler, H., 1985, DNA sequence homology among members of the Cinl repetitive DNA family in maize and teosinte, Maydica, in press.

Honjo, T., 1983, Immunoglobulin genes, Annual Rev. Immuno., 1:499-528.

Hubby, J. L., and Lewontin, R. C., 1966, A molecular approach to the study of genic heterozygosity in natural populations. I. The number of alleles at different loci in Drosophila pseudoobscura, Genetics, 54:577-594.

Hudson, R. R., 1982, Estimating genetic variability with

restriction endonucleases, Genetics, 100:711-719.

Hunkapiller, T., Huang, H., Hood, L., and Campbell, J. H., 1983, Impact of modern genetics on evolutionary theory, in: "Perspectives on Evolution," R. Milkman, ed., Sinauer Associates, Sunderland, Mass., pp. 164-189.

Johns, M. A., Mottinger, J., and Freeling, M., 1985, A low copy number, copia-like transposon in the maize genome, Eur. Mol. Biol. Organ. J., under revisison.

Johns, M.A., Strommer, J. N., and Freeling, M., 1983, Exceptionally high levels of restriction site polymorphism in DNA near the maize Adhl gene, Genetics, 105:733-743.

Jordan, E., Saedler, H., and Starlinger, P., 1968, 0° and strong-polar mutations in the gal operon are insertions, Mol. Gen. Genet., 102:353-363.

Kermicle, J. L., 1984, Recombination between components of a mutable gene system in maize, Genetics, 107:489-500.

Lucotte, G., Gal, A., Nahon, J.-L, and Sala-Trepat, J. M., 1982, Eco-hl restriction-site polymorphism of the albumin gene in different inbred strains of rat, Biochem. Genet., 20:1105-1115.

Lewontin, R. C., and J. L. Hubby, 1966, A molecular approach to the study of genic heterozygosity in natural populations. II. Amount of variation and degree of heterozygosity in natural populations of Drosophila pseudoobscura, Genetics, 54:595-609.

McClintock, B., 1947, Cytogenetic studies of maize and Neurospora, Carnegie Inst. Wash. Year Book, 46:146-152.

McClintock, B., 1951, Chromosome organization and genetic expression, Cold spring Harbor Symp. Quant. Biol., 16:13-47.

McClintock, B., 1954, Mutations in maize and chromosomal aberrations in Neurospora, Carnegie Inst. Wash. Year Book, 53:254-260.

McClintock, B., 1961, Some parallels between gene control systems in maize and in bacteria, Am Nat., 95:265-277.

McClintock, B., 1965, The control of gene action in maize, Brookhaven Symp. Biol., 18:162.

McClintock, B., 1967, Genetic systems regulating gene expression during development, Dev. Biol. Suppl., 1:84-112.

McDonald, J. F., 1983, The molecular basis of adaptation:a critical review of relevant ideas and observations, Annu. Rev. Ecol. Syst., 14:77-102.

Nevers, P., Shepherd, N.S., and Saedler, H., 1985, Plant transposable elements, Advances in Botanical Research, in press.

Oberthur, E., and Peterson, P. A., 1984, Uq controlled mutable allele at the c locus, Maize Genet. Coop. Newsl., 58:4-5.

Ohta, T., 1984a, Some models of gene conversion for treating the

evolution of multigene families, Genetics, 106: 517-528.
Ohta, T., 1984b, Population genetics of transposable elements, IMA J. Math. Appl. Med. Biol., 1:17-29.
Ohta, T., and Dover, G. A., 1984, The cohesive population genetics of molecular drive, Genetics, 108:501-521.
Peacock, W. J., Dennis, E. S., Gerlach, W. L., Llewellyn, D., Lorz, H., Pryor, A. J., Sachs, M. M., Schwartz, D., and Sutton, W. D., 1983, Gene transfer in maize: Controlling elements and the alcohol dehydrogenase genes, in: "Proc. 15th Miami Winter Symp.," K. Downey, K. Vollmy, F. Ahmad, and J. Schulty, eds., Academic Press: New York, 20:311-320.
Penny, L. H., and Eberhart, S. A., 1971, Twenty years of reciprocal recurrent selection with two synthetic varieties of maize (Zea mays L.), Crop Sci., 11:900-903.
Pereira, A., Schwarz-Sommer, Zs., Gierl. A., Bertram, I., Peterson, P. A., and Saedler, H., 1985, Genetic and molecular analysis of the Enhancer (En) transposable element system of Zea mays, Eur. Mol. Biol. Organ. J., 4:17-23.
Perutz, M. F., 1983, Species adaptation in a protein molecule, Mol. Biol. Evol., 1: 1-28.
Peterson, P. A., 1953, The mutable pale green locus, Maize Genet. Coop. Newsl., 27:58.
Peterson, P. A., 1960, The pale green mutable system in maize, Genetics, 45:115-133.
Peterson, P. A., 1961, Mutable al of the En system in maize, Genetics, 46:759-771.
Peterson, P. A., 1966, Phase variation of regulatory elements in maize, Genetics, 54:249-266.
Peterson, P. A., 1967, A comparison of the action of regulatory systems in maize and lysogenic bacteria, Genetics, 56:581.
Peterson, P. A., 1970, Controlling elements and mutable loci in maize: Their relationship to bacterial episomes, Genetica, 41:33-56.
Peterson, P. A.. 1978, Controlling elements: The induction of mutability at the A2 and C loci in maize. in: "Maize Breeding and Genetics," D. B. Walden, ed., John Wiley & Sons, NY., pp. 601-635.
Peterson, P. A., 1981, Instability among the components of a regulatory element transposon in maize, Cold Spring Harbor Symp. Quant. Biol., 45: 447-455.
Peterson, P. A., 1983, Newly originated mutable alleles at the bz locus, Maize Genet. Coop. Newsl., 57:2.
Peterson, P. A., 1984, C-I changes, Maize Genet. Coop. Newsl., 58:2-3.
Peterson, P. A., 1985a, A dominant color allele C-m(r), responsive to a specific Uq, Maize Genet. Coop. Newsl., 59: in press.

Peterson, P. A., 1985b, The isolation of En-1 in the wx 84-4 allele, Maize Genet. Coop. Newsl., 59: in press.
Peterson, P. A., 1985, Plant mobile elements, CRC, (submitted).
Peterson, P. A., and Cormack, J., 1984, wx alleles newly originated from an En-containing plot, Maize Genet. Coop. Newsl.,58:3.
Pohlman, R. F., Fedoroff,N. V., and Messing, J., 1984, The nucleotide sequence of the maize controlling element activator, Cell,37:635-643.
Rhoades, M. M., 1936, The effect of varying gene dosage on aleurone colour in maize, J. Genet., 33:347-354.
Rhoades, M. M., and Dempsey, E., 1982, The induction of mutable systems in plants with the high-loss mechanism, Maize Genet. Coop. Newsl.,56:21-26.
Robertson, D. S., 1978, Characterization of a mutator system in maize, Mutation Research,51:21-28.
Robertson, D. S., and Stinard, P., 1984, Putative forward mutation frequencies at the yl and wx loci in the presence of Mu, Maize Genet. Coop. Newsl.,58:11-12.
Robinson, R. R., Germain, R. N., Mckean, D. J., Mescher, M., and Seidman, J. G., 1983, Extensive polymorphism surrounding the murine Ia abeta chain gene, Jour. of Immuno., 131:2025-2031.
Russell, W. A., 1974, Comparative performance for maize hybrids representing different eras of maize breeding, Proc. 29th Annu. Corn Sorghum Res. Conf., Chicago, pp. 81-101.
Russell, W. A., 1984, Agronomic performance of maize cultivars representing different eras of breeding, Maydica,29:375-390.
Russell, W. A., 1985, Comparisons of the hybrid performance of maize inbred lines developed from the original and improved cycles of BSSS, Maydica, in press.
Russell, W. A., and Vega, U. A., 1973, Genetic stability of quantitative characters in successive generations in maize inbred lines, Euphytica,22:172-180.
Sachs, M. M., Peacock, W. J., Dennis, E. S., and Gerlach, W. S., 1983, Maize Ac/Ds controlling elements, a molecular viewpoint, Maydica,28:289-302.
Saedler, H., Bonas, U., Deumling, B., Gupta, M., Hahlbrock, K., Harrison, B. J., Kreuzaler, F., Peterson, P. A., Reif, J., Schwarz-Sommer, Zs., Shepherd, N., Sommer, H., Ubben, D., and Wienand, U., 1983, Transposable elements in plants, in: "Genetic Rearrangement," K. F. Chater, C. A. Cullis, D. A. Hopwood, A. A. W. B. Johnston, and H. W. Woolhouse, eds., The Fifth John Innes Symposium, Croom Helm Ltd., London and Canberra, pp. 107-116.
Saedler, H., and Nevers, P., 1985, Transposition in plants: A molecular model, Eur. Mol. Biol. Organ. J.,4:585-590.

Salamini, F., 1981, Controlling elements at the Opaque-2 locus of maize: their involvement in the origin of spontaneous mutation, Cold Spring Harbor Symp. Quant. Biol., 45:467-476.

Schnable, P. and Peterson, P. A., 1984, bz-m805137-A bronze mutable of the Cy system, Maize Genet. Coop. Newsl., 58:9.

Schnable, P., and Peterson, P. A., 1985, System relationships of the Cy transposable element system including tests against Robertson's mutator system, Maize Genet. Coop. Newsl., 59:4.

Schwarz-Sommer, Zs., Gierl. A., Cuypers, H., Peterson, P. A., and Saedler, H., 1985, Plant transposable elements generate the DNA sequence diversity needed in evolution, Eur. Mol. Biol. Organ. J., 4:591-597.

Schwarz-Sommer, Zs., Gierl, A., Klosgen, R. B., Wienand, U., Peterson, P. A., and Saedler, H., 1984, The Spm (En) transposable element controls the excision of a 2-kb DNA insert at the wx^{m-8} allele of Zea mays, Eur. Mol. Biol. Organ. J., 3:1021-1028.

Shaw, C. R., 1965, Electrophoretic variation in enzymes, Science, 149:936-43.

Shepherd, N. S., Schwarz-Sommer, Zs., Blumberg vel Spalve, J., Gupta, M., Wienand, U., and Saedler, H., 1984, Similarity of the Cinl repetitive family of Zea mays to eukaryotic transposable elements, Nature, 307: 185-187.

Shepherd, N.S., Schwarz-Sommer, Zs., Wienand, U., Sommer, H., Deumling, B., Peterson, P. A., and Saedler, H., 1982, Cloning of a genomic fragment carrying the insertion element Cinl of Zea mays, Mol. Gen. Genet., 188:266-271.

Smith, O. S., 1983, Evaluation of recurrent selection in BSSS, BSCBl, and BS13 maize poulations, Crop Sci., 23:35-40.

Sober E, ed., 1984, Conceptual Issues In Evolutionary Biology, The MIT Press, Cambridge, Mass. 725 pp.

Sprague, G. F., 1946, Early testing of inbred lines of corn, J. Am. Soc. Agron., 38:108-117.

Stuber, C. W., and Goodman, M. M., 1984, Inheritance, intracellular location and genetic variation of 6-phosphogluconate dehydrogenase isozymes in maize, Maydica, 29:453-471.

Stuber, C. W., Goodman, M. M., and Moll, R. H., 1982, Improvement of yield and ear number resulting from selection at allozyme loci in a maize population, Crop Sci., 22:737-740.

Stuber, C. W., Moll, E. H., Goodman, M. M., Schaffer, H. E., and Weir, B. S., 1980, Allozyme frequency changes associated with selection for increased grain yield in maize (Zea mays L.), Genetics, 95:225-236.

Sturtevant, A. H., 1913, The linear arrangement of six sex-linked factors in *Drosophila*, as shown by their mode of association, *J. Exp. Zool.*, 43-59.

Syvanen, M., 1984. The evolutionary implications of mobile genetic elements, *Ann. Rev. Genet.*, 18:271-293.

Taylor, A. L., 1963, Bacteriophage-induced mutation in *Escherichia Coli*, *Proc. Natl. Acad. Sci. USA*, 50:1043-1051.

Weck, E., Courage, U., Döring, H.-P., Fedoroff, N., and Starlinger, P., 1984, Analysis of *sh-m6233*, a mutation induced by the transposable element *Ds* in the sucrose synthase gene of *Zea mays*, *Eur. Mol. Biol. Organ. J.*, 3:1713-1716.

Wickler, W., 1968, "Mimicry", Weidenfeld and Nicolson, London, 255 pp.

MUTATION, APICAL MERISTEMS AND DEVELOPMENTAL SELECTION IN PLANTS

Edward J. Klekowski, Jr., Hans Mohr,*
and Nina Kazarinova-Fukshansky*

Botany Department
University of Massachusetts
Amherst, MA 01003 U.S.A.

*Biologisches Institut II
Universität Freiburg
Schänzlestrasse 1
D-7800 Freiburg im Breisgau
Federal Republic of Germany

ABSTRACT

Since in plants Weismann's doctrine of the separation of soma and germ is invalid, somatic mutations may occur in cell lines which in turn give rise to meiocytes. Although some mutations are adaptive, the majority of mutations are disadvantageous and ultimately reduce the fitness of a ramet (or genet). Many anatomical and developmental characteristics may reduce the mutation frequency within ramets. Apical meristems may influence mutation frequency by either reducing the mutation rate or enhancing diplontic selection. The former is possible in meristems with permanent apical initials (structured meristems) through nonrandom DNA strand segregation and in meristems with impermanent apical initials (stochastic meristems) which have a méristème d'attente. Diplontic selection is maximized in apical meristems in which the initials and derivative cells are least determined ontogenetically, thus allowing maximum intercellular competition (stochastic meristems). In spite of these mechanisms, the individual ramet meristems in long-lived genets still may be expected to diverge genetically through the fixation of mutations in the small cell pools within these meristems by mechanisms similar to Muller's Ratchet.

The continued accumulation of disadvantageous mutations within the apical initials of ramet apical meristems should ultimately reduce the reproductive capacity (sexual) of ramets. Many plant characteristics may be viewed as mechanisms which purge such mutational load without proportionate decreases in reproductive capacity. Thus characteristics such as pollen competition, low seed-ovule ratios and selective seed or fruit abortion may all be aspects of a soft selection sieve whereby mutational load is eliminated with little sacrifice in overall reproductive capacity. In long-lived plants, outbreeding mechanisms also may represent adaptations to increase fitness by covering recessive lethal and other kinds of disadvantageous alleles (mutational load). Plant genetic systems therefore may <u>not</u> represent a compromise "between the conflicting requirements of fitness and flexibility" but rather a means of negating or repairing recurrent mutational load. Thus genetic systems may be a means of achieving maximum immediate fitness rather than future evolutionary change.

INTRODUCTION

As has been recently emphasized by Whitham and Slobodchikoff (1981) and Buss (1983), in plants Weismann's doctrine of the separation of soma and germ is invalid, thus in some species within individual selection may be as important as selection at the level of the individual. The extent to which evolutionarily significant within individual selection is possible is dependent upon the degree of genetic variation present within a population of ramets. (Following White (1979) a genet is the entire plant, of whatever size or however subdivided or propagated, which is derived from a single zygote or meiospore. In contrast a ramet is a single module or shoot complex of a genet that is conveniently enumerated. All sib ramets have a mitotic origin and barring somatic mutation, should have identical genotypes.) Any theoretical discussion of the possible genetic variation within a population of ramets (or branches of a tree for that matter) must begin with an understanding of the dynamics of the loss or persistence of somatic mutations within apical meristems.

Apical meristems in vascular plants include a great variety of organizations and patterns of cell division ranging from simple meristems within a single apical cell as is characteristic of some pteridophytes to complex tunica-corpus systems found in most angiosperms (Popham, 1951). Within this spectrum of anatomical and ontogenetic diversity are many features which may either change the frequency of mutations which have accumulated in an apical initial or influence the effectiveness of diplontic selection within the population of cells which constitute the apical meristem. As will be shown in our discussion, features such as the stability or permanence of apical initials, the number of apical initials, the size of the cell pool in the apical meristem, patterns of cell division in the meristematic cell pool, presence of a quiescent area and patterns of DNA strand segregation are among the many facets of apical meristem development which have relevance in the generation of genetic diversity among sib ramets.

STOCHASTIC VERSUS STRUCTURED MERISTEMS

From the viewpoint of somatic genetic variability, the primary feature of vascular plant apical meristems influencing the loss or persistence of somatic mutations is whether the meristem is composed of permanent or impermanent apical initials. Meristems based upon permanent apical initials may maintain mutations indefinitely whereas impermanent initials allow the operation of diplontic or stochastic drift (Balkema, 1972) and diplontic selection to modify the frequency of mutant initials in the meristem.

Structured meristems have been defined as those based upon permanent apical initials (Klekowski and Kazarinova-Fukshansky, 1984a). The division of such apical initials results in one daughter cell remaining an initial and the other becoming a new body cell (Esau, 1977). In contrast, stochastic meristems are those based upon impermanent apical initials. To quote Esau (1977), "A cell functions as an initial not because of any inherent properties but because of its position. At the time of division of an initial it is impossible to predict which of the two daughter cells will 'inherit' the function of the initial and which will become the derivative. It is also known that a given initial may be replaced by a cell which through prior history would be classified as a derivative of an initial." "The concept of meristematic initials, therefore, implies that a cell is an initial, not because of its inherent characteristics, but simply because of its particular position in the meristem, a position that cannot be considered permanent," (Esau, 1953). Thus in this view, apical initials divide to give rise to a pool of cells which subsequently is sampled for the next set of initials. This pool of cells is analogous to the "continuing meristematic residue" of Newman (1965).

STRUCTURED MERISTEMS - GENERAL PROPERTIES

All structured meristems are based (by definition) upon permanent apical initials and thus may be viewed as stem cell multiplication systems in which the tangential system of cell renewal or multiplication occurs (Kay, 1965). Kay has argued that the frequency of mutations in a stem cell (or in the present case an apical initial) is a function of the number of mitotic divisions since the cell originated from the zygote. This argument is based upon the assumption that the division process is the most error prone portion of the cell cycle. Mutations arise both from disjunctive accidents and errors in DNA replication. With sufficient cycles of division mutations not only accumulate but the DNA synthesis and repair mechanism may become less effective due to mutation. As a consequence mutation rates would be expected to increase as age (in terms of numbers of cell divisions) of the stem cell increases. The tangential system of cell renewal maximizes the number of cell divisions a stem cell or initial must make to give rise to a pool of derivative cells.

Unstratified Structured Meristems

The classic examples of unstratified structured meristems are the apical cell based meristems found in the shoot apices of many pteridophytes (the psilopsids, some lycopsids, the sphenopsids, some eusporangiate ferns and the leptosporangiate ferns) (Bierhorst, 1971). Although there has been debate as to whether the apical cell is meristematic in such apical meristems (McAlpin and White, 1974), the pattern of cell walls in the apex can be accounted for only by assuming that the apical cell is meristematic (Lintilhac and Green, 1976; Bierhorst, 1977). Because of the relatively rigid developmental and structural restraints on patterns of cell division in these meristems, diplontic selection or competition between potential initial cells is minimized. Mutation of the apical cell generally results in the formation of mutant derivative cells and the ultimate transmission of the mutant to all subsequently formed organs (Klekowski, 1984).

Although apical cell based meristems would be expected to retain disadvantageous mutations more readily than other kinds of meristems, perhaps this is in part compensated for by mechanisms which reduced mutation rates. Recent studies by Kuligowski-Andres and Tourte suggest that cell division in apical cells in *Marsilea vestita* embryos has been modified to reduce the mutation accumulation in these cells (Kuligowski-Andres et al., 1979; Kuligowski-Andres and Tourte, 1980; Tourte et al., 1980). Using tritiated thymidine labeled gametes, two patterns of DNA synthesis in apical cells were noted. When the zygote was the result of the fertilization of a radioactively labeled egg and a nonlabeled spermatozoid, analysis of silver grain counts of autoradiographs of different aged embryos showed a continuous diminution of grains per cell with mitosis. In contrast, when a labeled spermatozoid fertilized a nonlabeled egg, radioactivity was retained selectively by the apical cells up to at least the 5,000 cell stage of the embryo. The French workers have suggested that their data support the hypothesis of nonrandom distribution of paternal chromatids in the mitotic divisions of the apical cells (Tourte et al., 1980). Thus the paternal copy of genetic information may function as a master copy retained within an apical cell. Cairns (1975) has suggested that during DNA replication, the newly synthesized daughter polynucleotide chain is more error prone than the original template chain and that stem cells which continuously divide may retain the master template chains. The systematic orientation of chromosomes so that the chromatids with the template polynucleotide chain are drawn to the pole which remains the apical cell would reduce the mutation frequency.

Figures 1 and 2 diagrammatically illustrate how nonrandom orientation of chromatids results in the loss of mutant cells to the somatic tissues. In contrast, random chromatid orientation can result in newly replicated error prone polynucleotide chains being incorporated into the apical cell genotype. Once the apical cell has a mutant genotype, all subsequently derived cells and tissues will have the mutant genotype.

These results are provocative to say the least and, if substantiated, will have far reaching implications. Apical cells are ubiquitous in both embryonic and adult meristems of many pteridophytes. If paternal template conservation occurs in these apical cells, mutations of the maternal genome should be more common than those of the paternal genome within a clone or population of sib ramets.

Unstratified meristems occur in some pteridophytes and most gymnosperms. Popham (1951) has reviewed evidence which suggests that some *Selaginella* species have 2-5 apical cells in their apical meristems. In the gymnosperms the majority of researchers indicate that permanent initials or apical cells are absent (Johnson, 1951). It is important to note that the reduction in mutation accumulation through nonrandom orientation of chromatids as described in *Marsilea* is only possible in apical meristems with permanent apical initials.

Stratified Structured Meristems

Apical meristems which are organized into one or more tunica layers overlaying an internal corpus will be considered as stratified meristems. The tunica is characterized by anticlinal cell divisions resulting in a sheet or sheets of cells clothing the corpus in which cell wall orientations appear random (anticlinal and periclinal) (Esau, 1977). In the present communication, the individual tunica layers as well as the corpus are considered as relatively discrete component meristems of a stratified system. Evidence for the autonomy of such component meristems is to be found in the numerous examples of stable periclinal chimeras documented in plants with stratified meristems (Neilson-Jones, 1969).

Stratified meristems appear to be restricted to the seed plants. In the gymnosperms, based upon anatomical data, stratified apical meristems have been documented in the *Ephedraceae*, *Gnetaceae*, and *Araucariaceae* (Johnson, 1951). Stratified meristems probably also occur in the Cupressaceae where numerous examples of relatively

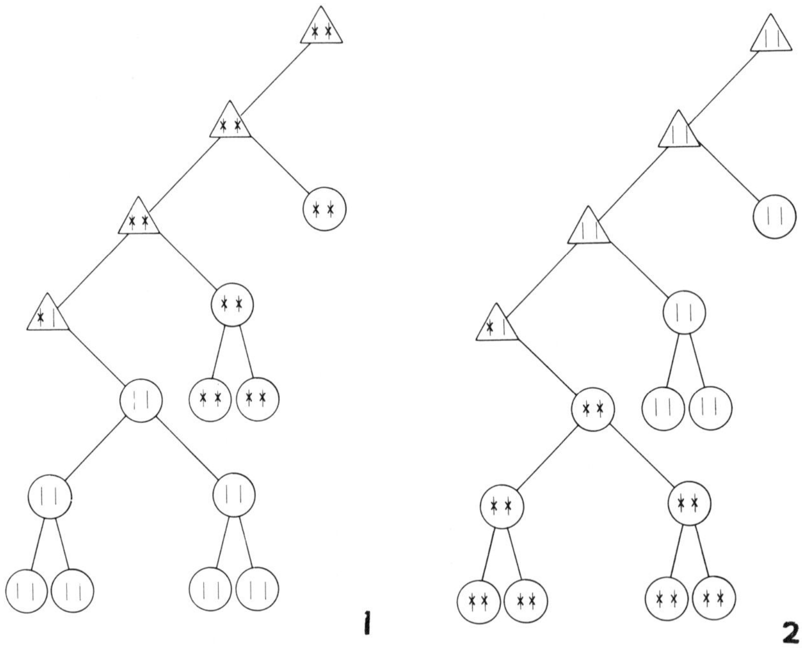

Fig. 1-2. Diagrammatic representation of chromatid segregation in a structured meristem based upon an apical cell (△). The (⚹) marks the newly synthesized DNA strand. This strand is believed to be more mutation prone. Fig. 1. Random chromatid segregation, in this case the newly synthesized strand was segregated into the initial during the first mitotic division. Note how all subsequent somatic cells carry copies of the mutation prone strand. Fig. 2. Nonrandom chromatid segregation, in this case the newly synthesized mutation prone strand is immediately eliminated to the somatic cells.

stable periclinal chimeras have been found (Pohlheim, 1980). In the angiosperms, stratified apical meristems seem to be the norm for practically all groups (Gifford, 1971). The individual component meristems in a species with two tunica layers and a corpus are termed the LI, LII, LIII. In the dicots the sporogenous tissue generally forms from the LII whereas in the monocots the LI and LII may contribute cells (Stewart and Dermen, 1979) to this tissue.

Stratified apical meristems which are structured have permanent apical initials in the individual component meristems. There is considerable debate in the literature concerning the permanence of initials in the tunica and corpus. Stewart and Dermen (1970), using data from the length of persistence of mericlinal chimeras, argued that permanent apical initials exist in the component meristems of stratified meristems in gymnosperms and angiosperms. More recently (Ruth et al., 1985) analyzed chimeras in *Juniperus* and concluded that the component meristems behaved stochastically. In angiosperms anatomical and ontogenetic studies support the hypothesis that initials are impermanent (Soma and Ball, 1964; Newman, 1965; Chen and Ross, 1965) whereas studies of mericlinal or sectorial chimeras suggest permanent initials (Stewart, 1978) (see however Balkema, 1972 for another view). The persistence of such chimeras may well be due to permanent initials, but it should be noted that such chimeras can persist also in stochastic apices (Klekowski and Kazarinova-Fukshansky, 1984a). The persistence of mericlinal or sectorial chimeras in stochastic meristems is a function of the number of apical initials, cell pool sizes, number of nodes per division of the initials, diplontic selection and the degree to which a quiescent center is present in the apex.

Where permanent apical initials are present in the component meristems of a stratified meristem, mutation of an initial will result in the long term persistence (semi-permanent?) of mericlinal or sectorial chimeras. Periclinal chimeras can result from buds formed in the mutant sectors of the original stem.

STOCHASTIC MERISTEMS - GENERAL PROPERTIES

Stochastic meristems are based upon impermanent apical initials. In such meristems, initials are sampled recurrently from a pool of cells in the apical meristem. This pool of cells may be sampled randomly or with some biases. In Fig. 3 is shown the operation of a stochastic meristem. In this case three original initials underwent two cycles of mitosis to generate a pool of 12 cells from which the next subsequent set of three initials was selected. In stochastic meristems the number of apical initials (α) and the number of cell divisions between selection of new initials (r) may vary; both of these parameters may greatly affect the loss or persistence of mutations in the meristem (Klekowski and Kazarinova-Fukshansky, 1984b).

Stochastic meristems which follow the above pattern of cell division with a continuous feed of cells to the soma as well as the regeneration of initials may be viewed as stem cell multiplication systems in which the tangential system of cell renewal occurs (in this respect they are identical to structured meristems). As has been discussed previously, this system of cell renewal maximizes the number of cell divisions between zygote and meiosis and consequently results in higher mutation frequencies in the initials (Kay, 1965).

Not all plants with stochastic meristems follow the tangential

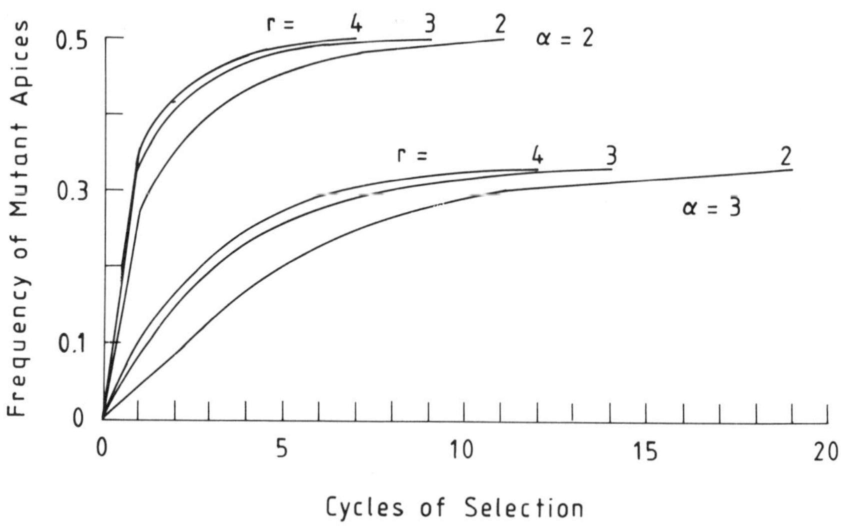

Fig. 3. Diagrammatic representation of α and r in terms of apical meristem growth. The beginning state has three initials, one of which is heterozygous for a new mutation (X). All initials go through two cycles of mitosis (r = 2) resulting in a pool of $\alpha 2^r$ cells from which new initials are chosen (from Klekowski and Kazarinova-Fukshansky, 1984a).

pattern of cell renewal. In the past 30 years botanists have come to realize that the shoot apices of many plants are characterized by considerably reduced mitotic activity than would be expected for meristematic tissues. According to a hypothesis formulated by French botanists, the summit of the apex in a vegetative shoot is occupied by a distinctive zone, the _méristème d'attente_, in which there is an absence or near absence of mitotic activity (Buvat, 1952). The _méristème d'attente_ is surrounded laterally by the _anneau initial_ and is subtended by the _méristème medullaire_, in both of which mitosis is common. In more recent interpretations of this hypothesis, the _méristème d'attente_ is characterized by some mitotic activity, but this is much more restricted than the rest of the apex (see Steeves and Sussex, 1972 for further discussion). From the viewpoint of mutation theory, the most interesting feature of this hypothesis is that although the _méristème d'attente_ is inactive mitotically during vegetative growth, with the onset of floral or inflorescence development this situation changes radically. Abundant mitoses may be detected in the _méristème d'attente_ at this time; and the previously inactive region gives rise to most or all of the reproductive structures. Such a pattern of development has been demonstrated clearly in the apical meristems of _Nicotiana_ (Sussex and Rosenthal, 1973) and _Helianthus_ (Langenauer and Davis, 1973).

From the viewpoint of developmental biology, the function of this population of potentially meristematic cells which "waits" throughout vegetative development is unknown (Sussex and Rosenthal, 1973). Although mutation buffering may not be the primary factor responsible for the presence of the _méristème d'attente_, the genetic consequence of such a population of cells destined for reproductive tissue is considerable. Such a pattern of development follows either the asynchronous logarithmic or skewed asynchronous system of cell renewal as described by Kay (1965). Consequently the number of cell generations between zygote and meiocyte is considerably less in plants with a _méristème d'attente_ than in plants lacking such a subpopulation of cells in their apical meristems. Based upon the previous discussion of tangential systems of cell renewal, the mutation frequency in the meiocytes is in part a function of their biological age (measured in cell generations). Thus assuming all other factors as similar, the mutational load in species with a _méristème d'attente_ should be less than in species which lack this type of apical organization.

The asynchronous logarithmic and skewed asynchronous systems of stem cell renewal described by Kay also can promote the resistance of meristems to damage by physical and chemical mutagens. Grodzinsky and Gudkov (1982) have argued that a cell population which is variable with reference to cell cycle lengths as well as is highly asynchronomous in terms of metabolism will present a broad array of cell sensitivities at any point in time. Thus the exposure of such a population to a genetic insult (ionizing radiation, chemical mutagen, or any other mutationally active environmental influence) will result in some cells less damaged than others. Such less-damaged cells may proliferate and restore the meristem and consequently enhance the probability of plant survival.

Unstratified Stochastic Meristems

Apical meristems which lack a tunica-corpus organization and in which permanent apical initials are not identifiable occur primarily in the gymnosperms (Johnson, 1951), although a similar apical organization has been reported in the angiosperm *Saccharum officinarum* (Thielke, 1959). Klekowski and Kazarinova-Fukshansky (1984a and b) have analyzed the mathematical consequences of stochastic growth in unstratified meristems with regard to the persistence or loss of neutral and selectively disadvantageous mutations. In Fig. 3 the mathematical parameters used in the analysis of stochastic growth are illustrated. The number of apical initials (α), the number of mitotic divisions (r) before the next selection of subsequent initials and the size of the cell pool from which initials are selected ($\alpha 2^r$) are important parameters which influence the fate of somatic mutations in apical meristems.

Let us first consider mutations which are selectively neutral within the apical meristem. Assuming a starting condition of α-1 nonmutant initials and one initial heterozygous for a mutation, stochastic growth will ultimately result in apical meristems which consist of either all mutant or all nonmutant initials. Klekowski and Kazarinova-Fukshansky (1984a) have shown that with continued stochastic growth the fixation asymptotics for mutants and nonmutants (i.e., the probability of apices which consist of only mutant initials or only nonmutant initials) is $1/\alpha$ for apices with only mutant initials and $(\alpha-1)/\alpha$ for nonmutant initials. As is shown in Fig. 4 these asymptotics are independent of the cell pool sizes ($\alpha 2^r$) from which initials are selected (note that stochastic growth is a mathematical formulation of diplontic drift as described by Balkema (1972)).

For selectively neutral mutations (again only within the context of the meristem) it is easy to show that the number of somatic mutations fixed in the plant is independent of the number of apical initials. If R is the mutation rate per initial per unit of biological time (e.g., cell generations, ramet generations, or sexual generations in annuals) then:

$$\text{Number of somatic mutations fixed per genet} = \alpha R\,(\text{Number of biological time units})1/\alpha$$

$$\text{Number of somatic mutations fixed per genet} = R\,(\text{Number of biological time units}).$$

Thus the number of apical initials only determines the proportion of ramets or buds which have fixed a given mutation, whereas the total number of separate mutations which have been fixed in various ramets of a genet is a function of the mutation rate and genet age.

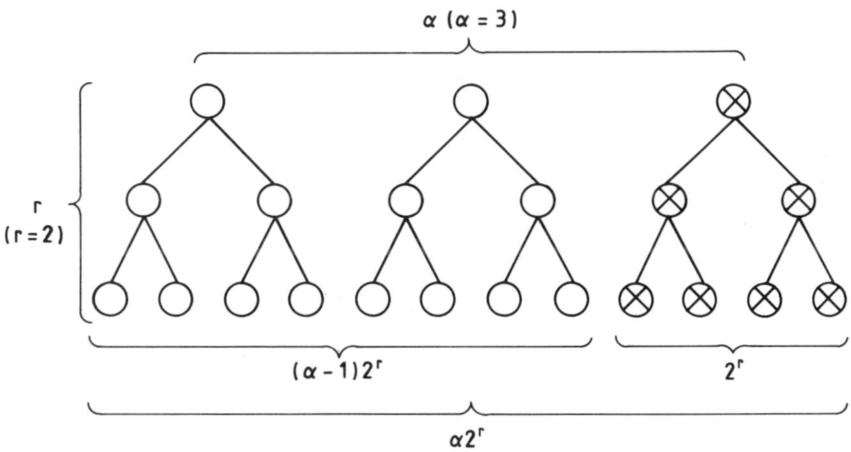

Fig. 4. The fixation of a neutral mutation in unstratified apices with different α and r values as a function of the number of apical initial selection cycles. Fixation is defined as having occurred when all of the initials of a meristem are of the mutant genotype (from Klekowski and Kazarinova-Fukshansky, 1984a).

Another interesting aspect of stochastic growth is that certain combinations of α and r promote the persistence of sectorial or mericlinal chimeras. The persistence of such chimeras may be measured in terms of the number of mitotic cycles of initials and daughter cells giving rise to $\alpha 2^r$, the number of selections of apical initials, or the number of nodes of growth. The number of nodes of stem growth is a function of the number of cell divisions of initials and daughter cells or the number of selection cycles of apical initials from the cell pool $\alpha 2^r$. Stewart and Dermen (1970) suggested that 3-5 nodes per division of the apical initials is reasonable in plants. In Table 1 are shown the probabilities of the persistence of a sectorial or mericlinal chimera in apices with different α and r values under conditions of stochastic growth. The number of nodes of persistence is four times the number of cell divisions of the apical initials and their immediate daughter cells from which new initials are selected. The persistence of sectorial chimeras is related to the number of apical initials (α) and the number of divisions (r) generating the cell pool ($\alpha 2^r$) from which apical initials are selected. For example when $\alpha = 2$ and $r = 1$ chimeras will exceed 20 nodes 10% of the time, when $\alpha = 2$ and $r = 2$ chimeras will exceed 32 nodes 10% of the time, when $\alpha = 4$ and $r = 1$ chimeras will exceed 56 nodes 10% of the time and when $\alpha = 4$ and $r = 2$ chimeras will exceed 112 nodes 10% of the time. Stewart and Dermen (1970) have stated that if every cell in the apical meristem had an equal probability of functioning as an initial (stochastic growth), then there could exist no sectorial chimeras which would show any length or permanence. The data in Table 1 show that in this regard Stewart and Dermen are in error; stochastic growth can give rise to sectorial chimeras which may persist for many nodes.

Experimental data which is consistent with the stochastic growth model is available in the extensive literature of flowering plant mutagenesis. In such studies the seeds are exposed to a mutagen, the resulting seedlings (the M_1 generation) are grown to sexual maturity and the flowers selfed. The embryos resulting from this selfing comprise the M_2 generation and may exhibit mutant phenotypes. Thus apical initials heterozygous for mutants undergo many cycles of mitosis which ultimately result in an M_1 plant which is a chimera. The mutants are made homozygous in the M_2 embryo and thus detected. Since the mutants are passed through the sporogenous tissue of the M_1, the mutant initials which occur in only one of the tunica layers (LII) are being studied. Thus although flowering plants have stratified apical meristems, the mutagenesis data deals primarily with only one of the component meristems (LII) and thus represents a single stochastic system.

The instability of mericlinal chimeras is routinely documented in studies of plant mutagenesis (Grinikh et al., 1974). For example Muller (1965) studied 255 mericlinal chimeras in _Arabidopsis_ M_1

plants for 20 consecutive plastochrons and found that in 34% the mutant sector had been displaced from the apex, in 28% the non-mutant sector was displaced, and in 38% the mericlinal chimera persisted but the sector size changed. Since the number of apical initials in the seed of this species is thought to be two (α = 2), the frequency of apices which lose the mutant should equal the frequency of apices which fix the mutant. If one assumes that α = 2, r = 2 and that one division of the initials or daughter cells occurs every 5 plastochrons then based upon Table 1, one would predict 33.7% of the apices to lose the mutant, 33.7% to have fixed the mutant and 32.7% chimeras. The correspondence between these expected values and the observed values of 34%, 28% and 38% is notable.

Experimental systems based upon somatic mutations resulting in discrete patches or sectors which are not lost with continued growth of the organ (petal or leaf) have documented an inverse relationship between areas of the mutant sectors and mutation frequency (Cuany et al., 1957; Buiatti et al., 1970). Thus for example flowers that have small mutant sectors generally have more sectors than flowers with large sectors. This relationship is explicable as a simple consequence of changing the number of initials (α). The mutation frequency in such systems is

$$\text{Number of somatic mutations in the organ} = \alpha R \text{ (Number of biological time units).}$$

Note this differs from the previous formula for mutation frequency by the omission of the fixation asymptotic, $1/\alpha$. In experimental systems in which all mutant patches or sectors are preserved, the fixation asymptotic is not relevant. Since the mutagen exposure is usually an acute single exposure, the number of biological time units is unity and therefore mutation frequency is simply the product of the mutation rate (R) and number of initials (α). The size of the mutant sector is a function of the fixation asymptotic $1/\alpha$. Thus increasing the number of apical initials (α), increases the product αR and decreases the reciprocal $1/\alpha$.

Diplontic or somatic selection is possible between mutant and nonmutant cells within the cell pool of the meristem ($\alpha 2^r$) (Gaul, 1965). Langridge (1958) proposed that in the case of deleterious mutations, diplontic selection would be lessened for genes inactive in the apical meristem and for genes responsible for diffusible metabolites. Mutations of the latter could be maintained by cross feeding from nonmutant cells within the meristem. Meinke and Sussex (1979a and b) reported that cells heterozygous for embryo-lethal mutants did not appear to have a growth disadvantage in apices which were a chimera of mutant and wild type initials. These results suggest that either the mutants were completely recessive, the genes

Table 1. Persistence of mericlinal and sectorial chimeras in stochastic meristems. The probability (or frequency) of a chimera (M) is given in terms of the number of cell generations of the apical initials and immediate derivatives, the number of cycles of apical initial selections, and the number of nodes of persistence (where four nodes occur every cell generation). The starting condition was α-1 nonmutant initials and one mutant initial and the mutant initials are selectively neutral.

	Cell Generations	Selection Cycles	Nodes	M
α=2, r=1				
	4	4	16	0.198
	8	8	32	0.039
	12	12	48	0.007
α=2, r=2				
	4	2	16	0.327
	8	4	32	0.107
	12	6	48	0.035
	16	8	64	0.011
	20	10	80	0.004
α=3, r=1				
	4	4	16	0.410
	8	8	32	0.168
	12	12	48	0.068
	16	16	64	0.028
	20	20	80	0.012
	24	24	96	0.004
α=3, r=2				
	4	2	16	0.529
	8	4	32	0.279
	12	6	48	0.148
	16	8	64	0.078
	20	10	80	0.042
	24	12	96	0.024
	28	14	112	0.012
	32	16	128	0.006
α=4, r=1				
	4	4	16	0.479
	8	8	32	0.257
	12	12	48	0.140
	16	16	64	0.75
	20	20	80	0.041
	24	24	96	0.022
	28	28	112	0.013

	Cell Generations	Selection Cycles	Nodes	M
$\alpha=4,\ r=2$				
	4	2	16	0.659
	8	4	32	0.478
	12	6	48	0.352
	16	8	64	0.263
	20	10	80	0.190
	24	12	96	0.141
	28	14	112	0.102
	32	16	128	0.075
	36	18	144	0.055
	40	20	160	0.041
	44	22	176	0.029
	48	24	192	0.022
	52	26	208	0.016
	56	28	224	0.012
	60	30	240	0.009

were not expressed within the apical meristem or the cells survived by cross feeding. For mutations which do not fulfill Langridge's requirements and have some expression in heterozygotes, diplontic selection must be considered in modeling the fate of somatic mutations within apical meristems. Klekowski and Kazarinova-Fukshansky (1984b) analyzed the fate of selectively disadvantageous mutations during stochastic growth of the apical meristem. The ability of an apical meristem to be buffered against mutation (i.e., to lose mutations through diplontic or somatic selection) was studied as a function of α and r. Given a starting condition of α-1 nonmutant initials and one initial heterozygous for a mutation, the viability of the nonmutant initial within the context of the meristem is unity ($V = 1$) whereas for the initial heterozygous for the mutation the viability is less than one ($V < 1$). In Table 2 are shown the fixation asymptotics or the frequency of apices which consist of only initials heterozygous for the mutation after repeated cycles of stochastic growth. Such apices consist of cells with disadvantageous genotypes in contrast to the apices with the original nonmutant initials. If we start with apical meristems with one initial heterozygous for a mutation and α-1 nonmutant initials, the frequency of resulting ramets which have fixed the mutation after repeated cycles of stochastic growth is related inversely to the values of α and r. Thus the number of buds or ramets which have only mutant initials decreases as α and/or r increase.

Table 2. Fixation asymptotics for buds with only mutant initials for the case where cells heterozygous for mutations have viabilities of less than one ($V < 1$).

α^*	r^*	V^*=0.95	0.90	0.85	0.80	0.75
2	1	0.449	0.397	0.346	0.296	0.250
	2	0.427	0.357	0.291	0.232	0.181
3	1	0.268	0.207	0.154	0.109	0.075
	2	0.245	0.170	0.111	0.069	0.041
4	1	0.179	0.120	0.075	0.044	0.024
	2	0.157	0.089	0.046	0.022	0.009

*α = The number of apical initials
r = The number of mitotic divisions between selections of apical initials
V = Cell viability

The total number of disadvantageous mutations fixed by a growing genet (tree, clone, etc.) is given by the equation,

$$(R_b)\ (\alpha)\ (\text{Age})\ (\text{Fixation Asymptotic}) = \text{Total Number of Disadvantageous Mutations Fixed},$$

where R_b is the mutation rate for disadvantageous mutations per initial per unit of biological time and age is the number of biological time units (ramet generations, cell cycles, etc.). The product of (α) (Fixation Asymptotic) is a useful index of the propensity of an apical meristem system to accumulate mutations. In Table 3 these products are given for different α, r and V values. It is clear that the total number of disadvantageous mutations fixed in a growing genet is related inversely to the α and r values. Thus all things being equal, genets with apices with high α and/or r values would be expected to accumulate fewer disadvantageous mutations and any given mutant would be present in a smaller proportion of sib ramets.

The within-individual evolution envisioned by Whitham and Slobodchikoff (1981) assumes the occurrence of somatic mutations in the apical meristem which have greater viabilities than the original wildtype cells. Since such somatic mutations would arise as heterozygotes in potential apical initials, some dominance must also be assumed. In an effort to model the dynamics of such diplontic selection and evolution we have assigned viabilities

Table 3. (Fixation asymptotics) α products for the fixation asymptotics in Table 2.

α^*	r^*	V^*=0.95	0.90	0.85	0.80	0.75
2	1	0.898	0.794	0.692	0.592	0.500
	2	0.8540	0.714	0.582	0.464	0.362
3	1	0.8040	0.621	0.462	0.327	0.225
	2	0.7350	0.510	0.333	0.207	0.123
4	1	0.7160	0.480	0.225	0.176	0.096
	2	0.628	0.356	0.184	0.088	0.036

*See Table 2.

of less than one to the nonmutant wildtype cells (V < 1) and a viability of one for the cells heterozygous for a given somatic mutation (V = 1). These different viabilities were substituted into the previously described equations of stochastic growth (Klekowski and Kazarinova-Fukshansky, 1984b). If one begins with meristems with α-1 nonmutant initials and one mutant initial, the fixation asymptotics after many cycles of stochastic growth may be calculated. In Table 4 are shown the fixation asymptotics for apical meristems with only nonmutant initials. Such apices are selectively disadvantaged in contrast to apices which have only mutant initials. It is apparent that the proportion of buds (or ramets) which have disadvantageous genotypes (consisting of only nonmutant initials) increases as α increases and that for any α value, increasing r results in a smaller proportion of ramets with disadvantageous (nonmutant) genotypes.

The total number of selectively advantageous mutations fixed by a growing genet is given by,

$$\text{Rg}\ (\alpha)\ (\text{Age})\ (\text{Fixation Asymptotic}) = \text{Total Number of Advantageous Mutations Fixed},$$

where Rg is the mutation rate for advantageous (good) mutations and the fixation asymptotic is the probability of a meristem having only mutant initials after many cycles of stochastic growth. It should be noted that this asymptotic is one minus the fixation asymptotic in Table 4 for any combination of V, α, and r. Table 5 shows the (fixation asymptotic) α products; it should be noted also that the greater the product the higher the total number of advantageous mutations fixed by a growing genet (assuming age and Rg as constants).

Table 4. Fixation asymptotics for buds with only nonmutant initials for the case where cells heterozygous for a mutation have V = 1 and nonmutant cells have V < 1. Apices which have only nonmutant initials are selectively disadvantaged in comparison to apices which have only mutant initials.

α^*	r^*	V^*=0.95	0.90	0.85	0.80	0.75
2	1	0.449	0.397	0.346	0.296	0.250
	2	0.427	0.357	0.291	0.232	0.181
3	1	0.596	0.520	0.440	0.363	0.290
	2	0.568	0.464	0.364	0.274	0.199
4	1	0.669	0.576	0.478	0.382	0.294
	2	0.636	0.511	0.388	0.281	0.195

*See Table 2.

Table 5. (Fixation asymptotics) α products for the fixation asymptotics for meristems which consist of only mutant initials. Since the viability of mutant initials is one (V = 1) and the viability of nonmutants is less than one (V < 1), apical meristems which consist of only mutant initials are selectively advantaged over those meristems which have only nonmutant initials (see Table 4 for these asymptotics).

α^*	r^*	V^*=0.95	0.90	0.85	0.80	0.75
2	1	1.102	1.206	1.308	1.408	1.500
	2	1.146	1.286	1.418	1.536	1.638
3	1	1.212	1.440	1.680	1.911	2.130
	2	1.296	1.608	1.908	2.178	2.403
4	1	1.324	1.696	2.088	2.472	2.824
	2	1.456	1.956	2.448	2.876	3.220

*See Table 2.

The total number of advantageous mutations fixed by a growing genet increases as α and r increase.

A comparison of the dynamics of the occurrence and fixation of disadvantageous and advantageous somatic mutations reveals some interesting correlations associated with meristem organization. These are shown diagrammatically below,

Increased proportion of a genet which fixes a given advantageous mutation	Low	Apical Initials (α)	High	Increased buffering against disadvantageous mutations, higher total number of advantageous mutations fixed

The increased buffering ability against disadvantageous mutations is reflected in the decreased proportion of a genet which may fix a given mutation and the lower total number of mutations fixed per genet in meristems with higher numbers of apical initials. High numbers of apical initials also maximize the total number of advantageous mutations fixed per genet. In contrast meristems with low numbers of apical initials maximize the proportion of a genet (in terms of number of ramets) which fix any given advantageous mutation. To restate this relation in terms of a tree, decreasing the number of apical initials per meristem increases the proportion of branches which have fixed the same advantageous mutation. Increasing the number of apical initials per meristem increases the number of different advantageous mutations fixed in the various branches, although each mutation is present in fewer branches.

With regard to the number of cell divisions (r) which generate the cell pool ($\alpha 2^r$) from which apical initials are selected, the situation is clearer. Increasing r increases the buffering capacity against disadvantageous mutations as well as the retention and multiplication of advantageous mutations. In terms of our analysis there seems to be no obvious selection for low r values from the viewpoint of mutation.

If one accepts the possibility of diplontic selection within the apical meristem and the general principles of stochastic growth, apical meristem organization may represent patterns which maximize the above genetic relations within the constraints of species specific developmental and ontogenetic patterns. If the mutation rate for advantageous mutations (Rg) is very low it may be that selection for apical systems which maximize the proportion of ramets that have fixed a rare advantageous mutation is more significant than selection to increase the frequency of such mutants. Thus the evolution of an apical meristem which even doubles the frequency of advantageous mutations fixed in a genet may still result in too insignificant a yield of advantageous mutations to generate a significant selection pressure for structural alteration to the meristem organization. In contrast if the rare advantageous mutation occurs, it may be more significant to alter meristem structure to promote the spread of the mutation in the genet (i.e., increase the proportion of ramets which have fixed the mutations). If these arguments are valid, an apical meristem may represent some form of an equilibrium between the following mutually opposing relations:

1.) Increase in the number of apical initials promotes the loss of disadvantageous mutations.

2.) Decrease in the number of apical initials promotes the spread of advantageous mutations.

3.) Increase in r promotes the loss of disadvantageous mutations as well as increasing the total number of advantageous mutations fixed per genet and increasing the proportion of ramets which have fixed a given advantageous mutation.

Stratified Stochastic Meristems

Apical meristems which consist of discrete tunica layers (one or more) with an internal corpus are stratified meristems. Such stratified meristems are composed of component meristems, the individual tunica layers and the internal corpus. The integrity of apical meristems with a tunica-corpus organization is maintained only if the cells of the tunica layer(s) divide primarily in an anticlinal plane. Such apical meristems therefore may exhibit both structured and stochastic properties, the former is a function of the anticlinal patterns of cell division in the tunica and whether permanent initials are present in the component meristems. Stratified meristems exhibit stochastic properties if impermanent apical initials occur in the component meristems and if occasional cell displacement occurs between adjacent component meristems. Such cell displacements have been documented in most periclinal chimeras, but with very low frequencies (10^{-2} - 10^{-3}) and are not simultaneously reciprocal events (Gifford, 1954; Bergann and Bergann, 1962; Tilney-Bassett, 1963; Dermen, 1969; and Stewart and Dermen, 1970). In apices with two tunica layers and an internal corpus, the component meristems are designated LI for the outer tunica, LII for the inner tunica and LIII for the corpus. Cell displacements commonly occur in the following pattern, LI⟶LII⟷LIII, although the origin of adventitious buds and wound tissue may further mix cells from different component meristems.

The individual component meristems may be either structured (with permanent initials) or stochastic (having impermanent initials). All of the variables that have been discussed previously concerning structured and stochastic unstratified meristems apply equally to the individual component meristems which comprise a stratified apical meristem. The mathematical modeling of stratified meristems composed of stochastic component meristems involves a modification of the models of Klekowski and Kazarinova-Fukshansky (1984a, b). Thus rather than having a single stochastic meristem per apical meristem with distinct α and $\alpha 2^r$ values (numbers of apical initials and size of meristematic cell pools), an apical meristem is assumed to have two adjacent stochastic component meristems with limited leakage of cells between component meristems. Each component meristem has the properties of a semi-independent stochastic meristem; these component meristems may be invaded via cell displacements (e.g., periclinal divisions in a tunica) by cells from the adjacent component meristem.

Recently Klekowski, Kazarinova-Fukshansky and Mohr (1985) completed a mathematical analysis of the properties of stratified apical meristems. An apical meristem was modeled which consisted of two adjacent component meristems with the possibility of cell displacement in either direction (this is analogous to the LII and LIII situation in some angiosperm apices). The two kinds of component meristems are designated I and II and both have the same number of apical initials ($\alpha_I = \alpha_{II} = \alpha$). The number of divisions between subsequent selections of initials within a component meristem is r where component meristems have identical r values. Cell displacement between component meristems was quantified and included in the model. Cell displacement probabilities of 0.01 and 0.001 were selected as these values approximate those reported in the literature (Dermen, 1969). Fig. 5 shows how cell displacement was modeled mathematically. If one considers two adjacent component meristems (I and II), the next set of initials for component meristem I are sampled from the cell pool of component meristem I with a probability of p and from the cell pool of meristem II with a probability of q, where $p + q = 1$. Thus for the 0.001 displacement value, $p = 0.999$ and $q = 0.001$. A similar pattern is used to select initials for component meristem II.

Using this model, Klekowski, Kazarinova-Fukshansky and Mohr (1985) studied the change in frequency of neutral and selectively disadvantageous mutations. Starting with α_I-1 nonmutant initials and one mutant initial in component meristem I and α_{II} nonmutant initials in component meristem II the fixation asymptotics were calculated. These fixation asymptotics represented the frequency of apical meristems in which the apical initials in both component meristems were mutant. Fixation asymptotics for various combinations of values of α, r, V, and for two displacement probabilities (0.01 and 0.001) were compared to fixation asymptotics for unstratified stochastic meristems with similar α, r and V values. Thus for example, a stratified meristem with $\alpha_I = 2$ and $\alpha_{II} = 2$ and $r = 2$ for both component meristems was compared to an unstratified meristem with $\alpha = 4$ and $r = 2$. In such comparisons, the fixation asymptotics for selectively neutral mutations were identical. For selectively disadvantageous or advantageous mutations, the fixation asymptotics for stratified meristems were essentially similar to unstratified meristems (+ 5%). Thus it was concluded that meristem

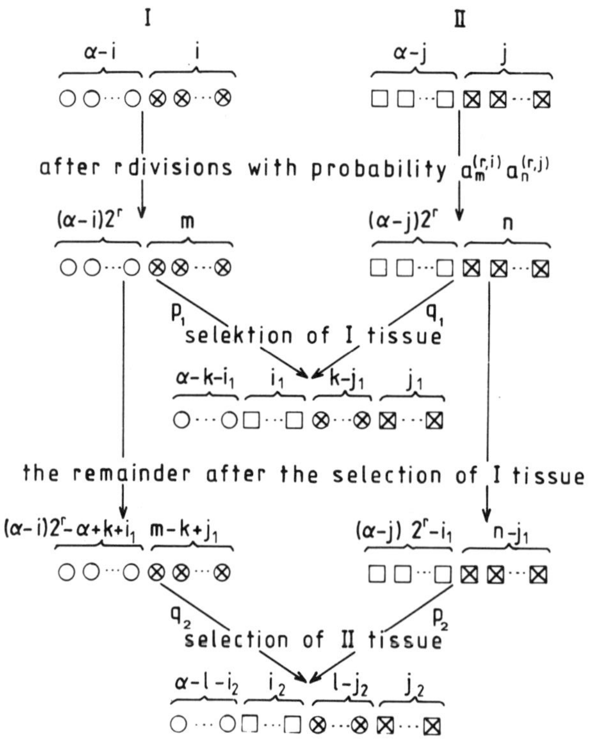

Fig. 5. Model for the mathematical analysis of a stratified stochastic apical meristem with two component meristems (I and II).

stratification has little influence on the frequency of apical meristems which consist solely of mutant initials.

Although the fixation asymptotics were similar, the length of biological time necessary to attain these asymptotics differed greatly between stratified and unstratified and stochastic meristems. The time necessary to approach the fixation asymptotics is orders of magnitude longer in stratified stochastic meristems in comparison to similar unstratified stochastic meristems. In Fig. 6 are shown curves comparing stochastic stratified and unstratified meristems. The frequency of apices with only mutant initials is plotted against number of mitotic cycles of the apical initials for selectively neutral mutations (V = 1). In all cases the fixation asymptotics are $1/\alpha$, (0.25 for α = 4 and 0.25 for $\alpha_I = \alpha_{II}$ = 2 or a total of four apical initials). Note that the most stratified meristem (p = 0.999) takes the greatest number of mitotic cycles to reach the fixation asymptotic.

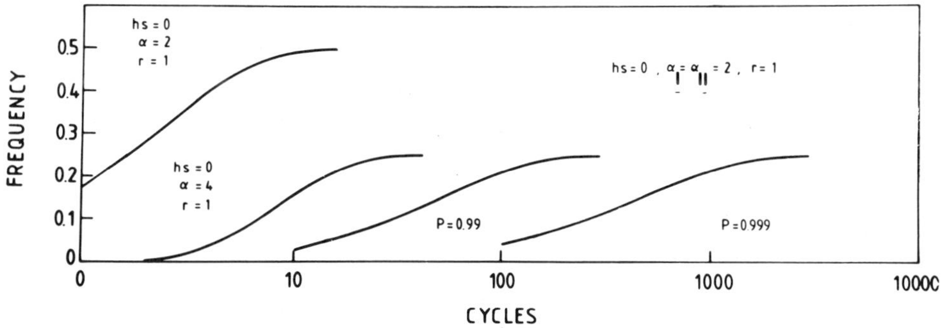

Fig. 6. The fixation of a neutral mutation in stratified and unstratified meristems as a function of the number of selection cycles of apical initials (log scale). Stratified meristems with $\alpha_I = \alpha_{II}$ = 2 and r = 1 and with two cell displacement patterns (P = 0.99 and 0.999) are compared to unstratified meristems with α = 2, r = 1 and α = 4, r = 1. Note that increasing stratification greatly retards the attainment of the fixation asymptotics.

For selectively disadvantageous mutations, meristem stratification results in a similar pattern. In Fig. 7 are plotted the frequency of apices which are a chimera as a function of mitotic cycles of the apical initials for mutant cells which have a lowered viability (V = 0.90). It is obvious that meristem stratification has one important consequence, such meristems promote the retention of cells which have disadvantageous mutant alleles.

Stratified apical meristems therefore appear maladaptive from the viewpoint of buffering against disadvantageous mutations. Although such meristems have comparable mutation asymptotics to similar nonstratified meristems, these asymptotics are achieved

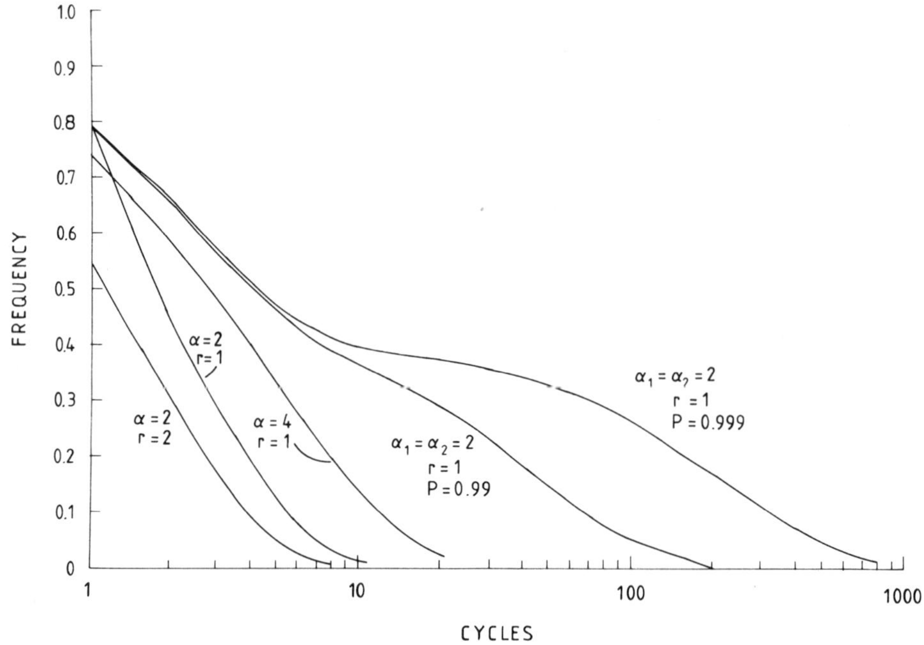

Fig. 7. The frequency of apices which are a chimera as a function of cycles of apical initial selections (log plot). Unstratified meristems (α = 2, r = 1; α = 2, r = 2; α = 4, r = 1) are compared to stratified meristems ($\alpha_I = \alpha_{II}$ = 2, r = 1, P = 0.99 or 0.999). The viability of the mutant cells was V = 0.90 and V = 1 for the nonmutants. Note the persistence of chimerism with increasing stratification.

very slowly. Disadvantageous mutations therefore may persist for long periods as periclinal chimeras within stratified meristems of long-lived perennials and are therefore more likely to give rise to sporogenous tissue and ultimately contribute to mutational load. Evidence for the long term maintenance of selectively disadvantageous mutations is to be found in the numerous examples of horticultural forms which are periclinal chimeras for nuclear or plastid determined chlorophyll deficiencies (Kirk and Tilney-Bassett, 1978). Stewart and Dermen (1979) report finding stable periclinal plastid chimeras in 60 genera within 10 families of monocots. In dicots such forms are even more common (see Hillier, 1972). Although such chlorophyll deficiencies are lethal mutations to an autotrophic plant, their persistence is due to the stratified nature of the apical meristems in these plants. In unstratified meristems such mutations would be fixed in a few ramets or buds and be lost eventually.

The computer analysis has shown that even cell genotypes which are selectively disadvantageous within the context of the apical meristem will persist for considerable periods as periclinal chimeras. Dermen (1969) has shown that plant species differ considerably in the frequency of periclinal divisions and cell displacements between the component meristems. In species with very low frequencies of cell displacements, the component meristems, although in one organism, would be expected to diverge with time due to recurrent mutation. Thus as a long-lived perennial grows, the component meristems could be expected to diverge due to the operation of Muller's Ratchet (Muller, 1964; Haigh, 1978) within the relatively small populations of meristematic cells.

In clonal plants, depending upon the origin of the ramets, different component meristems may be sampled. Reichardt (1955) and Rives (1961) suggested that sampling different component meristems could be a source of genetic variation in cultivated grapes (see also Thompson and Olmo, 1963). For example, assuming three component meristems (LI, LII, LIII), clonal plants reproduced by modified stems (rhizomes, tubers, corms, or bulbs) may reproduce the periclinal chimera in the parent plant unless an environmental insult disturbs normal ontogeny and adventitious buds from the LI form or if buds are regenerated from mixtures of cells from the component meristems at a wound. Clones which reproduce by modified roots (storage roots, root suckers) will represent primarily the LIII in the ramets. Clones in which the leaves are vegetative organs (e.g., *Bryophyllum*) can give rise to a diversity of offspring since component meristems may vary considerably to the extent to which they contribute to leaf tissue. Finally in clones which reproduce via a modification of the sporogenous tissue (as in the apomict *Taraxacum officinale*, Lyman and Ellstrand, 1984) the LII genotype will be represented primarily in these ramets (although it should be noted that the LI also may contribute sporogenous tissue in monocots (Stewart and Dermen, 1979). In upright clones, trees, different branches may be composed of cells derived from different component meristems. For example axillary buds will perpetuate periclinal chimeras, adventitious buds may represent cells from the LI or LII or LIII and root suckers are derived from the LIII (see Pratt et al., 1967 for such variations in apple trees).

Considering mutations which are disadvantageous either within the apical meristem or to the plant in general, selection is much more effective in unstratified meristems than stratified meristems. In plants the selective loss of mutations may occur at the level of diplontic selection within the apical meristem or by selection at the ramet or bud level. Unstratified meristems maximize diplontic selection at the level of the apical meristem and also enhance selective discrimination between ramets. In stratified meristems, diplontic selection operates at the level of the component meristems. Selective discrimination between ramets is not as effective because of the buffered expression of mutants fixed in periclinal chimeras (see Vaughn, 1983 for a review of the interactions of mutant and nonmutant tissues in periclinal chimeras).

CONCLUSIONS

The genetic variance among sib ramets is in great part a consequence of the kind of meristems present in the apices of the plant. Apical meristems may influence mutation by either reducing the mutation rate in initials or by enhancing diplontic selection. The former is possible in structured meristems through nonrandom DNA strand segregation and in stochastic meristems which have a méristème d'attente. Diplontic selection is maximized in apical meristems in which the initials and derivative cells are least determined ontogenetically, thus allowing maximum intercellular competition (i.e., stochastic meristems).

The seed plants primarily have either stochastic unstratified (most gymnosperms) or stochastic stratified meristems (angiosperms). Although diplontic selection is possible in such meristems, such selection is probably not an all inclusive filter against the accumulation of disadvantageous mutations (to the species) during the growth of a long-lived genet. Two arguments may be proposed which suggest the inadequacy of diplontic selection in this regard:

1.) Diplontic selection is dependent upon measurable differences in some aspect of cell viability (survival, division rate, etc.) between nonmutant cells and those heterozygous for a somatic mutation. Since nonmutant and mutant cells are in intimate physical contact nonmutant cells may crossfeed mutant cells and enhance the latter's viability. Also genes that are not expressed in the apical meristem or mutations which are completely recessive will not be influenced by diplontic selection.

2.) Stratified meristems promote the retention of mutations in component meristems regardless of whether such mutations are selectively neutral or disadvantaged in terms of diplontic selection.

Given these two limitations of diplontic selection one would predict that a significant (and generally unrecognized) problem facing plants is the accumulation of mutational load during the growth of any long-lived genet. Such a load has been documented in long-lived fern clones (Klekowski, 1984) and in many gymnosperms (Park and Fowler, 1984) (in angiosperms genetic load studies are uncommon (Levin, 1984) but it is interesting to note that long-lived perennials have the highest levels of genetic variation (Hamrick, 1979)). A consequence of the accumulation of mutations (including recessive lethals) in a growing genet is the progressive increase in self-sterility of the sib ramets. For unlinked recessive lethals,

$$\text{self-sterility} = 1 - (3/4)^n$$

and n is equal to the number of genes heterozygous for lethals in a ramet.

In the past, the evolutionary consequences of somatic mutation during plant growth were generally disregarded. This was due to the way in which mutation rates have been discussed in classical plant evolutionary genetics with its emphasis on advantageous mutations. Thus when mutation rates for given genes in plants are from 10^{-4} - 10^{-6} mutations per gene per generation, the neglect of mutation as an immediate evolutionary force is understandable but incorrect. Many plants are clonal and individual genets may be extremely old (Cook, 1983) thus the mutation rate per gene per generation in an annual has little meaning in an organism 500 years old composed of hundreds of ramets. Fitness in any organism is multigenic, thus if the negative effects of mutation are considered mutation rates must be summed across all loci which can mutate to decrease fitness. In *Drosophila melanogaster* (2n = 8) such estimates yield values that are surprisingly high. Wallace (1968) and Simmons and Crow (1977) estimate 0.005 - 0.006 lethals per chromosome per generation for the second and third chromosomes, or an overall mutation rate per generation in excess of 1% for lethals. Dobzhansky et al. (1977) calculated that the mean rate of all viability mutations is no less than 0.7055 mutations per individual per generation! In *Zea mays* the spontaneous mutation rate for a number of quantitative traits has been estimated at 0.045 mutations per trait per gamete per generation (Sprague, Russel, and Penny, 1960). The mutation rate to the chlorophyll-deficient condition in corn has been estimated at 0.002 per gamete per generation (Crumpacker, 1967). Recent research in plant tissue culture has demonstrated dramatically the frequency of somatic mutations in plants. Protoplast culture and regeneration have documented high frequencies of variant phenotypes in regenerated plants of various asexually reproduced plants (Shepard et al., 1980; Meins, 1983). This "genetic" variation originates in both the cultural and regenerative process as well as being present within the intact plant (*in planta*, Meins, 1983). Lorz and Scowcroft (1983) showed that 25% of regenerates in *Nicotiana tabacum* represented genetic changes which were pre-existing in the intact plants (*in planta*).

Wallace (1975, 1981) considered the problem of decreasing fertility associated with the accumulation of alleles which reduce the viability of homozygotes. Wallace proposed that at least two forms of selection may operate within the context of genetic load, hard selection and soft selection. Hard selection results in a progressive loss of zygotes and reproductive capacity of the population with the increasing frequency of lethals. On the other hand soft selection uncouples (at least partially) the loss of zygotes due to genetic causes and the general decline in reproductive capacity. The concept of soft selection (within the present context) is based on the premise that a certain proportion of zygotes will always fail regardless of their genotype. This loss of zygotes may be broadly considered as due to the environment. Thus for example, if 45% of zygotes fail regardless of genotype, the remaining 55% constitute the reproductive capacity of the plant. If this plant carries two unlinked recessive lethals, then self-sterility in terms of genetically defective zygote genotypes is 43.75%. The reproductive capacity of the plant from the viewpoint of hard selection is (55%) (56.25%) = 30.94% if one assumes that the loss of zygotes due to environment is independent of their genotypes with regard to the lethals. In contrast if soft selection occurs, the genetically defective zygotes are also those that would normally fail regardless of genotype (45%), thus the reproductive capacity of the plant is undiminished in spite of the fact it is heterozygous for two lethals.

In vascular plants a certain proportion of the "environmental" loss of zygotes occurs within the reproductive tissues of the individual plant itself. Buchholz (1922) in a classic but neglected analysis of such selection in vascular plants has termed this the "internal environment" and coined the term developmental selection to describe this kind of soft selection. In Fig. 8 is shown Buchholz's analysis of where developmental selection may operate in vascular plant life cycles. His analysis is broader than our previous discussion in pointing out how developmental selection at the gametophytic and sporophytic levels may serve to uncouple lethals from lost reproductive capacity (see also Klekowski, 1982 for an analysis of one aspect of this kind of soft selection). Many other characteristics such as incompatibility systems (Mulcahy and Mulcahy, 1983), selective seed or fruit abortion (Stephenson

Selection between vegetatively branching parts of either the gametophyte or sporophyte; buds and branches of trees, which later give rise to reproductive parts.

DEVELOPMENTAL SELECTION

Occurring during early embryonic or gametophytic stages within tissues of parent plant, under conditions uniform for competing individuals

Interovular selection, between ovules within same ovary: (1) after fertilization, largely due to activities of contained embryos; (2) before fertilization, due in part to activities of contained female gametophytes, megaspores, or archesporial cells.

Embryonic selection, between embryos within the same ovule, or within tissues of parent gametophyte.

Gametophytic selection: (1) between male gametophytes, such as pollen tubes within carpellary and nucellar tissue; (2) between female gametophytes within the same ovule.

Gametic selection: (1) between male gametes or sperms; (2) between female gametes or eggs.

Fig. 8. Developmental selection in vascular plants (modified from Buchholz, 1922). This kind of soft selection can eliminate lethals without reducing reproductive capacity.

and Bertin, 1983), and low seed-ovule ratios (Wiens, 1984) may all be aspects of a soft-selection sieve whereby mutational load is eliminated with little sacrifice in overall reproductive capacity.

Mutational load also may be a significant feature of other aspects of plant reproductive biology. In long-lived plants, outbreeding mechanisms may represent adaptations to increase fitness by covering recessive lethal and other kinds of disadvantageous alleles. This may explain the association of various outbreeding mechanisms with perenniality, woodiness, and vegetative reproduction (Stebbins, 1950; Grant, 1975; Bawa, 1980; Lloyd, 1982). In the past the genetic systems present in plants have been assumed to be a compromise "between the conflicting requirements of fitness and flexibility" (Stebbins, 1950). This explanation has the same difficulties as explanations of the evolution of sex itself (see Bell, 1982 for a discussion of the rival theories of sexuality and recombination). Muller (1964) suggested that the function of sex is to facilitate the elimination of unfavorable mutations (see Bell, 1982 for review). Because many vascular plants have patterns of growth and development which allow the accumulation of somatic mutations in cell lines which give rise to meiocytes, genetic systems which promote recombination may have evolved for the immediate end of eliminating or concealing deleterious mutations. Charlesworth and Charlesworth (1978) have demonstrated mathematically the importance of severe inbreeding depression in the evolution of dioecy; one of the consequences of mutational load of course is inbreeding depression. Further modifications, refinements and specializations of these genetic systems may well reflect resource limitations (Bawa and Beach, 1981) and aspects of sexual selection (Willson, 1979), but our analysis indicates that the primary selective value of genetic systems in long-lived plants is negating and repairing recurrent mutational load. Thus the reproductive characteristics of long-lived plants must be considered at least in part a reflection of the anatomical, morphological, and ontogenetic constraints of form and growth peculiar to these organisms.

ACKNOWLEDGEMENTS

The senior author (EJK) thanks the University of Massachusetts for travel funds and for granting a six month leave of absence to complete these researches in Freiburg. He also would like to thank the Alexander von Humboldt Stiftung for their continued support of this project. NKF is obliged to the Deutshe Forschungsgemeinschaft (Scha 303) for financial support. We benefited from discussions with Drs. L. Fukshansky and E. Schäfer of the University of Freiburg and Drs. E. Davis, O. Stein, M. Marcotrigiano and D. Mulcahy of the University of Massachusetts. We also thank Professor L. Tikos of the University of Massachusetts for translating the Grodzinsky and Gudkov publication and Dr. John Cairns of Harvard University for bringing the Kay publication to our attention.

REFERENCES

Balkema, G. H., 1972, Diplontic drift in chimeric plants, Radiation Botany, 12:51-55.

Bawa, K. S., 1980, Evolution of dioecy in flowering plants, Annual Rev. Ecol. Syst., 11:15-39.

Bawa, K. S., 1984, The evolution of dioecy-concluding remarks, Ann. Missouri Bot. Gard., 71:294-296.

Bawa, K. S., and Beach, J. H., 1981, Evolution of sexual systems in flowering plants, Ann. Missouri Bot. Gard., 68:254-274.

Bell, G., 1982, "The Masterpiece of Nature," University of California Press, Berkeley.

Bergann, F., and Bergann, L., 1962, Uber Umschichtungen (Translokationen) an den Sprossscheiteln periklinaler Chimaren, Zuchter, 32:110-119.

Bierhorst, D. W., 1971, "Morphology of Vascular Plants," The Macmillan Co., New York.

Bierhorst, D. W., 1977, On the stem apex, leaf initiation and early leaf ontogeny in Filicalean ferns, Amer. J. Bot., 74:125-152.

Buchholz, J. T., 1922, Developmental selection in vascular plants, Bot. Gaz., 73:249-286.

Buiatti, M., Baroncelli, S., Testi, R., and Boscariol, P., 1970, The effect of environment on diplontic selection in irradiated Gladiolus corms, Radiation Botany, 10:531-538.

Buss, L. W., 1983, Evolution, development and the units of selection, Proc. Natl. Acad. Sci. USA, 80:1387-1391.

Buvat, R., 1952, Structure, evolution et functionnement du meristeme apical de quelques dicotyledones, Ann. Sci. Nat. Bot. Ser., 13:199-300.

Cairns, J., 1975, Mutation, selection and the natural history of cancer, Nature, 225:197-200.

Charlesworth, B., and Charlesworth, D., 1978, A model for the evolution of dioecy and gynodioecy, Amer. Naturalist, 112: 975-997.

Chen, C. H., and Ross, J. G., 1965, Colchicine-induced somatic chromosome reduction in sorghum. V. Diploidization of the stem apex after treatment of tetraploid seedlings, Can. J. Genet. Cytol., 7:21-30.

Cook, R. E., 1983, Clonal plant populations, American Scientist, 71:244-252.

Crumpacker, D. W., 1967, Genetic loads in maize and other cross-fertilized plants and annuals, Evol. Biol., 1:306-424.

Cuany, R. L., Sparrow, A. H., and Pond, V., 1958, Genetic response of Antirrhinum majus to acute and chronic plant irradiation, Z. Induktive Abslammuings. u. Vererbungolehre, 89:7-13.

Dermen, H., 1969, Directional cell division in shoot apices, Cytologia, 34:541-558.

Dobzhansky, T., Ayala, F. J., Stebbins, G. L., and Valentine, J. W., 1977, "Evolution," W. H. Freeman, San Francisco.

Esau, K., 1953, "Plant Anatomy," J. Wiley and Sons, New York.

Esau, K., 1977, "Anatomy of Seed Plants," J. Wiley and Sons, New York.

Gaul, H., 1965, Selection in M_1 generation after mutagenic treatment of barley seeds, in: "Induction of Mutations and the Mutation Process," J. Veleminsky and T. Gichner, eds., Czechoslovak Academy of Sciences, Prague, pp. 62-71.

Gifford, E. M., 1954, The shoot apex in angiosperms, Bot. Rev., 20:447-529.

Gifford, E. M., and Corson, G. E., 1971, The shoot apex in seed plants, Bot. Rev., 37:143-229.

Grant, V., 1975, "Genetics of Flowering Plants," Columbia University Press, New York.

Grinikh, L. I., Shevchenko, V. V., Grigor'eva, G. A., and Draginskaya, L. Ya., 1974, Study of chimerism in the reproductive tissue of Arabidopsis thaliana plants following irradiation of seeds, Genetika, 10:18-28.

Grodzinsky, D. M., and Gudkov, I. M., 1982, Heterogeneity of meristems - the basis of higher plant reliability. Academy of Sciences of the Ukrainian S. S. R., Physiology and Biochemistry of Cultivated Plants, 14:107-118 (translated from Russian).

Haigh, J., 1978, The accumulation of deleterious genes in a population - Muller's Ratchet, Theor. Pop. Biol., 14:251-267.

Hamrick, J. L., 1979, Genetic variation and longevity, in: "Topics in Plant Population Biology," O. T. Solbrig, S. Jain, G. B. Johnson and P. H. Raven, eds., Columbia University Press, New York, pp. 84-113.

Hillier, H. G., 1972, Hillier's Manual of Trees and Shrubs, A. S. Baines and Co., New York.

Johnson, M. A., 1951, The shoot apex in gymnosperms, Phytomorphology, 1:188-204.

Kay, H. E. M., 1965, How many cell-generations?, The Lancet, 2: 418-419.

Kirk, J. T. O., and Tilney-Bassett, R. A. E., 1978, "The Plastids," Elsevier/North-Holland Biomedical Press, New York.

Klekowski, E. J., Jr., 1984, Mutational load in clonal plants: A study of two fern species, Evolution, 38:417-426.

Klekowski, E. J., Jr., and Kazarinova-Fukshansky, N., 1984a, Shoot apical meristems and mutation: Fixation of selectively neutral cell genotypes, Amer. J. Bot., 71:22-27.

Klekowski, E. J., Jr., and Kazarinova-Fukshansky, N., 1984b, Shoot apical meristems and mutation: Selective loss of disadvantageous cell genotype, Amer. J. Bot., 71:28-34.

Klekowski, E. J., Jr., Kazarinova-Fukshansky, N., and Mohr, H., 1985, Shoot apical meristems and mutations: Stratified meristems and angiosperm evolution, Amer. J. Bot., (submitted).

Kuligowski-Andres, J., and Tourte, Y., 1979, La determination des cellules apicales chez une Pteridophyte; role particulier du genome paternel, Bull. Soc. bot. Fr., Lettres bot., 126: 491-505.

Kuligowski-Andres, J., Tourte, Y., and Faivre-Baron, M., 1979, La differenciation cellulaire de l'embryon: genome paternel et cellules organogenes chez une Pteridophyte, C. R. Acad. Sc. Paris, 289:1093-1096.

Langenauer, H., and Davis, E. L., 1973, Helianthus annuus responses to acute x-irradiation. I. Damage and recovery in the vegetative apex and effects on development, Bot. Gaz., 134:301-316.

Langridge, J., 1958, A hypothesis of developmental selection exemplified by lethal and semi-lethal mutants of Arabidopsis, Aust. J. Biol. Sci., 11:58-68.

Levin, D. A., 1984, Inbreeding depression and proximity-dependent crossing success in Pholx drummondii, Evolution 38:116-127.

Lintilhac, P. M., and Green, P. B., 1976, Patterns of microfibrillar order in a dormant fern apex, Amer. J. Bot., 63:726-728.

Lloyd, D. G., 1982, Selection of combined versus separate sexes in seed plants, Amer. Naturalist, 120:571-585.

Lorz, H., and Scowcroft, W. R., 1983, Variability among plants and their progeny regenerated from protoplasts of Su/su heterozygotes of Nicotiana tabacum, Theor. Appl. Genet., 66:67-75.

Lyman, J. C., and Ellstrand, N. C., 1984, Clonal diversity in Taraxacum officinale (Compositae) an apomict, Heredity, 53: 1-10.

McAlpin, B. W., and White, R. A., 1974, Shoot organization in the filicales: the promeristem, Amer. J. Bot., 61:562-579.

Meinke, D. W., and Sussex, I. M., 1979a, Embryo-lethal mutants of Arabidopsis thaliana: A model system for genetic analysis of plant embryo development, Develop. Biol., 72:50-61.

Meinke, D. W., and Sussex, I. M., 1979b, Isolation and characterization of six embryo-lethal mutants of Arabidopsis thaliana, Develop. Biol., 72:62-72.

Meins, F., Jr., 1983, Heritable variation in plant cell culture, Ann. Rev. Plant Physiol., 34:327-346.

Mulcahy, D. L., and Mulcahy, G. B., 1983, Gametophytic self-incompatibility reexamined. Science, 220:1247-1251.

Muller, A. J., 1965, The chimerical structure of M_1 plants and its bearing on the determination of mutation frequencies in Arabidopsis, in: "Introduction of Mutations and the Mutation Process," J. Veleminsky and T. Gichner, eds., Czechoslovak Academy of Sciences, Prague, pp. 46-52.

Muller, H. J., 1964, The relation of recombination to mutational advance, Mutat. Res., 1:2-9.

Neilson-Jones, W., 1969, "Plant Chimeras," (2nd ed.), Methuen and Co., London.

Newman, I. V., 1965, Pattern in the meristems of vascular plants. III. Pursuing the patterns where no cell is a permanent cell, J. Linn. Soc. London Bot., 59:185-214.

Park, Y. S., and Fowler, D. P., 1984, Inbreeding in black spruce (Picea mariana (Mill.) B.S.P.): self-fertility, genetic load, and performance, Can. J. For. Res., 14:17-21.

Pohlheim, F., 1980, Zur sprossvariation bei den Cupressaceae, Wiss. Z. d. Humboldt-Univ. zu Berlin, Math.-Nat. R., 39S:295-306.

Popham, R. A., 1951, Principal types of vegetative shoot apex organization in vascular plants, Ohio Jour. of Science, 51: 249-270.

Pratt, C., Ourecky, D. K., and Einset, J., 1967, Variation in apple cytochimeras, Amer. J. Bot., 54:1295-1301.

Reichardt, A., 1955, Experimentalle Untersuchungen uber den Effekt von Rontgenstrahlen in der vegetativen Vermehrung einer alten Rebensorte, Die Gartenbauwissenschaft, 20:355-413.

Rives, M., 1961, Bases genetiques de la selection clonale chez la vigne, Ann. Amelior. Plantes, 11:337-348.

Ruth, J., Klekowski, Jr., E. J., and Stein, O. L., 1985, Impermanent initials of the shoot apex and diplontic selection in a juniper chimera, Amer. J. Bot., (in press).

Shepard, J. F., Bidney, D., and Shakin, E., 1980, Potato protoplasts in plant improvement, Science, 28:17-24.

Simmons, F. C., and Crow, J. F., 1977, Mutations affecting fitness in Drosophila populations, Ann. Rev. Genet., 11:49-78.

Soma, K., and Ball, E., 1964, Studies of the surface growth of the shoot apex of Lupinus albus, Brookhaven Symp. Biol., 16:13-45.

Sprague, G. F., Russel, W. A., and Penny, L. H., 1960, Mutations affecting quantitative traits in the selfed progeny of doubled monoploid maize stocks, Genetics, 45:855-866.

Stebbins, G. L., 1950, "Variation and Evolution in Plants," Columbia University Press, New York.

Steeves, T. A., and Sussex, I. M., 1972, Patterns in Plant Development, Prentice-Hall, Inc., Englewood Cliffs, New Jersey.

Stephenson, A. G., and Bertin, R. I., 1983, Male competition, female choice, and sexual selection in plants, in: "Pollination Biology," L. Real, ed., Academic Press, New York, pp. 110-151.

Stewart, R. N., 1978, Ontogeny of the primary body in chimeral forms of higher plants, in: "The Clonal Basis of Development," S. Subtelny and I. M. Sussex, eds., Prentice-Hall, Inc., Englewood Cliffs, New Jersey, pp. 131-160.

Stewart, R. N., and Dermen, H., 1970, Determination of number and mitotic activity of shoot apical initial cells by analysis of mericlinal chimeras, Amer. J. Bot., 57:816-826.

Stewart, R. N., and Dermen, H., 1979, Ontogeny in monocotyledons as revealed by studies of the developmental anatomy of periclinal chloroplast chimeras, Amer. J. Bot., 66:47-58.

Sussex, I., and Rosenthal, D., 1973, Differential ^{3}H-Thymidine labeling of nuclei in the shoot apical meristem of Nicotiana, Bot. Gaz., 134:295-301.

Thielke, C., 1959, Der Sprossscheitel in der Gattung Saccharum, Naturwiss., 46:478-479.

Thompson, M. M., and Olmo, H. P., 1963, Cytohistological studies of cytochimeric and tetraploid grapes, Amer. J. Bot., 50: 901-906.

Tilney-Bassett, R. A. E., 1963, The structure of periclinal chimeras, Heredity, 18:265-285.

Tourte, Y., Kuligowski-Andres, J., and Barbier-Ramond, C., 1980, Comportement differentiel des chromatines paternelles et maternelles au cours de l'embryogenese d'une fougere: Le Marsilea, European Journal of Cell Biology, 21:28-36.

Vaughn, K. C., 1983, Chimeras and variegation: problems in propagation, Hort. Science, 18:845-848.

Wallace, B., 1968, "Topics in Population Genetics," Norton, New York.

Wallace, B., 1975, Hard and soft selection revisited, Evolution, 29:465-473.

Wallace, B., 1981, Basic Population Genetics, Columbia Press, New York.

White, J., 1979, The plant as a metapopulation, Ann. Rev. Ecol. Syst., 10:109-145.

Whitham, T. G., and Slobodchikoff, C. N., 1981, Evolution by individuals, plant-herbivore interactions, and mosaics of genetic variability: the adaptive significance of somatic mutations in plants, Oecologia (Berl.), 49:287-292.

Wiens, D., 1984, Ovule survivorship, brood size, life history, breeding systems, and reproductive success in plants, Oecologia (Berl.), (in press).

Willson, M. F., 1979, Sexual selection in plants, Amer. Naturalist, 113:777-790.

PROPERTIES OF MUTABLE ALLELES RECOVERED FROM MUTATOR STOCKS OF ZEA MAYS L.

Virginia Walbot, Catherine P. Briggs, and Vicki Chandler

Department of Biological Sciences
Stanford University
Stanford, CA 94305

INTRODUCTION

The study of transposable elements and the highly mutable alleles they create allows recognition of the fluid nature of both prokaryotic and eukaryotic genomes. Mutations at particular alleles have been instrumental in dissecting the spatial and temporal patterns of gene expression in maize (*Zea mays* L.) and *Drosophila*. And with the advent of molecular cloning, the mutable alleles containing a particular insertion element can be isolated by virtue of the transposon "tag" they contain. Recently we have begun studying the genetic and molecular properties of mutable alleles derived from maize lines containing Mutator activity. In this paper we review the properties of unstable alleles of maize and the genetic properties of Mutator lines of maize. A new transposable element family has been discovered and isolated from an unstable allele recovered from a Mutator line (Barker, et al., 1984). The molecular properties of this element called *Mu-1* will be reviewed.

The methods for recovering mutable alleles using Mutator stocks will be described as well as the properties of the mutable alleles we have recovered in genes of the anthocyanin biosynthetic pathway. In particular, we will report on changes in the frequency of somatic reversion occuring within a kernel, on an ear or within the whole plant. Finally we present evidence for the segregation of *Mu-1* elements during outcrosses and discuss the prospects for identifying and cloning mutable alleles containing *Mu-1* elements.

BACKGROUND

Mutable alleles, which by definition demonstrate a frequent reversion to a normal or near-normal phenotype, were reported shortly after maize became an organism of interest to geneticists at the turn of the century (Emerson, 1914). Like so many other phenomena described in maize, these mutable alleles were studied in genes controlling the biosynthesis of anthocyanin pigment in the pericarp and aleurone tissue of the kernel and in plant parts, particularly the cob (Coe and Neuffer, 1977). For one such mutable allele of the A1 locus ($a1^{m-1}$), Rhoades (1938) demonstrated that another gene which he termed Dotted (Dt) could act *in trans* to promote the frequent somatic reversion of the colorless $a1^{m-1}$ aleurone tissue to full pigmentation characteristic of the wild type A1 allele.

This important observation on trans-activation was followed by the illuminating experiments of McClintock (1956; reviewed by Federoff, 1983) who demonstrated the existence and transposition of other two element gene regulators including Ac-Ds and Spm, in which one component such as Ac acts *in trans* to regulate the activities of Ds elements dispersed in the genome.

In the two element systems, only one component is capable of directing the insertion and excision events controlling its own transposition and that of related but defective elements; in the Ac-Ds system, the Ac member is the autonomous component while the non-autonomous Ds elements contain sequences responding to the Ac-encoded activities. The element systems described by McClintock in the 1940's and 1950's and additional systems described by Peterson and his colleagues (reviewed by Freeling, 1984) are shown to be independent entities by demonstrating that the individual autonomous elements exert no control over the affected alleles regulated by another system. This genetic proof of non-identity is likely to be confirmed by molecular evidence of a lack of shared sequence between element systems; there is some homology within an element family as has already been shown for Ac and numerous sizes of Ds elements, all of which share the same, short terminal inverted repeats with Ac (Freeling, 1984; Döring and Starlinger, 1984).

Recent studies of the molecular basis of the activities of these controlling elements have, for the most part, confirmed McClintock's hypotheses regarding their mode of action. In brief, these elements are transposable and when they insert into or near a gene they disrupt the normal expression of that gene, often resulting in null or low expression. These elements also cause a short duplication of host sequence information at the site of insertion, and the length of this duplication is a characteristic of the type of element involved (Freeling, 1984). The reversion of the affected allele to normal activity occurs when the inserted element is excised, either precisely at its termini leaving intact the host sequence duplication created during insertion or imprecisely in which case parts of the host sequence duplication are altered or lost (Sutton et al., 1983; Döring and Starlinger, 1984).

McClintock (1978) proposed that previously cryptic controlling elements may be activated by chromosome breakage or other forms of stress (see also Burr and Burr, 1981). Indeed, *Ds* was first recognized as a point on chromosome 9^S at which there was a high frequency of chromosomal breakage. Recently Mottinger, et al., (1984) recovered new insertion-mediated mutants of *Adhl* at a frequency exceeding spontaneous mutation in a stock infected by barley stripe mosaic virus; the insertion elements which moved into the *Adhl* gene were shown to be a normal, but previously stable, part of the maize genome (M.R. Johns, pers. comm.). Thus, it remains a mystery at what frequency these elements move, whether they are specifically activated in virus-infected plants, and how their movement is regulated within the maize genome.

PHYSICAL AND GENETIC PROPERTIES OF *Mu-1*, A TRANSPOSABLE ELEMENT ISOLATED FROM MUTATOR STOCKS

About ten years ago Robertson (1978 and references therein) reported a line of maize with an abnormally high spontaneous mutation rate. This line was termed Mutator. The properties of this line were dominant and were transmitted equally well through both parents (Robertson, 1980). The properties were not influenced in tests with a number of known transposable elements (Robertson and Mascia, 1981) and no separation could be made between autonomous and non-autonomous components. The inheritance was non-Mendelian, however, because about 90% of the Fl progeny of a Mutator crossed to a normal line inherited the activity (Robertson, 1978). This last property tended to rule out the otherwise plausible hypothesis that the Mutator activity was due to an error-prone DNA replication process and strongly suggested the presence of a new family of transposable elements.

A mutable allele of *Adhl-S* was recovered from a plant population with Mutator activity after allyl alcohol selection against pollen carrying wild type levels of alcohol dehydrogenase enzyme activity; this mutant, *S3034*, produced only 40% of normal ADHI enzyme activity and was shown to contain an insertion of ~1.5 kb relative to the progenitor *S* allele (Strommer et al., 1982). Subsequent analysis demonstrated that this insertion event had occurred within the first intron of the gene and was accompanied by a 9 bp host sequence duplication (Bennetzen et al., 1984). The insertion element was subsequently sequenced (Barker et al., 1984) and shown to be 1,367 bp long with 213 and 215 bp inverted terminal repeats. This element is termed *Mu-1*.

Multiple copies of DNA sequences related to Mu-1 are found in Mutator lines of maize. When nuclear DNA from Mutator lines is digested with a restriction enzyme such as Hinf I which recognizes sites only within the terminal repeat, blotted to nitrocellulose, and hybridized with Mu-1 probes, primarily a 1.3 kb size class characteristic of the cloned Mu-1 element is observed. A second size class, approximately 300 bp larger than Mu-1 is also detected in many Mutator lines (Barker et al., 1984). This larger element, designated Mu-L, has been cloned and analyzed (Taylor and Walbot, unpublished data). It contains inverted terminal repeats similar to Mu-1 and retains many, but not all restriction site characteristics of Mu-1. The additional 300 bp "insert" is not homologous to any region of Mu-1. The restriction site polymorphisms suggest that Mu-L and Mu-1 are not simply related by a single insertion or deletion event. In the Mutator lines analyzed in the Adh1 experiments, about 10-30 copies of cross-hybridizing sequences were detected (Strommer et al. 1982), but we have identified Mutator lines with, in some cases, more than 100 copies (Chandler and Walbot, unpublished data). Interestingly, the total copy number of Mu-1 homologous elements and the ratio of Mu-1 and Mu-L can vary widely between Mutator lines (Chandler, Taylor, and Walbot, unpublished data).

Cloning Genes Using Transposable Elements as Tags

In addition to the intrinsic interest in mutable alleles, such material provides a means to clone the gene into which the element has inserted. The principle behind transposon "tagging" is simple and has found wide application in the cloning of bacterial genes and a few genes in *Drosophila*. First, a clone of the transposable element must be obtained. The next step is to screen or select for mutants from a stock containing active copies of the cloned transposable element. The overall mutation rate should be much higher than normal in such a stock, and many of the new mutants will be due to the insertion of the transposable element into the gene of interest. To isolate this mutant gene from a genomic library, the cloned transposable sequence is used for the screening process on the assumption that the mutation arose from the insertion of this new sequence.

In eukaryotes, even lower eukaryotes such as yeast which have a small genome with little repetitive sequence, the transposable elements are present in multiple copies in the genome. (For example, there are about 30 copies of the Ty1 element in haploid *Saccharomyces cerevisiae* [Roeder and Fink, 1983] and about 300 copies of the Tc1 element in the Bergerac strain of *Caenorhabditis elegans* that has an active mutator system [Emmons, et al., 1983]). With such a large number of elements resident in the genome, the major task is determining which element is closely linked to the mutation of interest. One exception to the high copy number of transposable elements is Ac, the autonomous partner of the Ac-Ds controlling element system of maize. Although there are at least 3 types of Ds in the genome each present in dozens of copies, there are only a few copies of Ac. The Ac and Ds elements share some sequence homology, yet there is an internal region of Ac totally missing in the Ds elements, and this sequence is a unique probe for Ac. Using such an internal clone of Ac Fedoroff, et al., (1984) were able to identify the Bronzel (Bzl) gene of maize by first identifying and cloning a DNA fragment containing a bzl mutable allele harboring an intact Ac element. Only a few copies of sequences hybridizing to Ac were present in their stock, and the bzl mutant had been rigorously demonstrated to contain Ac by showing that this allele was required and sufficient for Ds instability elsewhere in the genome. Consequently, only a few sequences had to be cloned and examined for correlation with the mutable phenotype at bzl to verify that a particular clone was the Ac insertion into that locus. The advantages to using a low copy number element such as Ac as a transposon tag are obvious; however, in a stock containing Ac-Ds most mutations result from Ds insertion. This probably occurs simply because there are many mobile Ds elements and only one or a few Ac elements. Thus, the difficulty with using Ac is obtaining the appropriate mutant initially; this is no difficulty for those genes for which Ac insertions have already been verified, but may prove to be a major task for the generation of new mutations at specific loci.

The Mutator system represents a compromise between the choice of low copy number for ease of cloning and high copy number for ease of recovery of appropriate mutants. In Mutator lines, the initial copy number of Mu-like elements is in the range of 10 - 50, although some lines have more copies than that. It is unknown whether some elements are non-autonomous, and whether all mutants recovered from a Mutator background contain the same (or a closely related) insertional sequence. These features make it more difficult to clone the mutable alleles in a Mutator plant. On the other hand, the mutation frequency is significantly higher in Mutator backgrounds than for the recovery of Ac mutations.

In this report we describe our use of Mutator lines to obtain mutable alleles of genes controlling anthocyanin synthesis and the properties of the mutant alleles we have recovered. The genetic study of these alleles should allow development of the appropriate tester lines to determine the autonomy or non-autonomy of individual insertion elements and will provide successive generations for demonstrating the linkage between a particular DNA fragment and the mutable phenotype. With >10 copies of the insertion sequence, cloning of a particular affected allele may be most efficiently done in conjunction with such genetic tests.

GENERATION OF MUTANTS IN A MUTATOR LINE

The purpose of these experiments was to generate mutable alleles in as many genes of the anthocyanin biosynthetic pathway as possible using a Mutator stock containing the Mu-1 transposable element. One long term goal was to clone such alleles using the Mu-1 element as a transposon tag. The other goal was to study the pattern of somatic reversion of the affected alleles to discover whether the Mu-1 elements underwent changes of state and to discover how such alterations in the timing or pattern of somatic reversion were genetically controlled.

In 1982 we received 53 kernels of the Mutator stock (a generous gift from D. S. Robertson). Because of the limited seed stock, our initial screen for mutable alleles utilized pollen from these plants that were homozygous for the dominant alleles required for full kernel color (A1 A2 Bz1 Bz2 C1 C2 Pr R) crossed onto to seven individual recessive testers (see Table 1). All of the F1 kernels should be fully purple; rarely a colorless kernel is produced due to contamination, a diploid endosperm (no pollen contribution) or a new null mutation. Mutable alleles are flecked with dots or streaks of color against a colorless background. Such spotted kernels produced during the initial screen were picked for further study.

The purple Mutator stock was sib or self-crossed to provide a larger supply so that in 1983 we could use the Mutator plants as the female parent crossed by the recessive tester (Table 2). In such crosses, contamination resulting in colorless kernels is impossible; colorless kernels can arise from a haploid endosperm (no maternal contribution) or a new null mutation.

In addition appropriate material for the selection of mutations at R and B^{peru} were developed from the original Mutator stock (see Tables 3 and 4). From the 53 kernels of the purple Mutator stock provided by D. S. Robertson, plants with three kinds of R constitution were also identified by crossing each plant as male to an r-g tester. Each plant was also selfed or sibcrossed to propagate the stock. R/R plants (37/53) give a mottled F1 due to the incomplete activation of male-transmitted standard R. $R^{self\ color}$/R (14/53) yielded ears with half fully colored kernels (R^{sc}/r-g) and half with very infrequent patches of color (R′/r-g) due to the paramutation of standard R to an R′ allele of reduced activity in the presence of R^{sc} in the previous generation. R^{sc}/R^{sc} (2/53) plants were identified as producing a fully purple heterozygote. The Mutator line R^{sc} homozygotes were crossed to R^{sc}/R^{scm} stock to create a mutable R self color line which was used as a male parent onto r-g tester. Alternatively, R standard Mutator plants were used as female and pollinated by r-g in controlled crosses (1983) or in an isolation plot (1984).

Table 1. Maize Stocks and Strategies Used in the Screen for Mutable Alleles at A1, A2, Bz1, Bz2, C1, C2, and Pr in 1982

Locus	Number of Kernels Scored	Number of Mutable Mutants	Mutants per 10^6 Gametes
A1	23,250	0	-
A2	62,250	3	48
Bz1	49,500	0	-
Bz2	60,000	1	17
C1	82,250	0	-
C2	89,500	1	11
Pr	86,250	0	-
Total	453,000	5	11

Notes. Robertson's Mutator purple stock was homozygous for all of the dominant factors required for accumulation of purple anthocyanin pigment in the aleurone tissue. The type of R allele present was determined during the course of the experiment by crossing this stock onto r-g tester. The original Robertson stock was shown to be a mixture of R, R/R^{scm}, R^{scm}. Each tester was homozygous for all of the dominant factors required for the accumulation of purple anthocyanin pigment in the aleurone tissue (A1, A2, Bz1, Bz2, C1, C2, Pr, and R) with the exception of the one locus examined in each experiment. In most stocks the recessive gene of the anthocyanin biosynthetic pathway was flanked by other recessive markers: a1sh2, a2btpr, sh1bz1wx, bz2, c1sh1wx, c2su, and pr.

Purple Mutator stock was used as the male parent onto blocks of detasseled tester plants in controlled crosses. Because the Mutator parent was used as male all null kernels were discarded as possible contaminants and no attempt was made to calculate a mutation frequency except for the mutable alleles. All of the mutable alleles listed above have been propagated for several plant generations and continue to display a mutable phenotype.

Table 2. Maize Stocks and Strategies Used in the Screen for Mutable Alleles at Al, Bzl, and Cl in 1983

Locus	Number of Kernels Scored	Number of Mutable Mutants	Mutants per 10^6 Gametes
Al	124,250	0	-
Bzl	102,750	2	20
Cl	93,000	0	-
Total	320,000	2	6

Notes. Sib and self progeny of the Robertson Mutator purple stock with a standard R allele were used. Recessive testers were as described in Table 1 for Al, Bzl, and Cl. Blocks of 300 - 400 detasseled Mutator plants were pollinated by each tester in controlled crosses. About 25 Mutator plants were individually used as male onto 100 tester plants of each genotype. The two bzl mutable alleles were recovered from two plants in a family of 10 Mutator plants derived from a sib cross of the original Mutator stock in which one parent had given rise to a putative mutable event in the previous generation. This family, used as female, was crossed by the recessive bzl tester.

Table 3. Screen for Mutables Alleles of R from a Mutator Line

Year	Number of Kernels Scored	Number of Mutable Mutants	Mutants per 10^6 Gametes
1983	186,250	0	-
1984	72,000	6	84

Notes. As noted in Table 1, each of the original 53 purple Mutator stock were used as male on r-g tester to determine the R allele present. Paramutation, a reduction in pigmentation regulated by the R allele after exposure to certain self-color alleles of R (see discussion by Coe and Neuffer, 1977), would confound the identification of mutable alleles resulting from insertional mutations. Consequently care was taken to select homozygous standard R plants or homozygous self-color R plants from the original population, r-g tester was used throughout.

In 1983 a mixture of sib and self kernels from R plants were used as female in controlled crosses by the r-g tester producing fully purple kernels in the progeny. The families chosen had, in the previous generation, given rise to mutable alleles. Mutator stock containing a self-color R allele was used as male in controlled crosses onto the tester.

In 1984 the same Mutator sources were planted in an isolation plot, detasseled and open-pollinated by tester. The mutants obtained in 1984 have not yet been verified by linkage and transmission tests so no total mutation frequency is calculated. Interestingly some of the mutants identified appear to have arisen early in plant development because both ears from the plant in which the event occurred contained spotted kernels.

Subsequently, a B^{peru} Mutator line was derived by crossing the Mutator R/r-g (b/b) x B^{peru} r-g. When planted, the seedlings inheriting R were identified by their red leaf margins and were culled from the field to leave B^{peru}/b r-g Mutator plants. These were pollinated by b r-g individuals yielding ears with 50% fully purple B^{peru}/b r-g kernels (B^{peru} results in full kernel pigmentation as would an R^{sc} allele) and 50% colorless kernels b r-g. These colorless kernels were carefully scanned to identify any spotted kernels. A complete description of the strategies used is given in Table 4.

The first two years in which we screened for mutable alleles we recovered eight verified alleles. In 1984, three additional B^{peru} mutants with the appropriate phenotype were recovered. The mutant phenotypes were heritable, but their linkage must still be tested. Six putative R mutants were recovered, but their transmission and linkage must be tested. The mutation rate considering the data only for the eight verified mutable alleles is 1.2×10^{-5}. As reported by Robertson (1978), the frequency is much higher than the spontaneous mutation frequency of $\sim 10^{-6}$ for

Table 4. Screen for Mutable Alleles of *Bperu* from a Mutator Background

Year	Number of Kernels Scored	Number of Mutable Mutants	Mutants per 10^6 Gametes
1983	178,000	1	6
1984	275,000	3	11

Notes. One of the original purple stock plants (B139-17) gave rise to *a2*$^{mu-2}$ in 1982; this plant had also been crossed onto *r-g* tester and shown to carry the standard *R* allele. In winter 1983, the *R/r-g* hybrid progeny, carrying an unknown plant color *B* allele but not *B*^{peru}, were crossed to and by homozygous *B*^{peru} *r-g* stock. In summer 1983 the progeny were planted. The 50% of progeny with red seedling color (plants containing the R allele) were culled from the field leaving *B*^{peru} */b r-g* plants containing Mutator activity. For further stock construction these plants were crossed to or by *B*^{peru} to produce progeny in which half of the plants will be homozygous for *B*^{peru} and half heterozygotes. In 1983 the heterozygous *B*^{peru}*/b* plants were detasseled and pollinated by testers in controlled crosses; a few of the Mutator stock were used as male parents onto 50 tester plants. *b r-g* with either *pl* or *Pl* were used as tester; *Pl* allows anthocyanin pigmentation in the absence of sunlight and serves to intensify plant color. The ears from both of these crosses segregated 1:1 for purple:colorless kernels, and these ears were carefully examined for kernels with intermediate levels of pigment or a spotted phenotype.

In 1984 the *B*^{peru} Mutator stock, now a mixture of homozygous and heterozygous plants, was planted in an isolation plot, detasseled and pollinated by tester. One half of the resulting ears were fully purple and one half segregated 1:1 for purple:colorless kernels. On ears segregating

1:1 for purple:colorless, only one-half of the kernels could be readily scored. Thus the actual number of kernels generated in the experiment is twice the 1983 figure, and 1.5 times the 1984 figure.

most loci of maize (Coe and Neuffer, 1977). Furthermore, we have ignored putative null mutations which account for a significant proportion of the total mutants obtained.

One consequence of our current understanding of transposable elements is that recognition of an unstable mutant by its propensity for reversion is most likely to occur when the insertion event occurs in a noncoding region of the gene. Insertions in coding regions initially disrupt the reading frame of the gene and excision is unlikely to restore the normal reading frame if the host sequence duplication remains. Thus, the phenotype of alleles disrupted in coding regions by insertion of transposable elements would likely be null both at the time of insertion and after excision. Consequently, insertions into coding regions would often be scored as stable null mutations despite their inherent instability at the molecular level.

The nomenclature for the eight verified mutable alleles is listed in Table 5. The mu-# designation refers to the numerical order of discovery and the fact that each allele was from a Mutator background. It is not yet established whether these mutants contain an insertion element and, if so, whether the element is homologous to *Mu-1*. Complete characterization of the mutations at the molecular level will require cloning the affected allele and comparing it to the progenitor allele. In preliminary experiments with the two *bz1* mutable alleles, we have utilized the cloned Bz1 sequences (Federoff et al., 1984) as a hybridization probe on Southern blots to demonstrate that each mutable allele contains an insertion relative to the progenitor *Bz1* allele (Chandler and Walbot, unpublished).

The phenotypes of the original mutable kernels are depicted in Fig. 1. Most of the initial isolates show small somatic reversions uniformly distributed over the surface of the aleurone. The only exception was the *c2* mutable which had two star-shaped patches of revertant phenotype, one of which is shown in Fig. 1H. The original $b^{peru\ mu-1}$ kernel had very small, rare sectors. Among the revertant sectors on the $a2^{mu-3}$ kernel is one odd ring-shaped sector marked with an arrow in Fig. 1C.

The original isolates were planted in the greenhouse and propagated by selfing and outcrossing to an appropriate tester.

Table 5. Mutable Allele Nomenclature

Locus	Allele	Progenitor Allele
A2	$a2^{mu-1}$	A2 in Robertson Mu Purple
	$a2^{mu-2}$	"
	$a2^{mu-3}$	"
B	$b^{peru\ mu-1}$	B^{peru} from Neuffer stock
Bz1	$bz1^{mu-1}$	Bz1 in Robertson Mu Purple
	$bz1^{mu-2}$	"
Bz2	$bz2^{mu-1}$	Bz2 in Robertson Mu Purple
C2	$c2^{mu-2}$	C2 in Robertson Mu Purple

In subsequent generations, some lineages have maintained their original phenotype but considerable variation has also arisen. Fig. 2 depicts typical kernels of $a2^{mu-1}$ and $bz1^{mu-1}$ both of which continue to show late reversion events encompassing 1-20 cells. This phenotype is also typical of $a2^{mu-2}$ and $bz1^{mu-2}$ (data not shown).

The $c2^{mu-1}$ lineage has shown great variation in the size of the revertant sectors. The original kernel had two large (>500 cells each) sectors, but initial crosses yielded kernels with numerous but extremely faint, very small (1-4 cell) sectors as shown in Fig. 2C. However, when the c2 mutable was crossed into a C2 R^{scm} stock, and then this stock selfed to yield one-quarter spotted $c2^{mu-1}$ progeny, the revertant sectors were very rare, faint and typically involved only a single cell. However, when these kernels demonstrating very late, infrequent somatic reversion events were planted and crossed onto a c2 tester, very large frequent sectors were found (Fig. 2D). Within these large sectors there are several grades of pigmentation ranging from the initial faint pigmentation of the stock to full pigmentation.

These dramatic changes in the phenotypes of the mutable alleles are similar in scope to those described for *Ac* and *Spm*-regulated alleles. Heritable alterations in the timing of somatic reversion events are manifested as changes in the size of the sectors with large sectors indicating early events and small sectors very late events during the development of the aleurone. Similarly alterations in the frequency of reversion result in kernels with few to many sectors; alterations in the level of expression of the revertant allele result in variations in pigment intensity from pale to deeply colored.

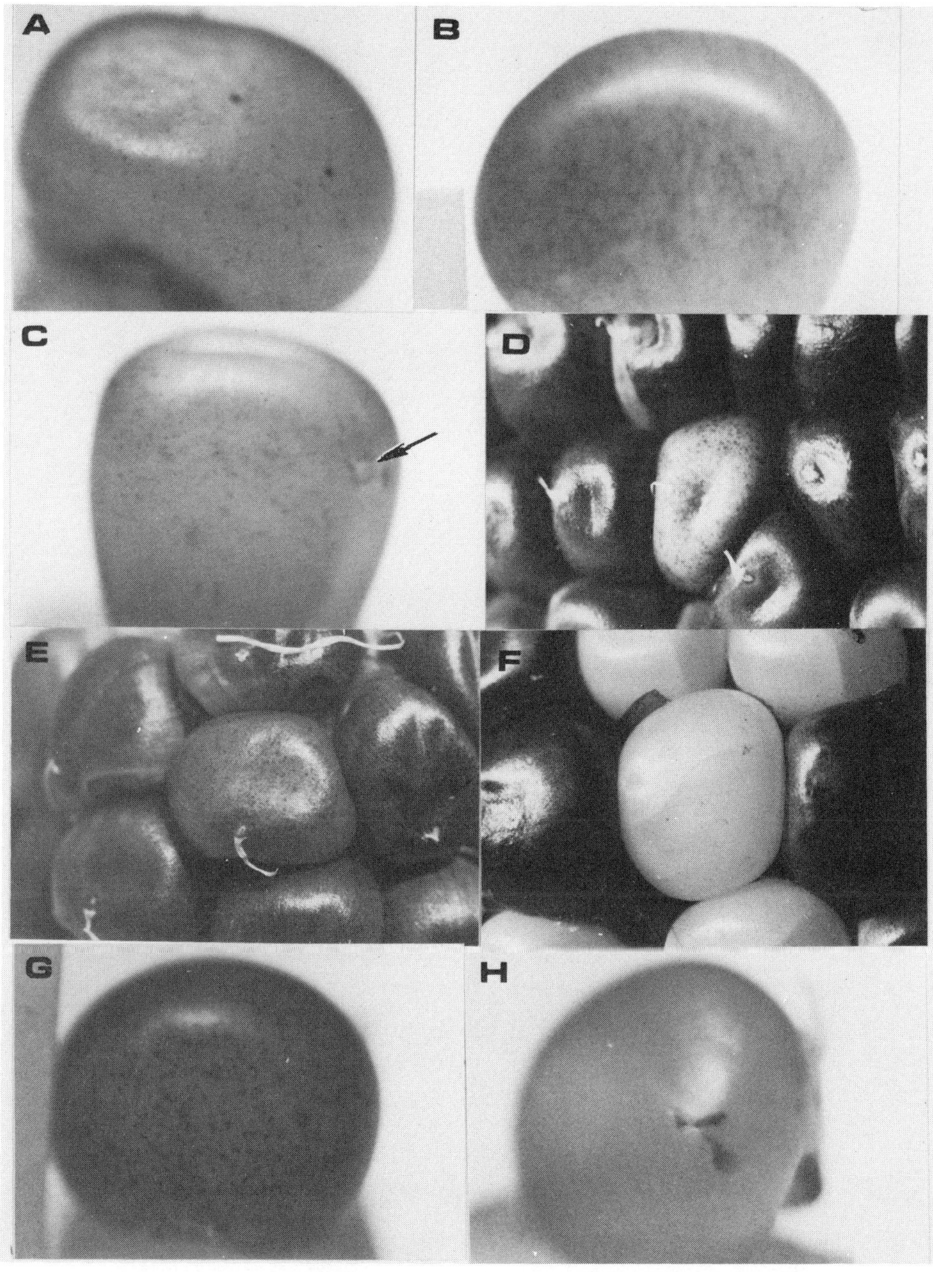

Fig. 1. Phenotype of the original isolates of the mutable alleles. Panel A, $\underline{a2}^{mu-1}$; B, $\underline{a2}^{mu-2}$; C, $\underline{a2}^{mu-3}$; D, $\underline{bz1}^{mu-1}$ among the fully pigmented sibling kernels; E, $\underline{bz1}^{mu-2}$ among its sibling kernels; F, $\underline{b}^{peru}$ among the kernels segregating 1:1 for $\underline{B}^{peru}$ pigmentation; G, $\underline{bz2}^{mu-1}$; and H, $\underline{c2}^{mu-1}$.

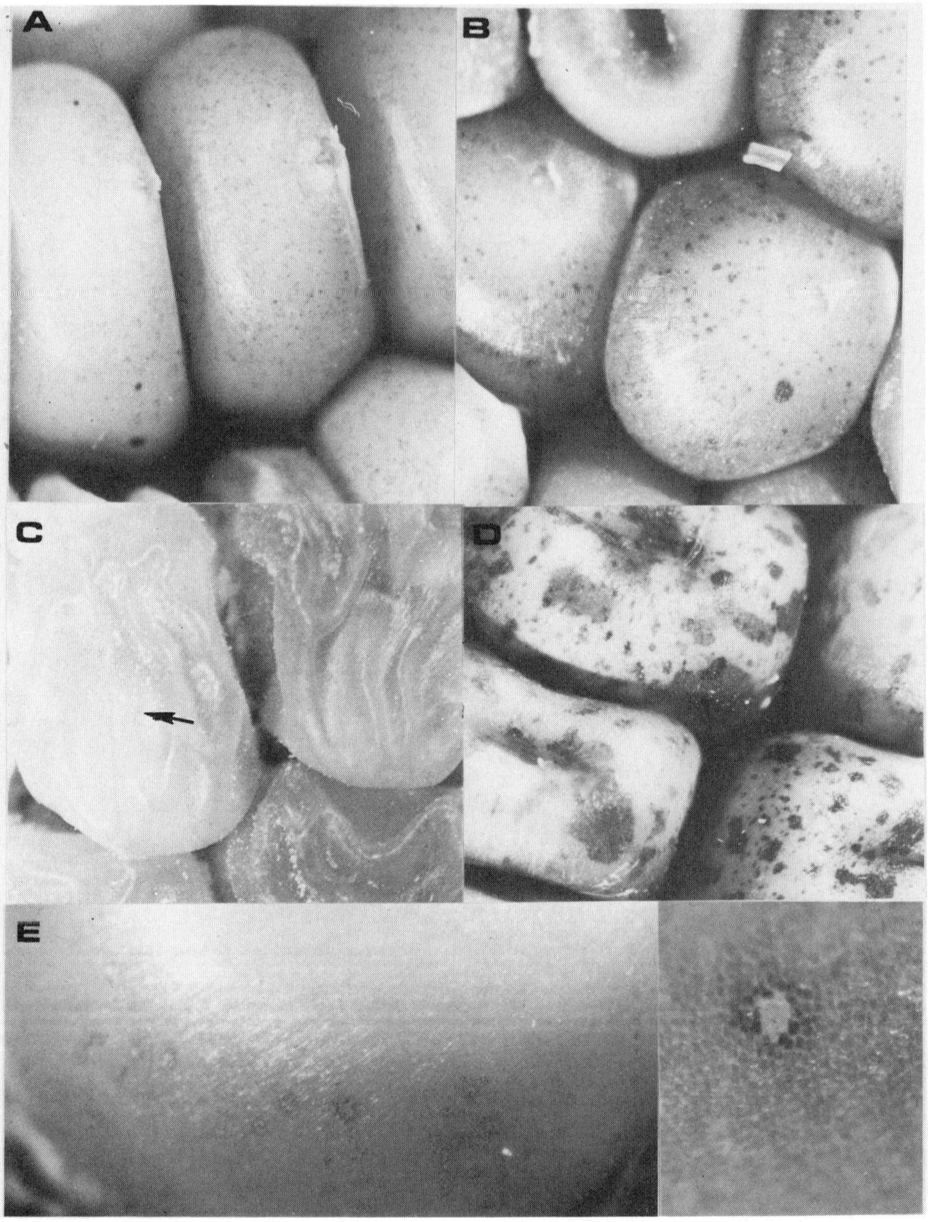

Fig 2. Subsequent expression of some mutable alleles. Panel A shows typical expression of *a2*$^{mu-1}$, B shows *bz1*$^{mu-1}$, C shows the very faint, small sectors (arrow) of *c2*$^{mu-1}$ in outcrosses to the *c2* tester in a W23 background, D shows *c2*$^{mu-1}$ sectors in a *R*^{scm} background, and E shows the ring-shaped sectors of the *a2*$^{mu-2}$ lineage. The inset in E is a close-up view of a ring-shaped sector.

The expression of the original $\underline{a2}^{mu-3}$ kernel was numerous small deeply pigmented sectors although one unusual ring-shaped sector encompassing about 500 cells was also observed (Fig. 1C, arrow). In subsequent generations the expression of this allele became predominantly this ring-spot type sector, and the revertant events became limited to a zone about midway between the crown and base of the kernel (Fig. 2E). The inset in Fig. 2E shows a close-up view of one of these ring-spot sectors in which it is clear that there are some fully pigmented cells at the periphery of the ring as well as cells of intermediate phenotype. Occasionally a pigmented cell is seen within the ring (data not shown). The events underlying this unusual phenotype are obscure. It is possible that expression of the A2 allele within the field of mutant cells is lethal so that only cells on the periphery of a revertant sector survive. It is also puzzling that revertant sectors themselves are restricted to a narrow zone in the middle of each kernel.

As shown in Fig. 3, the $\underline{bz2}^{mu-1}$ lineage has produced kernels with a variety of phenotypes. For example, kernels can have a very high density of very small one and two cell sectors (Fig. 3A) or a high density of much larger spots (Fig. 3B). In a few cases, the revertant sectors are restricted in a novel pattern on the kernel; as shown in Fig. 3C pigmented sectors are restricted to a narrow zone around the silk scar and to the base of the kernel with an unpigmented zone in between. This pattern violates the developmental history of the aleurone in which the early cell divisions establish wedge-shaped areas running from the base of the kernel to the center of the crown (not the silk scar) and later cell divisions produce progressively smaller pie-shaped, then tiny 2-16 cell rectagular blocks of sister cells within the aleurone (Coe, 1978).

Many kernels do show large regions with novel phenotypes, and these regions generally follow the developmental history of the aleurone, i.e. they arise as one-quarter or one-eighth sectors spanning the base to crown of the kernel. The $\underline{bz2}^{mu-1}$ kernel shown in Fig. 3D has numerous small, faint revertant spots except in one-eighth of the kernel. In this region the revertant events occurred earlier during aleurone development and encompass 50-100 cells, and the pigmentation is more intense. Fig. 3E shows a close-up view of the border of this region with a "change of state" compared to the basal pattern. Fig. 3E shows a kernel with numerous large revertant sectors, but a one-eighth region lacking spots; this phenotype most likely results from a "change of state" to an off condition during kernel development. However, another possibility is that this region of the kernel has lost the part of chromosome 1L carrrying the bz2 mutable allele. The changes of state within a kernel as shown here arise extremely frequently in the bz2 mutable lineage; approximately 5% of the kernels show a one-eighth or larger sector of a novel phenotype.

The mutable alleles do express in a tissue-specific fashion depending on the overall genetic constitution of the host plant. For example in an R-g background the aleurone, but not the scutellum contains anthocyanin, and we find that the $bz2^{mu-1}$ mutable phenotype is restricted to the aleurone in such kernels. In contrast, in a "self color" background (R^{sc}, R^{scm}, or B^{peru}) both the scutellum and the aleurone are pigmented, and the mutable alleles demonstrate this pigment distribution as well (data not shown).

The $b^{peru\ mu-1}$ is the most variable allele we have studied. As shown in Fig. 4 this allele displays many changes of state within a single outcross ear. Some kernels contain one-quarter sectors of full pigmentation contiguous to sectors with a fine spotted appearance. This allele produces a striking spectrum of pigmentation intensities with numerous pale and variegated pale derivatives.

We have also found instances of "changes of state" during the development of the ear; these result in groups of kernels with a novel phenotype. For example, in a cross between a $bz2^{mu-1}$/bz2 plant and bz2 tester most of the resulting ear contained no spotted kernels, but a small region at the base of the ear showed the expected 1:1 segregation of spotted:colorless kernels (Fig. 5A). Conversely, in a cross between a bz^{mu-1}/bz1 plant and tester the progeny ear showed primarily the expected 1:1 segregation pattern for spotted:colorless kernels, however, a large sector of about 75 kernels at the tip of the ear had no spotted kernels (Fig. 5B). Because individual ears are derived from just a few, typically two, cells present in the apical meristem of the seedling (Coe and Neuffer, 1978) with each cell forming 5-95% of the mature ear (Johri and Coe, 1983), our observations can both be interpreted as showing a very early change in state affecting one of the two apical cells destined to form the ear.

We also have examples of changes in the capacity of somatic reversion which depend on the kind of cross performed. In a few plants grown from kernels of the first outcross ear of the original stock of the $bz2^{mu-1}$ mutation, we found dramatic deviations from the expected Mendelian segregation ratios. When a heterozygous $bz2^{mu-1}$/bz2 plant is self-pollinated, a 3:1 spotted:colorless ratio should be obtained; when the second ear on this plant is test-crossed, a 1:1 ratio should be obtained. As shown in Fig. 5C, however, the C230-1 plant gave a 3:1 ratio in the self, but the outcross ear had only ~ 1% spotted kernels on it. The normal segregation ratios of spotted:colorless kernels on the selfed ear demonstrates that the mutable allele is still present in the stock and suggests the lack of spotted kernels in the testcross ear is significant. Subsequent crosses have shown that the testcross progeny, half of which should contain a mutable bz2 allele, can be "reactivated" to restore the spotted phenotype when these lines are crossed to active Mutator lines (Chandler and Walbot, unpublished).

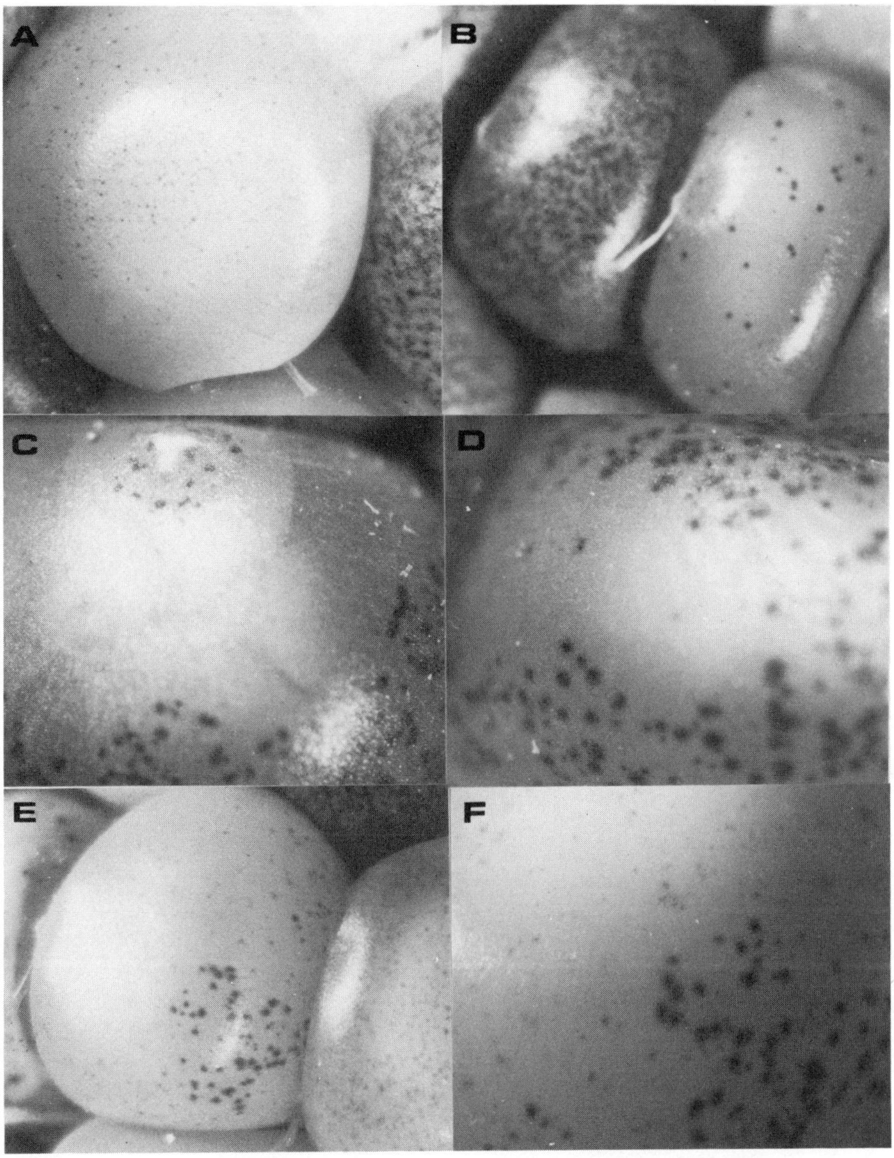

Fig 3. Variation in expression of the $bz2^{mu-1}$ allele. Panel A shows variation among kernels in the frequency and intensity of spots. Panel B shows a kernel with very infrequent but large spots. Panel C illustrates a "crown" pattern with revertant sectors restricted to a zone near the silk scar and to the bottom of the kernel. Panels D and E show sectors with alterations in revertant spot frequency on a single kernel; panel F is a close-up view of the sector in E.

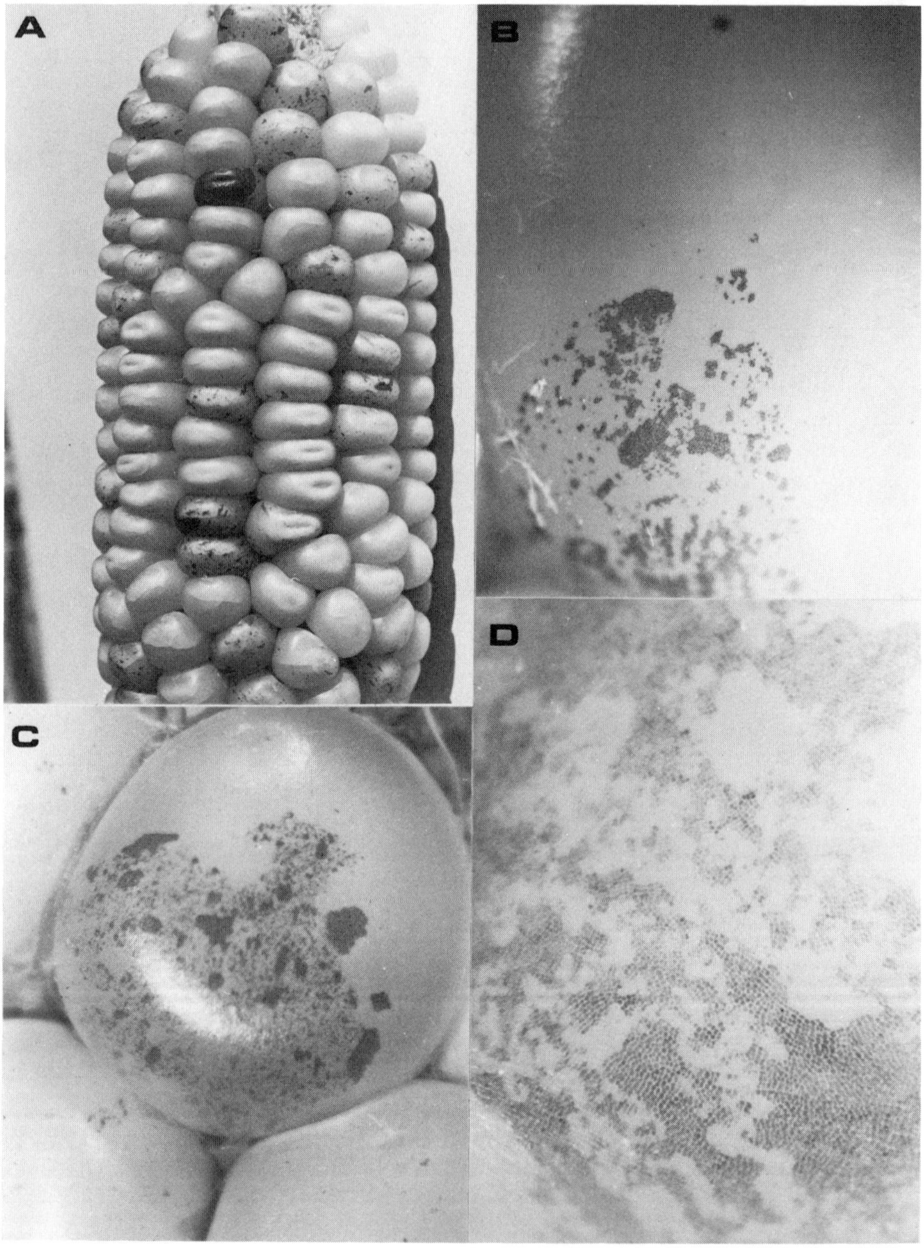

Fig. 4. Variation in expression of the $\underline{b}^{peru\ mu\text{-}1}$ allele. Panel A shows an ear segregating 1:1 for the mutable allele in which the great variety of revertant phenotypes is evident. Panels B and C show variant kernel sectors, and D shows a close-up view of the variegated, pale phenotype.

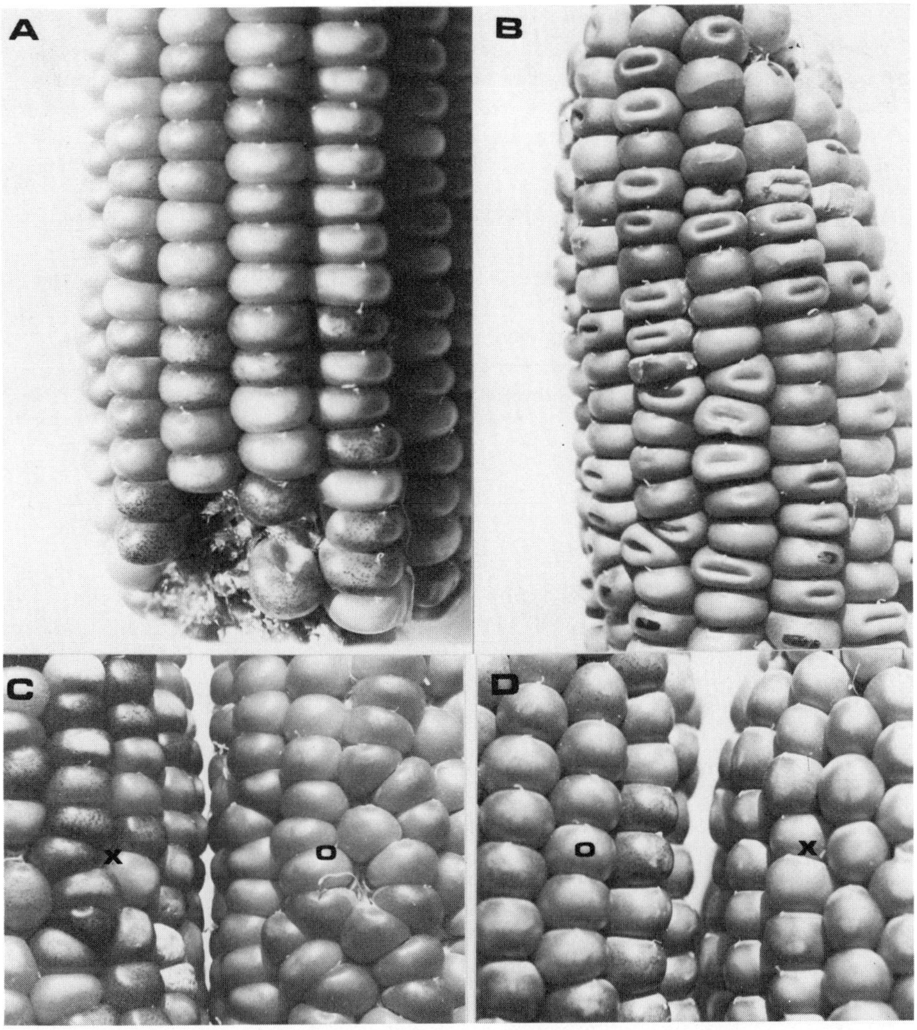

Fig. 5. Deviation from expected segregation ratios within or among ears. Panel A shows the cross of mutable parent C230-2 ($bz2^{mu-1}$/bz2) x tester in which half of the progeny kernels should be spotted; however, note that only a ~50 kernel sector at the base has the expected 1:1 ratio of spotted: colorless kernels. Panel B shows cross of D412-10 $bz1^{mu-1}$/bz1 by tester pollen in which there is a substantial sector lacking any spotted kernels (black lines indicate spotted kernels). Panel C has the outcross (O) and self (X) $bz2^{mu-1}$/bz2 plant C230-1 in which the selfed ear shows a 3:1 ratio of spotted:colorless kernels but the outcross ear does not show the expected 1:1 ratio. Panel D depicts the outcross and self ears of a plant from the same family, C230-3, in which the reciprocal result was obtained.

A second class of non-Mendelian inheritance occurred with plant C230-3 which segregated 50% spotted kernels when testcrossed, but the selfed ear had no spotted kernels (Fig. 5D). The testcross result demonstrates that the mutable allele is still present in the stock and suggests that the lack of mutable kernels on the selfed ear is highly significant. Similar results were obtained with several kernels taken from the selfed ear of the original *bz2*$^{mu-1}$ when they were again selfed in the subsequent generation. The significance of selfing in turning off the Mutator system is not precisely known, but Robertson (1983) reported a loss of the forward mutation rate in stocks subjected to several generations of self-pollination. Thus far, no crossing program whether to active Mutator lines or to inbred lines lacking Mutator activity has restored the mutable *bz2* phenotype to the "off" progeny of C230-3 or other lines in which the selfed ear of the plant switched "off."

CORRELATING PARTICULAR *MU*-CONTAINING DNA FRAGMENTS WITH A MUTABLE ALLELE

Individual *Mu-1* homologous fragments visible on a Southern blot can be treated as genetic traits that segregate due to an element's insertion into specific sites in the genome. A *Mu-1* insertion into a particular gene should produce a restriction fragment that segregates with the mutant phenotype.

If the mutable alleles described in the previous section are due to the insertion of an *Mu* element (either *Mu-1* or a related sequence) into the gene then it should be possible to clone the affected allele by virtue of homology to the *Mu* insert. To accomplish this a particular *Mu* insertion must be correlated with the mutable phenotype. To display the multiple *Mu* elements, each inserted in a different genomic location, total DNA is digested with an enzyme with no recognition sequence within the *Mu* element; consequently, cleavage occurs in the flanking DNA. The resulting fragments are electrophoresed on an agarose gel; a Southern blot is prepared and hybridized to a *Mu*-specific probe to identify fragments containing these fragments. To simplify the restriction pattern, the different Mutator restriction fragments should be separated as much as possible. In our experience, the restriction enzyme *Eco* RI tends to yield the widest range of fragments when maize DNA is digested, with fragment sizes ranging from ~3-20 kb (i.e. Fig. 6A). This is usually sufficient if the line contains less than 10 copies of *Mu-1* homologous elements. However, even with Eco RI digestions there are frequently too many fragments larger than 10 kb to be individually resolved on a Southern blot. Restricting the DNA with two different enzymes tends to decrease the number of large restriction fragments (>10 kb), resulting in better resolution of the restriction fragments.

To identify the *Mu*-containing fragment most likely to be in the gene of interest, the segregation of the fragments is followed through test crosses. All spotted kernels (for example, $bz2^{mu-1}$/*bz2*) should contain the relevant fragment, and barring a rare change of state, all sibling colorless kernels (*bz2*) should lack this fragment. At each generation segregation will decrease the number of shared fragments among the spotted kernels; in addition, some of the *Mu* elements may transpose at each generation further reducing the number of shared fragments but maintaining copy number. Consequently, comparison of lines containing the mutable allele should eventually indicate only one shared fragment.

The rapidity with which the appropriate fragment can be identified depends on the total number of *Mu* fragments present; if there are >30 fragments some are likely to co-migrate obscuring the segregation analysis, and with >20 fragments several generations will be required for the segregation analysis. To maximize the likelihood of identifying the appropriate fragment, we survey *Mu* copy number at each generation and follow segregation in those lineages with the lowest copy number. If, for example, a single mutable plant contains 30 copies of *Mu* and is outcrossed to a tester line, progeny with 20-40 copies are likely. Fig. 6B shows a theoretical calculation of the co-segregation analysis for a line with 30 copies of *Mu* in which the fragment most likely to represent the mutant allele of interest is evident in the F2 population.

In our experience there can be a two-to-three fold variation in *Mu* copy number among the progeny of a single plant. One example, comparing a Mutator parent with two of its outcrossed progeny, is shown in Fig. 7. DNA was prepared from the immature second and third ears on the Mutator plant and the first ear was outcrossed by a non-Mutator line (*b r-g* tester). Sibling progeny kernels were germinated and DNA was prepared from five day etiolated seedlings. All DNA samples were digested with two enzymes that do not restrict in *Mu-1*, *BamHI* and *BclI*, electrophoresed, blotted to nitrocellulose and hybridized with *Mu-1*. The Mutator parent line contained ~18 restriction fragments (lane c), one of its progeny (lane d) contained ~12 restriction fragments, and the other progeny (lane e) contained ~30 restriction fragments. Five to seven fragments are shared by all three lines, some fragments are missing and some new restriction fragments are observed in the progeny.

The disappearance and appearance of new restriction fragments could have several explanations. The most likely reason for loss of a restriction fragment is segregation, and if no transposition is occurring the total copy number should decrease in outcrosses, to non-Mutator lines. However, an element could promote a rearrangement or deletion in adjacent DNA removing a restriction site which would result in the loss of one fragment and the appearance of a new one, but no transposition would occur. Alternatively, an element

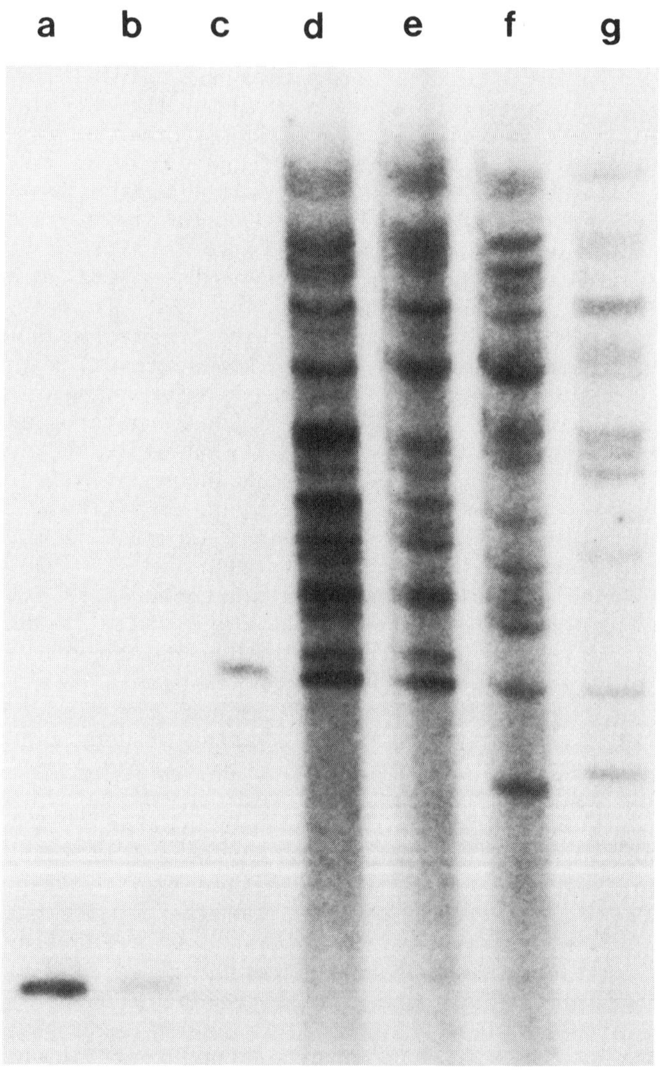
a
b
c
d
e
f
g

Fig. 6A. Multiple genomic restriction fragments are homologous to Mu-1. Four sibling kernels from an ear segregating 50% $bz2^{mu-1}$/ bz2 and 50% bz2/bz2 were germinated and DNA prepared from 5-day etiolated seedlings. DNA was digested with Eco RI, electrophoresed on a 0.5% agarose gel, blotted to nitrocellulose, and hybridized with nick-translated Mu-1 sequences. Lanes a and b are reconstructions with cloned Mu-1 sequences approximately 3 and 0.5 copies, respectively. Lane c is the non-Mutator bz2 tester DNA (negative control). The restriction fragment observed in lane c is the Mu-1 homologous sequences observed in this inbred line (Chandler, unpublished data). Lanes d-g are from a Mutator line. Lanes d and e are DNA samples from spotted kernels, $bz2^{mu-1}$/bz2; lanes f and g are DNA samples from bz2/bz2 kernels. Several restriction fragments, marked with arrows, are in the $bz2^{mu-1}$ lines, but not in bz2 Mutator lines.

6B. Sample calculation of the co-segregation of restriction fragments hybridizing to Mu-1 by chance alone. At each generation only kernels showing the mutable phenotype are selected, and the pattern of Eco RI fragments hybridizing to Mu is determined. For this calculation we will assume that the parental (initial) kernel contains an equal mixture homozygous and heterozygous Mu insertion sites. The Eco RI restriction fragments corresponding to the homozygous condition will not show segregation in the subsequent generation but the heterozygous fragments will. Thus, in the absence of transposition (loss of a fragment) or new insertion events, progeny will be expected to share 75% of their fragments in common, all of the homozygous fragments and half of the heterozygous fragments. In practice fragment loss and gain events reduce the total number of shared fragments.

Initial Mutant Kernel	30 Fragments
Compare 8 F1 Kernels	3 Fragments in Common (0.75^8)
Compare 8 F2 Kernels	0.3 Fragments in Common (0.1^8)

By comparing 8 siblings kernels from the same outcross ear in pairwise combinations to score the shared fragments only 10% of the fragments should be shared among the entire set by chance alone. By the F2 generation, generated by a random choice of F1 kernels for the subsequent outcross and then testing 8 kernels chosen from the outcross ears of that generation there is likely to be only one fragment shared among all of these progeny. This fragment will represent the Mu insertion into the gene of interest. The likelihood of identifying the appropriate fragment can be increased simply by increasing the number of kernels.

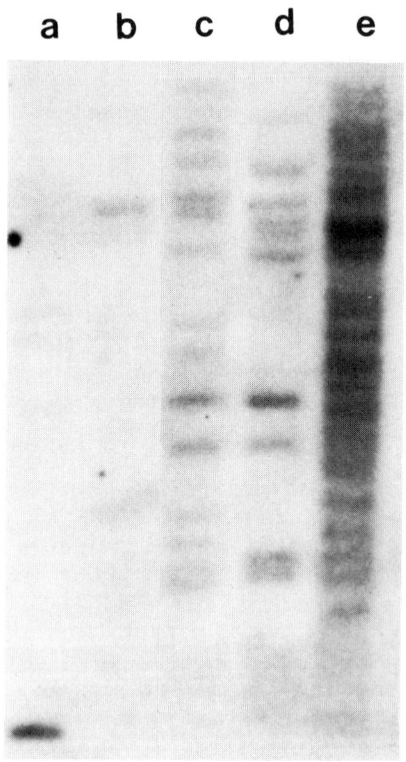

Fig. 7. Increase in the number of genomic restriction fragments homologous to Mu-1 in one plant generation. Lane a contains a reconstruction with cloned Mu-1 plasmid, ~3 copies. Lane b contains the non-Mutator inbred line that the Mutator lines in lanes c-e were crossed to. The band visible in lane b is the Mu-1 homologous sequence in this inbred line, the particular pattern observed is characteristic for each inbred tested (Chandler, unpublished data). Lane c is a DNA prep from a Mutator line that was subsequently outcrossed to the line, represented in lane b. Two F1 progeny from this cross were germinated and DNA prepared (lanes d, e).

could excise from one site and insert into a new site. Both of these alternatives would maintain the total copy number, but change the number of restriction fragments shared between a parent and its progeny. It would require probes specific for a particular gene to distinguish between rearrangement and transposition. However, both of these activities, in combination with segregation in outcrosses, would quickly decrease the number of restriction fragments in common between siblings. A third explanation for the appearance of new restriction fragments - the element remains at the original site, duplicates and inserts into a new site - would result in the retention of the original restriction fragment, the creation of a new fragment, and an increase in the total copy number.

The large increase of Mu-1 homologous sequences, ~2 fold, within one plant generation suggested that many of the elements duplicated or a few elements transposed many times within one life cycle (lanes c,e). It is not possible to determine if the new restriction fragments apparent in the plant shown in lane d result from rearrangement, deletion of restriction sites or new transposition events. However, duplication, and probably transposition, of elements has occured in the plant shown in lane e. This variation has an important impact on segregation analyses which would be possible in plants with copy numbers similar to that in plants shown in lanes c and d, but extremely difficult with plants like that in lane e. Therefore, it is important to identify the plants containing the lowest number of Mu elements, and this must be monitored every generation because the copy number can increase dramatically with one generation.

A major assumption in our molecular analysis is that the affected allele results from the insertion of an element homologous to the Mu-1 probe. So far three mutable alleles of Adh1 have been recovered from a mutator line and all contain a Mu-1 element. We have preliminary evidence consistent with a Mu insertion in both mutable bz1 alleles recovered from a Mutator background. Thus, Mu elements appear to be the major active transposable element family in Mutator lines.

The advantages of using a Mutator line as a transposon tag are the high forward mutation rate and the moderate copy number of the element. The major disadvantages with mutator lines include the inability to perform a genetic test to demonstrate that a Mu element is inserted in a particular allele and the lack of revertants of Mu-affected alleles. Part of the confirmation of insertion element involvement in creation of a mutant is a comparison of the structure of the progenitor, mutant, and revertant alleles. In the case of mutable alleles from Mutator lines only one genetically confirmed germline revertant has been reported (Robertson, pers. commun.). The fact that most somatic reversions in Mutator lines occur late in development may influence the low recovery of germinal revertants.

CONCLUSION

Mutator lines, defined by their high mutation rate and derivation from other Mutator lines, contain at least one active transposable element family at present represented by the *Mu-1* and *Mu-L* elements. Mutable alleles are readily recovered from such lines at a frequency about 10-fold above the spontaneous mutation rate; because null mutations are disregarded in the calculations, the true frequency of mutation is higher than the recovery of mutable alleles. The phenotype of many of the mutable alleles isolated in genes of the anthocyanin biosynthetic pathway are remarkably similar, namely late, frequent reversion in the somatic aleurone tissue. The distribution of pigment in plants containing these mutable alleles appears to depend on other genetic factors present in the background, indicating that these are not mutations in the tissue-specificity of anthocyanin biosynthesis. Exceptional kernels show sections with particularly large (early) or infrequent (even null) expression demonstrating that "changes of state" do occur in these mutable alleles during kernel development. Furthermore, clusters of kernels with novel phenotypes have also been noted demonstrating that changes of state in the expression of these alleles can occur early in plant development.

By following the segregation of the various genomic locations containing a member of the *Mu* family of sequences, it can be demonstrated that generally each outcross results in fewer *Mu* fragments shared in common among all of the spotted kernels. Among these shared fragment size classes should be the mutable allele corresponding to the phenotype under observation, provided that the mutation is caused by the insertion of an *Mu* element. However, it is not uncommon to identify plants that have a dramatically increased copy number of *Mu* elements. Although we have not yet cloned any allele by virtue of its containing a *Mu* element, our work with *bz2*$^{mu-1}$ and *b*$^{peru\ mu-1}$ has already narrowed the candidate fragment size classes to just a few. Consequently, we are confident that mutants generated in mutator backgrounds can be used within a few plant generations to clone the affected mutable alleles.

ACKNOWLEDGEMENTS

We wish to thank E.H. Coe, Jr., M.G. Neuffer, and especially D.S. Robertson for their generous gifts of seeds that made these studies possible. We are indebted to Drs. Coe and Neuffer, and their associates in the corn genetics community at the University of Missouri for their continuous and enthusiastic intellectual and material support for our research. We dedicate this paper to these two gentlemen and the warm spirit of sharing that makes participation in corn research so personally rewarding. Original research reported here was supported by a grant from the N.I.H.

(GM 32422 to V.W.). V.C. was supported by an NSF Postdoctoral Fellowship in Plant Biology.

REFERENCES

Barker, R. F., Thompson, D. V., Talbot, D. R., Swanson, J., and Bennetzen, J. L., 1984, Nucleotide sequence of the maize transposable element Mu1, Nucleic Acid Res., 12:5955-5967.

Bennetzen, J. L., Swanson, J., Taylor, W. C., and Freeling, M., 1984, DNA insertion in the first intron of maize Adh1 affects message levels: Cloning of progenitor and mutant Adh1 alleles, Proc. Natl. Acad. Sci. USA, 81:4125-4128.

Burr, B., and Burr, F., 1981, Transposable elements and genetic instabilities in crop plants, in: "Stadler Genet. Symp.," vol. 13, G. Redei, ed., University of Missouri, Ag. Exp. Stn., pp. 115-128.

Coe, E. H. Jr., 1973, The aleurone tissue of maize as a genetic tool, in: "Maize Breeding and Genetics," D. B. Walden, ed., Wiley-Interscience, pp. 447-459.

Coe, E. H. Jr., and Neuffer, M. G., 1977, The genetics of corn, in: "Corn and Corn Improvement," G. F. Sprague, ed., Amer. Soc. of Agronomy, Inc., Madison, pp. 111-223.

Coe, E. H. Jr., and Neuffer, M. G., 1978, Embryo cells and their destinies in the corn plant, in: "The Clonal Basis of Development," S. Subtelny and I. M. Sussex, eds., Academic Press, New York, pp. 113-129.

Döring, H.-P., and Starlinger, P., 1984, Barbara McClintock's controlling elements: Now at the DNA level, Cell, 39:253-259.

Emerson, R. A., 1914, The inheritance of a recurring somatic variation in variegated ears of maize, Am. Nat., 48:87-115.

Emmons, S. W., Yesner, L., Ruan, K. and Katzenberg, D., 1983, Evidence for a transposon in Caenorhabditis elegans, Cell, 32:55-65.

Federoff, N. V., 1983, Controlling elements in maize, in: "Mobile Genetic Elements," J. Shapiro, ed., Academic Press, New York, pp. 1-63.

Federoff, N. V., Furtek, D.B. and Nelson, O. E. Jr., 1984, Cloning of the bronze locus in maize by a simple and generalizable procedure using the transposable element Activator (Ac), Proc. Natl. Acad. Sci. USA, 81:3825-3829.

Freeling, M., 1984, Plant transposable elements, Annu. Rev. Plant Physiol., 35:277-298.

Johri, M. M., and Coe, E. H. Jr., 1983, Clonal analysis of corn plant development. I. The development of the tassel and the ear shoot, Develop. Biol. 97:154-172.

McClintock, B., 1956, Intranuclear systems controlling gene action and mutation, Brookhaven Symp. Biol., 8:58-74.

McClintock, B., 1978, Mechanisms that rapidly reorganize the genome, in: "Stadler Genet. Symp.," vol. 10, G. Redei, ed., University of Missouri, Ag. Exp. Stn., pp. 25-47.

Mottinger, J., Johns, M. A., and Freeling, M., 1984, Mutations of the Adh1 gene in maize following infection with barley stripe mosaic virus, Mol. gen. Gen., 195:367-369.

Rhoades, M. M., 1938, Effect of Dt on the mutability of the a1 allele in maize, Genetics, 23:377-397.

Robertson, D. S., 1978, Characterization of a mutator system in maize, Mutat. Res., 51:21-28.

Robertson, D. S., 1980, The timing of Mu activity in maize, Genetics, 94:969-978.

Robertson, D.S., 1983, A possible dose-dependent inactivation of mutator (Mu) in maize, Mol. Gen. Genet., 191:86-90.

Robertson, D. S. and Mascia, P. N., 1981, Tests of 4 controlling-element systems of maize for mutator activity and their interaction with Mu mutator, Mutat. Res., 84:283-289.

Roeder, G. S., and G. R. Fink, 1983, Transposable elements in yeast, in:, "Mobile Genetics Elements," J. Shapiro, ed., Academic Press, New York, pp. 300-328.

Strommer, J. N., Hake, S., Bennetzen, J., Taylor, W. C., and Freeling, M., 1982, Regulatory mutants of the maize Adh1 gene caused by DNA insertions, Nature, 300:542-544.

Sutton, W. D., Gerlach, W. L., Schwartz, D., and Peacock, W. J., 1984, Molecular analysis of Ds controlling element mutations at the Adh1 locus of maize, Science, 223:1265-1268.

PLANT DNA VARIATION AND STRESS

C. A. Cullis

John Innes Institute
Colney Lane
Norwich, U.K.

SUMMARY

The generation of quantitative DNA changes in the flax variety Stormont Cirrus in response to various stresses is described. Members of all the highly repetitive sequence families have been cloned and used as probes to characterise this variation. The variation occurs in a particular subset of the genome, irrespective of the stress applied, although the frequency with which a family of sequences varies may depend on the particular stress applied. In the case of the 5SRNA genes a specific labile set of these sequences which frequently undergoes deletion was detected. The evolutionary consequences of such an interaction between the genome and the environment are considered.

INTRODUCTION

Plant genomes contain a large proportion of repeated sequences. The study of the arrangement of these repeats in related species has provided evidence for amplification, deletion and transposition events which have resulted in repeated sequence turnover during evolution (Flavell, 1982). However these comparisons describe the end products of the processes and cannot provide evidence for the actual series of events by which a particular arrangement was produced. To characterise the cellular mechanisms by which the changes are generated a dynamic system in which variation occurs over a small number of, and ideally within a single, generation is required. This is also a necessary condition to determine whether or not the processes which bring about change act on the genome in a random manner or are localised to some fraction, and can be influenced in their frequency by any "shocks" or "stresses" to which the genome may be subjected. Genomic "shocks" have already been shown to affect the activity and frequency of transposition of transposable elements in maize (Zea mays L.) (McClintock, 1984). In a number of other systems external stresses have been shown to be associated with quantitative variation in DNA amount (Evans, 1968; Price et al., 1983; Cullis, 1983a).

Here the generation of rapid quantitative DNA changes, where there is no evidence for the involvement of transposable elements, will be considered. Attention will be focussed on flax (Linum usitatissimum L.) in which quantitative DNA variation has been generated in three ways. Firstly, in response to variations in the growth environment of the whole plant (Durrant, 1962; Evans, 1968). Secondly, by passage through a tissue culture cycle (Cullis, 1983a; Cullis and Cleary, 1985a). Finally, DNA variation has been generated in intraspecific crosses between genotrophs (Cullis, 1979). Intraspecific crosses have also been shown to result in DNA variation in Microseris species (Price et al., 1983) and maize (Rivin and Cullis, 1983). Three questions are considered. Are there distinct regions of the genome in which this variation occurs? Can specific subsets of the genome be defined by differing responses to various stresses? Finally, does this rapid quantitative variation cause any phenotypic changes and so have evolutionary consequences?

ENVIRONMENTALLY INDUCED HERITABLE CHANGES IN FLAX

Heritable changes can be observed in some flax varieties after they have been grown in different, characterised environments for a single generation (Durrant, 1962, 1971; Cullis 1981a). The progeny, with the stable heritably altered phenotype, from the selfed plants grown in these environments were termed genotrophs. These stable lines differed from one another and from the original variety (Stormont Cirrus, termed Pl) from which they were derived in a number of characters. These include plant weight and height, the total nuclear DNA amount (as determined by Feulgen staining) (Evans, 1968), the number of hairs on the capsule septa (Durrant and Nicholas, 1970) and the isozyme band patterns for peroxidase acid phosphatase and esterase (Cullis and Kolodynska, 1975).

The change in the peroxidase isozyme band pattern has been shown to be controlled at a single locus by a dominant and a recessive allele. In crosses between genotrophs a 3:1 segregation of the two parental patterns was obtained in the F2 (Cullis, 1979a).

The change in the nuclear DNA amount between the genotrophs

has been characterised in much greater detail and the sequences involved are described in subsequent sections.

There is no obvious relationship between the environmental conditions and any of the variation in the characters mentioned above. However, since only a small number of phenotypic characters have been studied it is possible that only a minor portion of the variation is of adaptive significance with respect to the stresses of these inducing environments. The remainder of the variation may be due to a general destabilisation of the genome and so any specific adaptive changes may appear insignificant among the wide range of variation observed.

THE FLAX GENOME

The flax genome size is 1.5pg/2C nucleus (Timmis and Ingle, 1973) and it has a diploid complement of 30 small, approximately equally sized chromosomes. One pair of chromosomes each have a nucleolar organiser region (Schweizer, 1979) and the majority, if not all the ribosomal RNA genes have been located on this pair of chromosomes by *in situ* hybridisation (Martin, pers. comm.).

The DNA from flax has been characterised by renaturation kinetics using the hydroxylapatite method (Britten et al. 1974). About half of the genome is made up of repeated sequence families with the remainder being low copy number, or single copy, sequences. The characterisation of the interspersion pattern of the repetitive and single copy sequences, by renaturation kinetics, has shown that the flax genome is organised in a long period intersperion pattern (Cullis, 1981). The minimum size of the blocks of repetitive and low copy number sequences from this analysis was about 3 kilobases.

Representative elements from all the highly repetitive sequence families of the flax genome have been cloned and characterised (Cullis and Cleary, 1985b). It has been shown that, in every case, these families exist in long tandem arrays of at least 50 kilobases. Thus the cloning of all these families has confirmed the flax genome organisation as a long period interspersion pattern. The 10% of the genome that is present as intermediately repetitive sequences may be interspersed to some degree with low copy number sequences but this has yet to be confirmed by cloning such fragments.

COMPARISON OF THE NUCLEAR DNAS FROM GENOTROPHS AND LINUM SPECIES

The DNAs from 12 genotrophs have been characterised by renaturation kinetics using the hydroxylapatite method with a DNA of an average single-stranded length of 300 base pairs (Cullis, 1983b). The Cot curves obtained were analysed by a non-linear regression computer program. A wide variation was found, for both rate constants and the proportion of the total DNA falling into different repetitive classes, when the DNAs from different genotrophs were analysed. The overall conclusions were that the DNAs of the genotrophs differed significantly from one another in both the highly repetitive and intermediately repetitive sequences. However these experiments gave no evidence for a change in the overall sequence information present in the genotrophs (Cullis, 1983b). These studies could only give information on relatively large changes occurring within the genome and could not detect small changes, especially those occurring in unique or low copy number sequences. For the determination of this type of variation specific cloned probes would be required. In addition this approach would only detect a net change in a particular repetitive class, so if both amplifications and deletions were occurring it is possible that no difference would be observed.

In order to characterise the changes in the highly repetitive fraction of the flax genome a number of cloned sequences were isolated which represent all the highly repetitive sequence families of the flax genome (Cullis and Cleary, 1985b). These sequences and their organisation within the genome, are listed in Table 1. The relative amounts of a number of these sequence families have been estimated in two ways; firstly by the hybridisation to southern blots of equal loadings of different DNAs (Cullis, 1983b) and secondly by a modified dot blot, termed slot-blot procedure (Cullis et al., 1984; Cullis and Cleary, 1985). Using the slot blot procedure the relative proportions, and

Table 1. The recombinant plasmids representing all of the highly repetitive sequence families of the flax genome.

Plasmid	Function/ Localisation	Size of insert (kilobases)	Size of repeat unit in genome (kilobases)	Fraction of P1 genome(%)
pBG35	rDNA	8.6	8.6	1.5
pBG13	5S DNA	3.5	0.35	3.0
pBG87	unknown	3.4	0.325	2.5
pCL13	satellite	0.6	0.5	2.2
pCL21	satellite	4.2	0.6	5.6
pCL2	unknown	0.4	0.57	4.2
pCL8	unknown	0.6	0.57	2.6
pCL53	unknown	0.55	0.55	2.3
pDC7	unknown	0.8	0.33	3.2

absolute amounts, of each of these sequence families was determined in 17 lines. These included 12 genotrophs, one flax variety (Liral Monarch) one linseed (Linum usitatissimum) variety (Royal, two L. bienne accessions (L. bienne is the supposed progenitor of flax) and one L. grandiflorum Desf. accession. The values obtained for a subset of these lines are given in Table 2.

Only one of the highly repetitive sequence families tested represented by the clone pCL21 and forming a major portion of the light satellite DNA, showed no significant differences between the lines. In all the other cases there was significant variation both between genotrophs, and between the genotrophs progenitor, P1, and the other cultivars as well as the related species. The range of variation between P1 and the related Linum species was not

Table 2. The relative amounts of different repeated sequences in leaf DNA from flax genotrophs and related Linum species. L6, LH, C1, C2, C3 were all derived from the flax variety Stormont Cirrus (Designated P1) (Cullis, 1977, 1981a). Liral Monarch was supplied by the Northern Ireland Agricultural Department, L. grandiflorum Desf. from the Botanical Garden, Aachen and L. bienne (106) from the Botanical Garden, Coimbra, Portugal.
*values are significantly different from P1 ($P \leq .01$). Values are given as the percentages of the genome homologous to the probe. Data are taken from Cullis and Cleary (1985b).

Source of DNA	Probe rDNA	5SDNA	pCL2	pCL8	pCL53	pBG87	pDC7	pCL13	pCL21
P1	1.5	3.0	4.2	2.6	2.3	2.5	3.2	2.2	5.6
L6	*1.0	*1.4	3.6	*2.1	1.9	2.3	*2.1	*1.5	4.8
LH	1.6	3.2	*5.0	*3.3	*1.1	2.3	2.8	*1.7	5.8
C1	1.5	*2.0	3.8	2.5	*1.5	*2.1	*2.1	*1.3	5.6
C2	1.6	2.9	4.2	2.7	2.2	2.4	3.1	*1.5	5.7
C3	*0.8	*2.2	*2.5	*2.1	*1.6	*1.9	*1.7	*1.2	5.6
Liral Monarch	*1.2	*2.2	*3.0	*2.1	*0.9	*2.0	*1.7	*1.2	4.7
Linum grandiflorum Desf.	1.6	*4.0	*2.1	2.5	2.2	*1.9	*2.6	1.9	5.5
Linum bienne (106)	1.5	*1.8	*1.8	*1.5	*1.2	*1.3	-	*1.1	5.0

significantly greater than that observed between the genotrophs themselves. Thus, quantitatively, the variation induced in the genotrophs was similar to that between species. However, as will be shown later there are qualitative differences between the species which are not present between the genotrophs.

Another point that emerges from all the data on the variation in the repetitive sequence families is that there is no obligatory association between the change in any one sequence and that occurring in any other. Thus each of these sequence families appeared to vary independently although some did vary at a higher frequency and to a greater extent than others. A more detailed consideration of the variation in a number of these sequences is given in the following sections.

5S RNA Genes

The 5S RNA genes in flax are arranged as tandem arrays of a 350 to 370 base pair repeating sequence (Goldsbrough et al., 1981). There is both length and sequence heterogeneity for this set of genes and the copy number can vary by more than two fold between the genotrophs (Table 2). In the genotrophs, the line with the highest number of copies of this sequence, has some 3% of its genome made up of 5S genes. One manifestation of the heterogeneity of the 5S gene tandem arrays is the lack of complete digestion to the monomer length by the restriction endonuclease Taq I (Figure 1). A subset of these genes is uncut by the enzyme and so can be distinguished from the remainder. In each case where the number of 5S genes has been reduced in the genotrophs it is this Taq I subset which has been preferentially deleted (Fig. 1; Cullis and Cleary, 1985). Thus within the 5S genes there is a subset which appears particularly labile and is readily deleted (Fig. 1 a,b,c,d).

When the 5S genes in another flax and a linseed cultivar were compared it was also found that the copy number was variable. This Taq I resistant set of sequences also varied between the two cultivars (Fig. 1 e,f).

However, when the variation in the 5S genes is considered in an evolutionary time frame the variability appears less precisely defined. The comparison of a number of accessions of two related *Linum* species shows a variation in the 5S gene number (Table 2). However a characterisation of the restriction digest pattern of those DNAs showed that variation could occur in other sets of sequences apart from the Taq I resistant subset. Thus with *L. grandiflorum* Desf., which had the highest 5S gene number, there were very few of the genes in Taq I resistant arrays. However there was s subset present in arrays of approximately 60 copies without a Taq I site, which was virtually absent from *L.*

usitatissimum (Fig. 1, g). Similarly in the accessions of L. bienne there is variation for the number of 5S genes that has no correlation with the amount of this sequence in Taq I resistant stretches (Fig. 1 h,i,j).

Thus in the case of the 5S genes, it would appear that in the

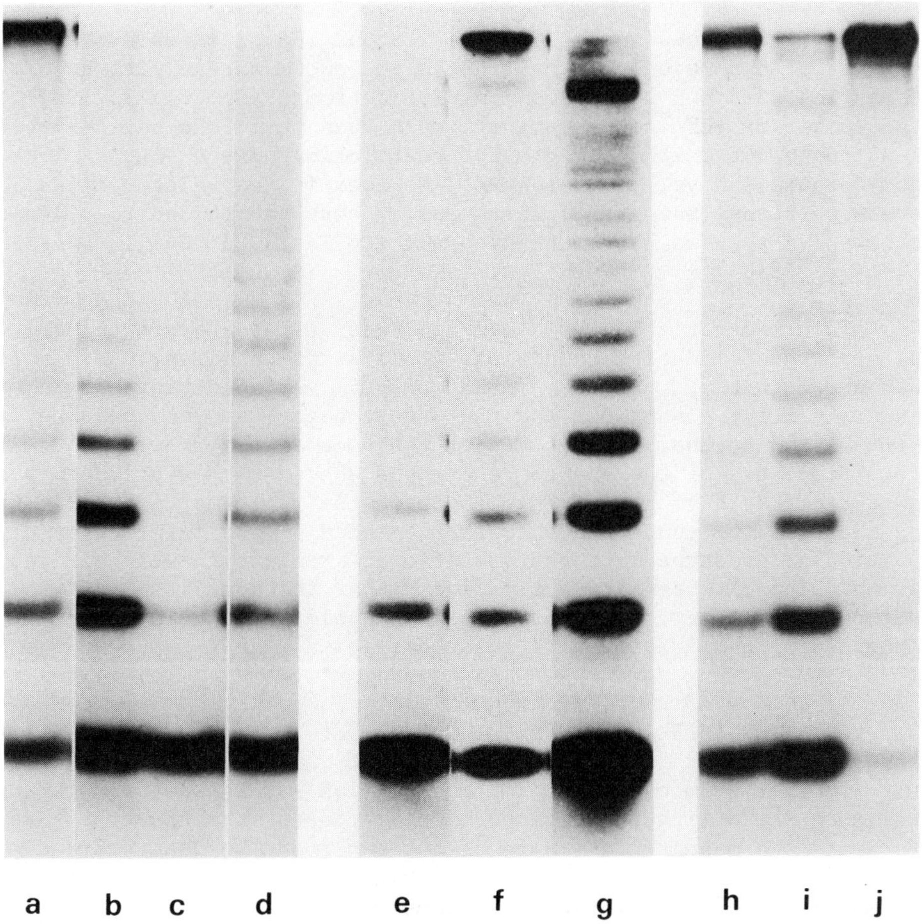

Fig. 1. TaqI digests of various DNAs run on a 1.4% agarose gel and transferred to a nitrocellulose filter. The filter was hybridised with nick translated cloned 5S DNA (pBG13). DNAs were extracted from (a) P1, (b) L6, (c) C1, (d) C3, (e) Liral Monarch, (f) Royal, (g) L. grandiflorum Desf. (h) L. bienne (106), (i) L. bienne (131), (j) L. bienne (111).

formation of the genotrophs, a particular subset of these genes was particularly susceptible to deletion. In every case where a reduction in the 5S gene number has been observed it occurred in this specific subset. However, only these rapid changes in the 5S genes are confined to this particular subset. In the longer term variation can occur in any part of the 5S gene array as demonstrated in the comparison between related species.

Ribosomal DNA variation

The flax genes coding for the 18S and 25S ribosomal RNAs consist of a homogenous set arranged in tandem arrays with a repeating unit 8.6 kilobases long (Goldsbrough and Cullis, 1981). The number of rRNA genes can very more than threefold between the genotrophs but neither length nor restriction enzyme site polymorphisms have been detected. However in the related *Linum* species (*bienne* and *grandiflorum* Desf.) both length and restriction site polymorphisms have been observed (Cullis 1983; Cullis and Cleary, 1985b).

The rDNA variation has been followed during the growth of plants, under inducing conditions (Cullis and Charlton, 1981). It was shown that, for Pl, the amount of rDNA varied during growth under inducing conditions and that the change observed could be transmitted to the progeny. Under the same set of growth conditions two of the stable genotrophs showed no variation in rDNA amount either during their growth or in their progeny. Thus in the case of Pl, the change in the rDNA depended on the interaction of a particular genotype with the specific environment in which it was grown. The change was quantitative rather than qualitative, although evidence for the potential of qualitative changes was obtained by comparisons with related species.

DNA VARIATION IN FLAX TISSUE CULTURE

Another form of stress imposed on flax was the passage through callus culture with the regeneration of plants from those cultures. Variants and mutant lines have often been found in plant tissue cultures and in plants regenerated from these cultures (Larkin and Scowcroft, 1983). The flax lines subjected to this passage included the original Pl line, which had been shown to respond to particular inducing environments as well as a number of the stable genotrophs, including the two shown not to respond to inducing environment in terms of changes in rDNA (Cullis and Charlton, 1981). DNA variation associated with the culture process was demonstrated in a number of lines (Table 3; Cullis and Cleary, 1985a).

DNAs from leaf and callus tissue from the same lines have been

Table 3. The relative amounts of different repeated sequences in leaf and callus DNAs. *values are for significant differences between leaf and callus DNAs from the same line ($P \leq .01$). Data from Cullis and Cleary 1985a. RT37/1 is an Agrobacterium tumefaciens transformed callus line from the S1 genotroph and was provided by Dr. A.G. Hepburn.

Source of DNA	rDNA	5SDNA	pCL2	pCL8	PCL53	pBG87	pDC7	pCL13	pCL21
P1 leaf	1.5	3.0	4.2	2.6	2.3	2.5	3.2	2.2	5.6
P1 callus	*0.9	2.6	4.0	2.4	*1.3	2.5	*2.5	*1.0	5.9
L3 leaf	1.5	2.8	4.8	2.6	2.0	2.3	2.6	1.6	4.9
L3 callus	*1.2	*2.3	*3.6	*2.1	1.9	2.3	2.7	1.7	5.0
S6 leaf	0.9	2.6	3.4	2.3	2.2	2.0	2.7	2.0	4.9
S6 callus	0.8	2.6	3.4	2.4	2.3	2.0	3.0	2.0	4.9
S1 leaf	1.1	3.0	3.9	2.9	2.4	2.3	2.8	1.9	5.2
FT37/1	*0.3	2.7	*2.1	*1.8	*1.3	*1.3	*1.1	1.6	5.0

compared using the set of cloned probes representing all the highly repetitive sequence families of the flax genome. It was observed that the various lines behaved differently in culture. In certain cases thare was no difference observed for any of the probes between the leaf and callus DNAs. In other cases differences were observed for the majority of the probes (Table 3). However for all the pairwise combinations of leaf and callus only one probe proved to be invariant. This was pCL21 which had already been found to be invariatn in the comparison of the genotroph DNA and with the related Linum species.

In only one case was a difference in the 5S gene number observed between callus and leaf. When the Taq I digests of these two DNA were hybridised with the 5S DNA it was found that the reduction in the 5S gene number could be completely accounted for in terms of the reduction in the Taq I resistant fraction (Cullis and Cleary, 1985a).

From this study it was apparent that the same range of DNA variation was appearing in tissue culture as that observed in the formation of the genotrophs. However this variation was also being found in genotrophs which have previously been shown not to respond to the various growth conditions which have been tried. Thus the "stable" genotrophs had not completely lost the ability to generate rapid genomic change, but had become unresponsive to those particular treatments by which they, themselves have been generated.

INSTABILITIES IN HYBRID PLANTS

McClintock (1978) suggested that wide species crosses were among the stresses that might trigger the reorganisation of parental genomes. Evidence for this comes from interspecific crosses in *Nicotiana* and *Microseris* species. However there is also evidence from *Microseris*, maize and flax that wide crosses may not be necessary to trigger rapid genomic change and that this may be initiated in intraspecific crosses.

The inheritance of nuclear 2C DNA amount in interspecific and intraspecific hybrids of *Microseris* species has been determined (Price et al., 1983). The values for F1 progeny from a cross between *M. douglassii* and *M. bigloveii* as well as the characterisation of 5 families of the F2 progeny indicated that the F1 plants produced were not of identical DNA content. The intraspecific cross between *M. douglassii* biotypes differeing in nuclear DNA content showed that the progeny could have unexpected values.

Analysis of crosses between flax genotrophs and the flax variety Liral Monarch gave an indication of a quantitative change in the nuclear DNA amount of the F1 (Durrant, 1981). Also the rDNA was shown to be significantly different from the expected value in the F1 of crosses between flax genotrophs (Cullis, 1979b).

By the use of the slot blotting technique, the quantitative inheritance of a number of repetitive sequence families was determined in F1 hybrids betwen maize inbreds (Rivin and Cullis, 1983). In a number of cases, including the rDNA and 5S genes the observed values differed significantly from their expected values.

In the examples above there was no striking evidence for the chromosomal location of the DNA sequence alterations. However in root tip, corolla and pollen mother cells of *N. tabacum* x *N. otophora* hybrids, chromosomes up to 15 times the normal length have been observed. The mechanism for the production of these megachromosomes is unknown. They may be the result of amplifications such as those which produce the homogeneously staining region in the chromosomes of some methotrexate resistant cell lines which have amplified copies of the dihydrofolate reductase gene (Dolnick et al., 1979).

DISCUSSION

It is clear that parts of the plant genome can be rapidly

modulated, particularly in response to "stress" or "shocks". Three different types of stress have been considered. The first was an environmental stress applied to the whole plant (the generation of the flax genotrophs), the second was the growing of plant cells in tissue culture, and the third was the combination of dissimilar genomes. In the application of all these stresses to flax it has emerged that the same set of genomic sequences appeared to be affected whenever variation was observed. Within this set of genomic sequences, in two cases it was the same subset of a particular family which was preferentially affected. Thus it is probable that the same basic mechanism was responsible for the changes in all three instances. However the nature of this mechanism is not known.

Another point to emerge was that not all species, nor all members of a single species had genomes which were equally susceptible to stress induced changes. Thus not all flax and linseed cultivars can undergo environmentally induced heritable changes. Not all interspecific hybrids give large, or even detectable changes in DNA. The basis for the variability in this response is not known but since it is restricted to certain lines it is possible to investigate the genetic bases of the modulating mechanism.

Thus the plant genome may be considered to have islands of instability in that only a subset of the DNA is susceptible to stress induced variation. However, how can this subset be distinguished within the genome? It appears unlikely that the recognition can be on the basis of primary nucleotide sequence since the subsets of 5S genes appear to have very few nucleotide differences. Thus it is likely to reside in the structure of the chromatin in which they are located. One possibility is that there are particular sequences which, because of their chromosomal location, are more susceptible to variation. This susceptibility may be dependent on the physiological status of the cell, so although the same sets of sequences may be affected with different stresses, the frequency with which any one sequence varies will be dependent on the particular perturbation applied.

Does this stress response have any special function? One possibility is that the genomic rearrangements provide a new set of novel variants which can overcome the threat to survival generated by the stress (McClintock, 1984). In order for this to be true, the generation of quantitative variation in these cases must have some effect on the phenotype. The most probable mechanism by which quantitative variation in repeated sequences may affect the phenotype is by a form of position-effect variegation which has been well documented in Drosophila (Rushlow and Chovnick, 1984).

Thus it appears that stress can be very important in the

mobilisation of the cell mechanisms involved in the restructuring of the genome. The mechanisms involved in the production of the changes in the genome by transposition, amplifications and deletions are already recognised. However there is nothing known about how the cell perceives the stress and translates this into a signal to bring about reorganisation of the genome that can be both radical but still limited.

REFERENCES

Britten, R.J., Graham, D.E. and Neufeld, B.R., 1974, Analysis of repeating sequences by reassociation kinetics. in "Methods in Enzymology," L. Grossman, and K. Moldave, eds, Academic Press, New York. pp 363-405.

Cullis, C.A., 1977, Molecular aspects of the environmental induction of heritable changes in flax, Heredity, 38:129-154.

Cullis, C.A., 1979a, Quantitative variation of ribosomal RNA genes in flax genotrophs, Heredity, 42:237-246.

Cullis, C.A., 1979b, Segregation of the isozymes of flax genotrophs, Biochem. Genet, 17:391-401.

Cullis, C.A., 1981a, Environmental induction of heritable changes in flax: defined environments inducing changes in rDNA and peroxidase isozyme band patterns, Heredity, 47:87-94.

Cullis, C.A., 1981b, DNA sequence organisation in the flax genome, Biochim. et Biophys. Acta, 652:1-15.

Cullis, C.A., 1983a, Environmentally induced DNA changes in plants, CRC Critical Reviews in Plant Sciences, 1:117-131.

Cullis, C.A., 1983b, Variable DNA sequences in flax, in: "Genetic Rearrangement," K.F. Chater, C.A. Cullis, D.A. Hopwood, A.W.B. Johnston and H.W. Woolhouse, eds, Croon Helm, U.K. pp253-264.

Cullis, C.A., and Charlton, L.M., 1981, The induction of ribosomal DNA changes in flax, Plant Sci. Letter., 20:213-217.

Cullis, C.A., and Cleary, W., 1985a, DNA variation in flax tissue culture,. Theoret. and Appl. Genet., submitted.

Cullis, C.A., and Cleary, W., 1985b, Rapidly modulated DNA sequences in flax, J. Molec. Biol., submitted.

Cullis, C.A., and Kolodynska, K., 1975, Variation in the isozymes of flax (linum usitatissimum) genotrophs, Biochem. Genet., 13: 687-697.

Cullis, C.A., Rivin, C.J., and Walbot, V., 1984, A rapid procedure for the determination of the copy number of repetitive sequences in eukaryotic genomes, Plant Mol. Biol. Reporter, 2:24-31.

Durrant, A., 1962, The environmental induction of heritable changes in Linum, Heredity, 17:27-61.

Durrant, A., 1971, Induction and growth of flax genotrophs, Heredity, 27:277-298.

Durrant, A., 1981, Unstable genotypes, Phil. Trans. R. Soc. Lond. B.,

292:467-474.

Durrant, A., and Nicholas, D.B., 1970, An unstable gene in flax, Heredity, 25:513-527.

Evans, G.M., 1968, Nuclear changes in flax, Heredity, 23:25-38.

Flavell, R.B., 1982, Sequence amplification, deletion and rearrangement: major sources of variation during species divergence, in: "Genome Evolution." R.B. Flavell, and G.A. Dover, eds., Academic Press, New York. pp 301-323.

Goldsbrough, P.B., and Cullis, C.A., 1981, Characterisation of the genes for ribosomal RNA in flax, Nuc. Acids Res., 1301-1309.

Larkin, P.J., and Scowcroft, W.R., 1983, Somaclonal variation and crop improvement, in: "Genetic Engineering in Plants," T. Kosuge, Meridith, C.P., and Hollander, A., eds., Plenum Press, Hew York, pp 289-314.

McClintock, B., 1978, Mechanisms that rapidly reorganise the genome. Stadler Genetics Symposium, 10:25-47.

McClintock, B., 1984, The significance of responses of the genome to challenge, Science, 226:792-801.

Price, H.J., Chambers, K.L., Bachman, K. and Riggs, J., 1983, Inheritance of nuclear 2C DNA content variation in intraspecific and interspecific hybrids of Microseris (Asteraceae), Amer. J. Bot., 70:1133-1138.

Rivin, C.J., and Cullis, C.A., 1983, Modulation of repetitive DNA in the maize genome, Genetics, 104:859-860.

Rushlow, C.A., and Chovnick, A., 1984, Heterochromatic position effect at the rosy locus of Drosophla Melanogaster: cytological, genetic and biochemical characterisation, Genetics, 108:589-602.

Schweizer, D., 1979, Fluorescent chromosome banding in plants: applications, mechanisms and implications for chromosome structure, in: "The Plant Genome," D.R. Davies, and D.A. Hopwood, eds, The John Innes Charity, Norwich, pp 61-72.

Timmis, J.N., and Ingle, J., 1973, Environmentally induced changes in rRNA gene redundancy, Nature New Biol., 244:235-236.

CONDITIO SINE QUA NON FOR *DE NOVO* EMERGENCE OF NEW GENES AND THE CONCEPT OF PRIMORDIAL BUILDING BLOCKS

Susumu Ohno, Nozomu Mori and Takeshi Matsunaga

Beckman Research Institute of The City of Hope

1450 E. Duarte Road, Duarte, CA 91010

INTRODUCTION

So long as a particular vital function is assigned to the single gene locus in the genome, all function-altering mutations are diligently eliminated by natural selection. Thus, the mechanism of gene duplication emerges as the prime means to create new genes with previously nonexistent functions (Ohno, 1970). Redundant copies of the existing gene created by this mechanism are largely ignored by natural selection, and while being ignored, they are free to accumulate function-altering mutations to emerge as new genes endowed with new functions. Indeed, the very fact that nearly all the existing genes can be considered as members of different families by propinquity of their descents reveals the prime role this mechanism played in evolution; e.g., the serine protease family, β_2-microglobulin family. Yet, this mechanism is very slow, inefficient and cumbersome as shall be pointed out shortly. Therefore, this mechanism is unfit to fulfill the organism's need to cope with rapidly changing environments. In this paper, we shall point out that under certain conditions, new genes with new roles can emerge in a flash by two novel means.

THREE SHORTCOMINGS OF THE MECHANISM OF GENE DUPLICATION

While relatively high spontaneous mutation rates are rather conductive of evolution of small organisms with proportionally small number of gene loci and very short generation times, such venturesomeness is not at all compatible with the life of modern vertebrates. Thus, DNA polymerase of modern vertebrates has become practically error-free, and multitudes of DNA repair mechanisms further reduces the spontaneous mutation rate of higher vertebrates. For example, the nucleotide substitution rate for *c-mos* oncogene of mammals has been computed to be 1.71 X 10^{-9} per site per year, whereas that of its viral counterpart *v-mos* was estimated as 1.31 X 10^{-3}; a million-fold difference (Gojobori and Yokoyama, 1985). This extremely slow mutation rate, then, is the first problem with the mechanism of gene duplication as it operates in higher vertebrates.

25 to 42 million years are required for the emergence of a new gene from a redundant copy of the preexisted gene

Fibrionopeptides A and B are discarded from fibrionogen when it is transformed to fibrin at the site of blood coagulation. Accordingly, there has been but little evolutionary constrain imposed upon their amino acid sequence changes. These coding segments must have been accumulating nearly all the randomly sustained mutational base changes. Yet, they have been undergoing evolutionary amino acid sequence changes at the rate of 1% per 1.0 and 1.7 million years, respectively (Dayhoff, 1972). While being ignored by natural selection, the product of a redundant gene copy would be undergoing amino acid sequence changes at about the similar rate. Under what seems a reasonable assumption that on the average, it takes 25% amino acid sequence divergence for the polypeptide to acquire a new function distinctly different from the ancestral one, the time required for a new gene to emerge from the redundant copy of the preexisted gene becomes 25 to 42.5 million years. Corollary to the tortoise-like pace of evolutionary changes noted above is the realization that many of the genes presently functioning in the vertebrate genome might prove themselves to be readily dispensable. By becoming tetraploid, an organism duplicates every gene locus in the genome simultaneously, and among modern teleost fish of the super order *teleostei*, there are a number of tetraploid species that have become as such in different times of the past (Ohno, 1970). This situation enabled Ferris and Whitt (1977) to estimate the average halflife of enzyme loci that have become redundant as a consequence of tetraploidy as 50 million years. Needless to say, after becoming degenerate as to their assigned functions as enzymes, these redundant genes may continue to encode functionless polypeptide chains for a long while yet.

The above reveals the vulnerability of the still widely held view that any gene which is expressed in precisely regulated manner must be essential to the well being of organisms. I shall give a few contrary evidences to refute this seemingly irrefutable view. *Slp* locus in the midst of MHC (major histocompatibility antigen) region of the mouse (*Mus musculus domesticus*) is a tandem duplicate of *Ss*. While *Ss* encodes C_4 complement, a degenerate polypeptide specified by *Slp* no longer functions as C_4. Yet, the expression of *Slp* is androgen-dependent in most mice, while it is constitutive in a mutant (Hansen and Shreffler, 1976). Serum albumin produced in liver is the major component of vertebrate serum and its expression is developmentally regulated in mammals, as it postnatally replaces its fetal counterpart; α-fetoprotein. Yet, analbuminemia is a harmless trait in man (*Homo sapiens*) (Bowman, Hermodson, Hammond, and Motulsky, 1976), the dog, (*Canis familiaris*) (Tanabe, 1980) and the rat (*Rattus norvegicus*) (Nagase, Simamune, and Shumiya, 1979).

salvandrum paucitas, damnandrum multitudo

Single base substitutions randomly sustained by coding sequences bring about three consequences; samesense (no alteration in the amino acid sequence), missense (an amino acid substitution) and premature chain-termination; e.g., tyrosine codon U A C changing into a terminator UAG. The expected ratio between the above three is 1 : 2.77 : 0.15. On the average, one out of every 26 randomly sustained single base substitutions would result in the production of shorter than original polypeptide chains. A new polypeptide chain representing only the amino terminal one-third or even half of the original is not likely to acquire a new function of any sort. Of randomly sustained deletions and insertions, small numbers of bases that are multiples of three are relatively harmless, for they merely delete or insert a few amino acids. The numbers that are not multiples of three, however, cause frameshifts which usually result in premature chain-termination in their immediate downstream. All in all, the chance of a redundant gene copy emerging triumphant as a new gene of previously nonexistent function appears exceedingly small, more often redundant copies becoming degenerate pseudogenes that abound in the genome. The consequence, thus, of hundreds of millions of years of continuous gene duplications should have been desertification of genomic DNA, distances between neighbouring functional genes becoming progressively long. Some years ago, one of us had estimated the average distance between neighboring mammalian genes as 45,000 base pairs (Ohno, 1972). Actually measured distances between six mouse immunoglobulin heavy-chain constant region genes (C_{δ}, $C_{\gamma 3}$, $C_{\gamma 1}$, $C_{\gamma 2b}$, $C_{\gamma 2a}$, C_{ε} and C_{α})ranged from 55,000 to 12,500 base pairs; the average distance being 25,300 base pairs (Nishida, Kataoka, Ishida, Nakai, Kishimoto, Bottcher, and Honjo, 1981; Shimizu, Takahashi, Yamawaki-Kataoka, Nishida, Kataoka, and Honjo, 1981). While this average distance is considerably less than the estimated 45,000, it should be remembered that there is an extremely long void in front of C_{μ}, C_{δ} gene complex separating groups of immunoglobulin variable region V_H genes from these constant region genes. Thus, the second shortcoming of the mechanism of gene duplication as the means to create new genes is its inefficacy, redundant copies far more often becoming degenerate pseudogenes than new genes of previously nonexistant functions.

Limitation on functional diversification

Perhaps the most serious shortcoming of the mechanism of gene duplication is that a truly novel function cannot be acquired by a new gene born by this mechanism. Instead, functions assumed by genes of the same family related by the propinquity of their descents are mere variations of the same theme. Recently, this inherent limitation imposed upon functional diversification was dramatically revealed by antigen-binding pockets of membrane-bound T-cell receptors. While $V_L + V_H$ dimeric antigen-binding pockets of immunoglobulins recognize individual antigenic determinants *per se*, $V_{\alpha T} + V_{\beta T}$ dimeric antigen-binding pockets of cytotoxic T-cell receptors are designed to recognize not alien (e.g., viral) antigens themselves but class I MHC antigens of the self (H-2D and H-2K of the mouse, HLA-A, HLA-B and HLA-C of man) altered by their association with alien antigens on target cell surface (Zinkernagel, and Doherty, 1974). In the case of *helper* T-cells, their membrane-bound receptors are designed to recognize on the surface of antigen-presenting macrophages, not alien antigens themselves but altered self class II MHC antigens (e.g., I_A antigens of the mouse, HLA-DW antigens of man). Thus, there was a great expectation that $V_{\alpha T} + V_{\beta T}$ antigen-binding pockets of T-cell receptors as a whole must show a unique configuration fundamentally different from $V_L + V_H$ antigen-binding pockets of immunoglobulin and that $V_{\alpha T} + V_{\beta T}$ antigen-binding pockets of cytotoxic T-cell receptors directed against class I MHC antigens might be significantly different from those of *helper* T-cells directed against class II MHC antigens. Recently, amino acid sequences of one human and two mouse $V_{\beta T}$ derived from cytotoxic and *helper* T-cells have been deduced by three groups (Chien, Gascoigne, Kavaler, Lee, and Davis, 1984; Saito, Kranz, Takagaki, Hayday, Eisen, and Tonegawa, 1984; Siu, Clark, Yoshikai, Malissen, Yanagi, Strauss, Mak, and Hood, 1984). Disappointingly, these $V_{\beta T}$ proved themselves to be but merely modified forms of immunoglobulin V_L as shown in Fig. 1 (Ohno, Matsunaga, and Lee, 1984), and that $V_{\beta T}$ of cytotoxic T-cell receptors and those of *helper* T-cell receptors were obviously derived from the same pool. Furthermore, one sequence identified as $V_{\alpha T}$ had no distinctive feature that separates it from $V_{\beta T}$ (Ohno, Matsunaga, and Lee, 1984). At the top left of Fig. 1, the amino acid sequence of human subgroup I $V_{\lambda L}$ NEW which is anti-vitamin K_1OH is shown in its realistic secondary structure configuration as revealed by the X-ray crystallographic study by Saul, Amzel, and Poljak (1978). The outer β-sheet structure shown on the right-hand side is comprised of two loops and one strand represented by residues, 1-26, 61-76 and 101 to 107. The inner β-sheet structure shown on the left-hand side is comprised of two loops representing residues 32-47, 84-100. It is the tryptophane loop (residues 35 to 47 shown in very large capital letters) of the inner β-sheet structure which forms the floor of $V_L + V_H$ antigen binding pockets by making contact with the corresponding tryptophane loop of V_H (Saul, Amzel, and Poljak, 1978). Six residues, 4 of which belonging to the tryptophane loop, that make contact with V_H are identified by being placed in squares. One residue (31st Asp) of CDR-1 region (residues

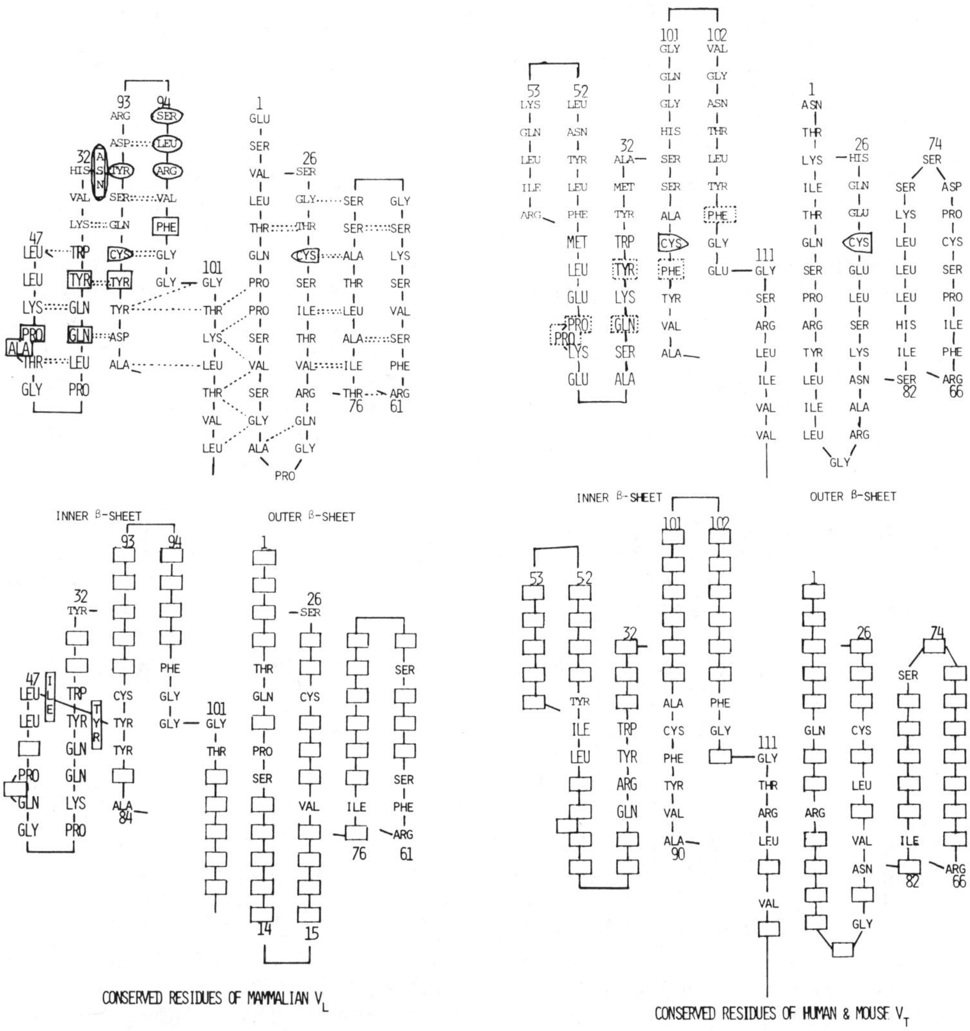

Fig. 1: Extreme similarities between the secondary structure configuration of V_L of immunoglobulin and that of V_T of T-cell receptor are illustrated.

24 to 34) and three residues of CDR-3 region (residues 89-97) that actually combine with vitamin K_1OH antigen are encircled (Twining, and Attassi, 1978). Residues of CDR-1,2,3 regions are shown in smallest capital letters. Residues of three pieces connecting outer and inner β-structure are not shown. They are residues 27-31, 48-62 and 77-83. At the bottom, the more general configuration applicable to both $V_{\kappa L}$ and $V_{\lambda L}$ of mammals is presented. Only residues conserved by majority of $V_{\kappa L}$'s as well as $V_{\lambda L}$'s of

several mammalian species listed by Kabat, Wu, Bilofsky, Reid-Miller, and Perry (1983) are identified. At the top right of Fig. 1, the deduced amino acid sequence of mouse 86T1 *helper* $V_{\beta T}$ (Chien, Gascoigne, Kavaler, Lee, and Davis, 1984) is shown in its assumed secondary configuration. At the bottom right, the general configuration applicable to both $V_{\alpha T}$ and $V_{\beta T}$ is presented. Only those residues conserved by three of the four published V_T sequences (one mouse $V_{\alpha T}$ and one human, two mouse $V_{\beta T}$) are identified (Ohno, Matsunaga, and Lee, 1984). While V_T and V_L are exceedingly similar, CDR-2 region of the latter appears to be longer thus forming 3rd loop within the inner β-sheet structure. Details on the functional significance of this feature have been discussed elsewhere (Ohno, Matsunaga, and Lee, 1984). At any rate, V_T for the antigen-binding pocket of T-cell receptor is but a mere variant of V_L for the antigen-binding pocket of immunoglobulin, $V_{\alpha T}$ + $V_{\beta T}$ dimeric antigen-binding pockets resembling artificial V_L + V_L dimer.

In spite of the three shortcomings noted above, the mechanism of gene duplication can be innovative on occassions. The case in point is the β_2-microglobulin family of genes to which V_L, V_H, $V_{\alpha T}$ and $V_{\beta T}$ genes just discussed belong.

INNOVATION OF ANTIGEN-BINDING DOMAINS FROM NONANTIGEN-BINDING DOMAINS WITHIN THE β_2-MICROGLOBULIN FAMILY

The membership of this family is vast, for included in it are single domain β_2-microglobulin and Thy-1 antigen as well as domains of Class I and Class II major histocompatibility antigens, immunoglobulins, T-cell receptors and of poly-Ig receptor for transepithelial transport. Yet, most of these domains are nonantigen-binding, the ability to recognize specific antigenic determinants is the unique property conferred only to V_L + V_H dimers of immunoglobulins and to $V_{\alpha T}$ + $V_{\beta T}$ dimers of membrane-bound T-cell receptors as already noted. Thus, there occured a true innovation within members of the β_2-microglobulin family. The nature of this innovation becomes evident when V_L and V_T shown in Fig. 1 is compared with nonantigen-binding $C_{H1\gamma1}$ constant domain of immunoglobulin $\gamma1$ heavy-chain shown at the top, left as well as with nonantigen-binding HLA-DR shown at the bottom, left of Fig. 2. The inner β-sheet structure shown on the left-hand side of each molecule of antigen-binding domains (Fig. 1) being made of two (V_L) or three (V_T as well as V_H) antiparallel loops is far more developed than corresponding structures of nonantigen-binding domains that are made of only one loop and a strand (Fig. 2, left). This difference in their inner β-sheet structures profoundly affect their manner of dimer formation. In the case of dimers formed between antigen-binding domains (e.g., between V_L and V_H and between $V_{\alpha T}$ and $V_{\beta T}$) the contact are mainly made between residues of each's respective trytophane loops. Thus, the inner β-sheet structure literally

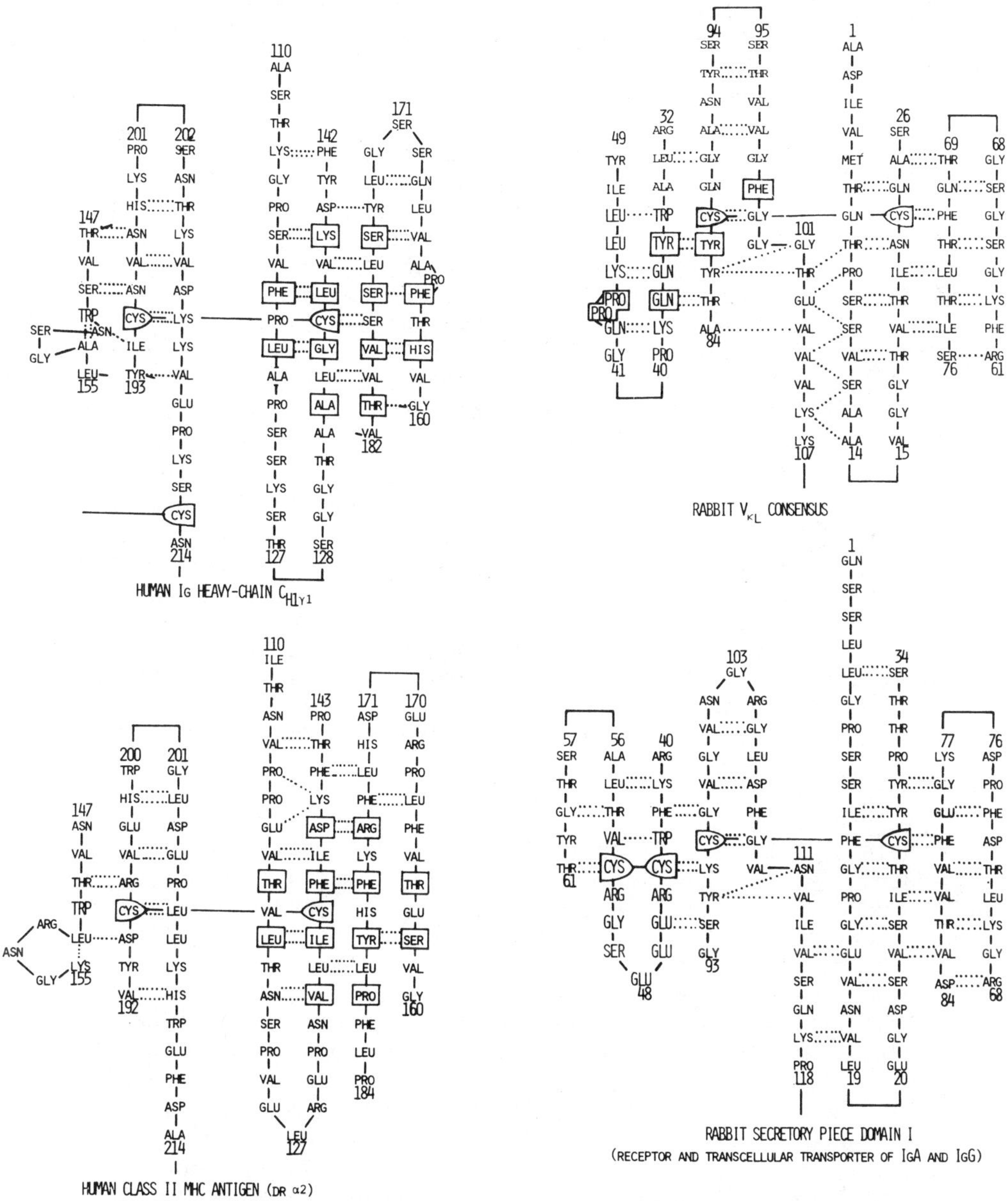

Fig. 2: The characteristic secondary structure configuration of nonantigen-binding domains of the β_2-microglobulin family is represented by human immunoglobulin γ1-heavy chain C_{H1} domain as well as by human class II antigen HLA-DR α2 domain at the left. At the right, antigen-binding domain-like configuration are represented by the rabbit consensus $V_{\kappa L}$ as well as by 1st domain of the rabbit poly-Ig receptor for transepitheial transport of poly-IgA and Poly-IgM.

occupies the inside position and the paired tryptophane loops form the floor of antigen-binding pockets. Hypervariable residues of CDR-1, CDR-2 and CDR-3 regions brought in the immediate vicinity of the paired tryptophane loops functioning as antigen-combining residues (Ohno, Matsunaga, and Lee, 1984). The situation is entirely different for dimers of nonantigen-binding domains. Contacts between, for example, C_{H1} and $C_{\kappa L}$ or $C_{\lambda L}$ of immunoglobulins or between $\alpha 2$ and $\beta 2$ of class II major histocompatibility antigens are made between residues of outer β-sheet structures as shown at the left of Fig. 2, thus, the outer β-sheet structure becomes a misnomer. At the top left of Fig. 2, the amino acid sequence of C_{H1} domain of human immunoglobulin $\gamma 1$-heavy-chain is shown in its realistic secondary structure configuration as revealed by the X-ray crystallographic study by Saul, Amzel, and Poljak (1978). As before, residues of three connecting pieces are ommitted. Only 150th TRP corresponding to 35th or 36th TRP of antigen-binding domains is shown in large capital letters. This TRP is present in nearly all the domains of the β_2-microglobulin family; only exception being β_2-microglobulin and Thy-1 antigen as well as 2nd and 3rd domains of the six-domain-long poly-Ig receptor for transepithelial transport (Mostov, Friedlander, and Blobel, 1984). Hydrogen bonds that stabilize β-sheet structure are shown as broken lines. Three cysteine residues, two for the intrachain and one for the interchain disulfide bridge are encased in bullet-shaped boxes. 12 residues of the outer β-sheet structure (right-hand side) encased in squares are those that make contact with corresponding residues of $C_{\kappa L}$ or $C_{\lambda L}$ to form the dimer.

At the bottom left, the amino acid sequence of $\alpha 2$ domain of α-chain of human HLA-DR antigen is shown in similar configuration deduced from various coor-dinates (e.g., Chou, and Fasman, 1978). The deduced configuration shown here proved nearly identical to that predicted from computer graphics techniques by Travers, Blundell, Sternberg, and Bodmer (1984).

Thus, it would appear that the key to the transformation from nonantigen-binding domains to antigen-binding domains within the β_2-microglobulin family was the formation of the tryptophane loop within their inner β-sheet structure. This is easier said than done, for amino acid sequences of this family are rich in Ser, Thr, Val, Leu residues that can readily form hydrogen bonds not only with each other but also with many other residues to stabilize the existing outer and inner β-sheet structures. Thus, the particular secondary structure configuration once assumed would not readily change even by rather extensive amino acid substitutions. The hint of how the formation of the tryptophane loop in the inner β-sheet structure was accomplished was given by the recently deducted amino acid sequence of the six-domain-long rabbit poly-Ig receptor for transepithelial transport of polymeric IgA and IgM to saliva, guts etc, (Mostov, Friedlander, and Blobel,

1984). Thus, here is the member of the β_2-microglobulin family that specifically binds to polymeric forms of IgA and IgM, and the amino acid sequences of its six domains was shown to share an average of 28% identical residues with $V_{\kappa L}$; the homology between 1st domain and $V_{\kappa L}$ rising to 36% (Mostov, Friedlander, and Blobel, 1984).As shown at the bottom right of Fig. 2, the 43rd Trp of the 1st domain which corresponds to 35th and 36th Trp of V_L and V_H is immediately followed by Cys and that there is another Cys 7 residues further downstream. The disulfide bridge formation between these two cysteine residues would in one stroke have created the tryptophane loop in its inner β-sheet structure. Accordingly, the transformation of the formerly nonantigen-binding configuration to an antigen-binding domain-like configuration was accomplished at once. Comparing the rabbit (*Oryctolagus cuniculus*) consensus $V_{\kappa L}$ sequence shown at the top right of Fig. 2 with the sequence of rabbit poly-Ig receptor domain 1 presented in the similar secondary configuration at the bottom right of Fig. 2, it would be noted that identical residues occupy the positions critical to the maintenance of the assumed configuration. Of five other domains of poly-Ig receptor, we found only 5th domain to be capable of assuming the similar secondary structure configuration as domain 1. Our prediction is that the binding of poly-Ig receptor to polymeric IgA and IgM involves only its 1st and 5th domains.

Needless to say, the above should not be construed to mean that 1st or 5th domain of poly-Ig receptor for transepithelial transport of polymeric IgA and IgM was the ultimate ancestor of V_L, V_H and V_T, for such a receptor should have been the late comer in evolution arising after the establishment of adaptive immune system. Rather what we contend here is that the initial transformation from nonantigen-binding to antigen-binding domains within the β_2-microglobulin family must have been accomplished by the similar means; i.e., the birth of a tryptophane loop by the disulfide bridge formation involving a pair of strategically placed cysteine residues.

DE NOVO EMERGENCE OF CODING SEQUENCES THAT ARE REPEATS OF BASE OLIGOMERS

Although the mechanism of gene duplication may occasionally be innovative as discussed above, the concept of evolution by gene duplication is inherently incomplete in that the ultimate origin of the ancestral gene of each family remains unexplained. It is likely that Rudolph Virchow's immortal dictum, *omnis cellula a cellula* does not apply to the origin of genes, for the simple reason that the first living creature on this earth could not possibly have started its life with one gene and its duplicates. There ought to have been several different kinds of genes at least; some specifying α-helical polypeptide chains, while others encoding β-sheet forming polypeptide chains. The multiple origin of genes immediately suggests that the transformation of noncoding sequences in the genome to new coding sequences ought to occur every now and then even in modern organisms.

Inasmuch as only one (A U G) of the 64 base triplets functions is the chain initiator, while three (U A A, U A G and U G A) are chain-terminators, the probability of 300-base-long randomly generated base sequences being able to specify a 100 amino acid residues long polypeptide chain is $(1/64) \times (61/64)^{99}$ which amounts to only 1.35×10^{-4}. Thus, it would appear that noncoding unique sequences, although abundant in the genome, have a slim chance of yielding a new gene i.e., a new coding sequence. It is granted that since AT/GC ratio of three chain-terminators (T A A, T A G, T G A) is 7-to-2 in favor of AT, GC-rich unique sequences have a far better chance of containing long open reading frames than those of more balanced AT/GC ratios. However, it should be noted that various signal sequences, not only the initiating codon but also the promotor and the poly-A attachment signal, that a new coding sequence has to acquire are all AT rich, thus, negating an apparent advantage enjoyed by GC-rich unique sequences as the likely candidtae to emerge as a coding sequence. Repeats of base oligomers, on the other hand, have a very good chance of emerging as a new gene. For example, 82.5% which is $(61/64)^4$ of the duodecameric repeats has a chance of having an open reading frame of indefinite lengths. Since such repeats in the genome tend to be very long indeed, all they have to do to become a new coding sequence is to wait for the emergence of various signal sequences in more or less appropriate places. It was on this basis that the notion of oligomeric primordial building blocks in the construction of various coding sequences was proposed by Ohno (1981). Earlier still Ycas (1972) has realized that periodical polypeptide chains that would be encoded by oligomeric repeats have a far better chance of assuming either α-helical or β-sheet secondary structures, when compared to randomly generated amino acid sequences. It was on this basis, he proposed that all the polypeptide chains of modern world must have been derived from primitive polypeptides of exact periodicities (Ycas, 1972).

Indeed, primarily α-helical modern polypeptide chains such as collagen as well as primarily β-sheet forming polypeptide chains such as serum albumin, immunoglobulins and class I major histocompatibility antigens still demonstrate discernible periodicity in their amino acid sequences. Accordingly, the base sequence of each's primordial building block (oligomeric unit) have been deduced with variable certainty in the collagen gene (Yamada, Avvedimento, Murdryj, Ohkubo, Vogeli, Irani, Pastan, and Crombrugge, 1980), serum albumin coding sequence (Alexander, Young, and Tilghman, 1984; Ohno, 1981), immunoglobulin and class I major histocompatibility antigen coding sequences (Ohno, Matsunaga, and Wallace, 1982; Ohno, Matsunaga, Epplen, Itakura, and Wallace, 1982;

Yazaki, and Ohno, 1983). The question which arises is: how periodical is a new polypeptide chain encoded by a newly emerged coding sequence directly derived from noncoding oligomeric repeats? We shall give a few example of newly emerged polypeptide chains that demonstrate exact periodicities of various types. Cods and hakes (*Gadus* and *Microgadus*, respectively) of the family *Gadidae* and flounders and halibuts (*Pleunectes species*) of the order *Heterosomata* mostly inhabit temperate waters of the world's oceans, but a few species of each live in freezing Arctic and Antarctic waters. These few have evolved to produce antifreeze proteins to be contained in their blood. The prototype antifreeze protein of cods are tripeptidic periodical glycoproteins of variable lengths: Ala-Thr-Ala tripeptide repeating many times. For each Thr residue, a disaccharide, galactosamino-galactose, is attached (DeVries, 1982). Thus, it would appear that Arctic and Antarctic cods such as *Gadus ogac, Boreogadus saida* and *Dissostichus mawsoni* managed to meet the new need to produce antifreeze device by recruiting a certain of the nanomeric repeat base sequences in the genome as a new coding sequence or sequences. Antifreeze proteins possessed by north-temperate and Arctic flounders are of entirely different origin and are not glycoproteins. The hexadecapeptidic periodicity of the antifreeze protein of the Alaskan plaicie *Pleuronectes quadritaberulatus* was determined as Asp-Thr-Ser-asp-Ala-Ala-Ala-Ala-Ala-Ala-Ala-Ala-Thr-Ala-Ala-Asp (DeVries, 1982). For the same purpose of evolving antifreeze proteins, flounders have recruited a different repeat sequence, the unit sequence being 48-base-long, of the genome as a new coding sequence.

Proteins synthesized by larval salivary glands of dipteran insects are of special interest for genes synthesizing these proteins can be visualized as Balbiani rings or RNA puffs on giant polytene chromosomes. Furthermore, these proteins are new proteins that fullfill specific demands of particular life styles adapted by larvae of different species. For example, the salivary glands of *Drosophila melanogaster* larvae synthesize large amounts of a small set of polypeptides that are secreted at the end of larval life to form a glue which attaches the pupal case to dry surfaceses during metamorphosis to the adult. One such encoded by Sgs4 gene was deduced to have the Heptapeptidic periodicity; the primordial heptapeptide being: Thr-Glu-Pro-Pro-Arg-Cys-Glu (Muskavitch, and Hogness, 1982). The Balbiani ring 2 gene of *Chironomus tentans* active in the larval salivary gland encodes a polypeptide chain of an enormous length which is used to spin tubes housing the aquatic larva. This Balbiani ring 2 gene is repetitious in a rather complex way probably suggesting that it has been in existence for quite awhile. Two types of related repeating units-the α and β types - are tandemly arranged in separate blocks. Every repeat unit has one constant region and one subrepeat region. The unit 30-residue-long sequence specified by the above noted constant region is: Lys-Cys-Gly-Ser-Lys-Met-Arg-Arg-Val-Leu-Ala-Glu-Lys-Cys-

Ala-Ala-Arg-Lys-Gly-Arg-Phe-Ser-Ala-Ser-Lys-Cys-Arg-Cys-Phe-Ser (Hoog, and Wieslander, 1984). The ultimate oligomeric unit of this coding sequence apparently was 18-base-long something like: A G C A A G C C A A G C A A A C A C specifying a hexapeptide Ser-Lys-pro-Ser-Lys-His (Sumegi, Wieslander, and Daneholt, 1982).

All in all, it would appear that most, if not all, of the coding sequences at their inception were oligomeric repeats, thus, encoding periodical polypeptide chains that tend to assume either α-helical or β-sheet secondary structure, and that even in modern organisms, new coding sequences are arising *de novo* on rare occassions from oligomeric repeats in the genome.

4) A SUDDEN BIRTH OF A NEW ENZYME FROM AN ALTERNATIVE OPEN READING FRAME OF THE PREEXISTED CODING SEQUENCE

Newly arisen polypeptides of exact periodicities noted above may be immediately useful as secreted proteins as we have seen or even as structural components of the cell. In order for them to acquire more sophisticated functions such as catalytic activities of enzymes, however, the accumulation through 50 million years or so of mutational amino acid sequence changes would be required. Thus, oligomeric repeats in the genome as such cannot be the source of suddenly appearing new enzymes. Yet, such does happen as we shall shortly see. What if the number of bases in the unit sequence of olygomeric repeats was not a multiple of three? When transcribed and translated, the unit periodicity to a polypeptide chain was given by three consecutive copies of the oligomeric unit base sequence translated in three different reading frames. Thus, while the repeats of the base nanomer could have given only a tri-peptidic periodicity to its polypeptide chain, the repeats of the base decamer should have given a decapaptidic periodicity to its polypeptide chain. Furthermore, the inherent property of this type of internally repetitious coding sequence is that not one but all three reading frames should have remained open initially encoding three polypeptide chains of the identical amino acid sequence as shown in the decameric repeats below as an example (Ohno, and Epplen, 1983):

```
  Arg  Thr  SER  Thr  Tyr Gln  His  Val  Pro  Ala
 C G T A C C A G C A/C G T A C C A G C A/C G T A C C A G C A
Thr  Tyr  Gln  His  Val  Pro  Ala  Arg  Thr  Ser
   Val  Pro  Ala  Arg  Thr  Ser  Thr  Tyr  Gln  His
```

The coding sequence started as the above type of repeats should have maintained one or the other alternative reading frame open for a long time, in spite of base substitutions, insertions, and deletions. Yet, with time, differences began to outweigh similarities between the amino acid sequence of the currently produced polypeptide chain and that of a polypeptide chain specifiable by its open alternative reading frame. At this stage, such an alternative open reading frame becomes a potential source of a suddenly appearing, entirely new polypeptide chain. This type of potential is characteristically possessed by viral as well as host genomic oncogenes (Ohno, and Yazaki, 1983) and viral nucleic acid polymerases (Ohno, 1984).

The successful synthesis of the linear polyamides, Nylon, was announced by W.H. Carothers of E.I. DuPont de Nemours & Co. in 1938, and the first Nylon factory was not built in Japan until 1950. Yet as early as 1975, Kinoshita, Negoro, Murayama, Bisaria, Sawada, and Okada (1977) isolated *Flavobacterium* Sp. K172 from waste water of a Nylon factory that can grow in the culture medium containing 6-aminohexanoic acid cyclic dimer as the sole source of carbon and nitrogen. Subsequently, they have shown that this unique property is conferred to *Flavobacterium* Sp. K172 by the possession of two plasmid encoded enzymes: 6-aminohexanoic acid cyclic dimer hydrolase and 6-aminohexanoic acid linear oligomer hydrolase. Further, they have established the high substrate specificities of these two enzymes in that they ignored a variety of natural amido bonds offered (Kinoshita, Negoro, Murayama, Bisaria, Sawada, and Ikada, 1977; Kinoshita, Terada, Taniguchi, Takene, Masuda, Matsunaga, and Okada, 1981). Apparently, these two enzymes were not preexisted amidohydrolases of broad substrate specificities. Thus, the swiftness with which *Flavobacterium* Sp. K172 acquired these two enzymes appeared to have broken all the rules of evolution. When the coding base sequence for the second enzyme was published by Okada, Negoro, Kumura, and Nakamura (1983), however, it was noted that this coding sequence at its inception must have been the decameric repeats of the type shown immediately above, and that in addition to the one open reading frame that encoded 6-aminohexanoic acid linear oligomer hydrolase, there existed yet a longer open alternative reading frame (Ohno, 1984). On this basis, it was proposed that this coding sequence originally was encoding a 427-residue-long arginine-rich polypeptide chain of an unknown function. The insertion of a single base T at the position noted below silenced this original reading frame, at the same time creating the chain-initiator A T G to start off an alternative open reading frame that happened to have encoded a 392-residue-long polypeptide with the 6-aminohexanoic acid linear oligomer hydrolase acitivity (Ohno, 1984).

```
 29                 33
Phe  Gly  Asp  Thr  Arg  Ter
T T C G G A G A C A C T C G A(T)G A A C G C A C G T T C C A C C
                            Met  Asn  Ala  Arg  Ser  Thr
                             1

G G C C A G C A C C C C G C C A G G T A T C C C G G A G C C G C G
 Gly  Gln  His  Pro  Ala  Arg  Tyr  Pro  Gly  Ala  Ala
            9
```

```
G C C G G G G A G C C G
 Ala  Gly   Glu   Pro
                  21
```

In the above, the reading frame shown above the base sequence was thought to be the original reading frame encoding the 427-residuelong arginine-rich (7.65%) polypeptide chain. The insertion of a single base T between 33rd Arg codon C G A and 34th Glu codon G A A silenced this original reading frame by creating the chain-terminator T G A. At the same time, the alternative open reading frame was activated by the creation of the chain-initiator A T G which is shown below the base sequence. This activated alternative reading frame now encodes 392-residue-long 6-aminohexanoic acid linear oligomer hydrolase. Details are described in Ohno (1984).

Whether or not this has actually happened in the plasmid harbored by *Flavobacterium* Sp. K172 is rather irrelevant to the present discussion. The point to be made here is that provided that the number of bases in the primordial oligomeric unit was not a multiple of 3, coding sequences descended from oligomeric repeats maintain for a long time in evolution one or the other alternative reading frame open, and such an alternative long open reading frame is the potential source of suddenly appearing new polypeptide chains with truely novel function. A new polypeptide encoded by such an alternative open reading frame share many identical oligopeptidic sequences with the polypeptide chain encoded by the original reading frame (Ohno, 1984a and b), yet its amino acid sequences is sufficiently different to be endowed with a novel function.

SUMMARY

Inasmuch as it is extremely unlikely that all the multitudes of genes of today have descended from the single primordial gene of eons ago, the theory of evolution by gene duplication is inherently incomplete in that the origin of the ultimate ancestor of each family or superfamily of genes remains unexplained. For various reasons, it was concluded that each coding sequence that arises *de novo* almost has to be repeated base oligomers, thus, encoding a polypeptide chain of an exact periodicity. Periodical polypeptide chains thus produced tend to assume either α-helical or β-sheet secondary structure.

Even modern organisms on rare occasions recruite from the silent part of their genomes new coding sequences that as expected encode polypeptide chains of exact periodicities. Provided that polypeptide chains encoded by them are either mainly α-helix or mainly β-sheet formers, their ultimately primordial oligomeric unit base sequence can still be deduced on a number of modern genes, although they have existed for hundreds of millions of years. Such oligomeric unit base sequences have been defined as primordial building blocks of individual coding sequences.

Those genes whose primordial building blocks consisted of numbers of bases that were not multiples of 3 tend to maintain for a long time, one or the other alternative reading frame open for long stretches. Such long alternative open reading frames are the potential source of suddenly appearing new polypeptide chains with truly novel functions.

Although the mechanism of gene duplication is inefficient (most of redundant copies becoming degenerate pseudogenes rather than new genes) and slow (the transformation of a redundant copy to a new gene requiring 25 to 50 million years), this mechanism too can be innovative on rare occasions. Thus, within the β_2-microglobulin family, the acquisition of the tryptophane loop in their inner β-sheet structure transformed nonantigen-binding domains to antigen-binding domains of immunoglobulins as well as T-cell receptors.

REFERENCES

Alexander, F., Young, P.R., and Tilghman, S.M., 1984, Evolution of the albumin:α-fetoprotein ancestral gene from the amplification of a 27 nucleotide sequence, J. of Mol. Biol., 173:159-174

Bowman, H., Hermodson, M., Hammond, C.A. and Motulsky, A.G., 1976, Analbuminemia in an American Indian girl, Clin. Genet., 9:513-526.

Chien, Y., Gascoigne, N.R.J., Kavaler, J., Lee, N.E., and Davis, M.M., 1984, Somatic recombination in a murine T-cell receptor gene, Nature, 309:322-326.

Chou, P.Y., and Fasman, G.D., 1978, Empirical predictions on protein conformation, Ann. Rev. Biochem., 47:251-276.

Dayhoff, M.O., ed., 1972, "Atlas of Protein Sequence and Structure," National Biomedical Research Foundation, Silver Springs, Maryland.

DeVries, A.L., 1982, Biological antifreeze agents in cold water fishes, Comp. Biochem. Biophysiol., 73A:627-640.

Ferris, S.D., and Whitt, G.S., 1977, Loss of duplicate gene expression after polyploidisation, Nature, 265:258-260.

Gojobori, T., and Yokoyama, S., 1985, Evolutionary rates of *c*- and *v-mos* genes, Proc. Natl. Acad. Sci., USA, in press.

Hansen, T.H., and Shreffler, D.C., 1976, Characterization of a constitutive variant of the murine serum protein allotype, *Slp*., J. Immunol., 117:1507-1513.

Hoog, C., and Wieslander, L., 1984, Different evolutionary behavior of structually related, repetitive sequences occuring in the same Balbiani ring gene of *Chironomus tentans*, Proc. Natl. Acad. Sci. USA, 81:5165-5169.

Kabat, E.A., Wu, T.T., Bilofsky, H., Reid-Miller, M., and Perry, H., 1983, Sequences of proteins of immunological interest., U.S. Dept. of Health and Human Serv.'s, Natl. Inst. of Health, Bethesda, MD.

Kinoshita, S., Negoro, S., Murayama, M., Bisaria, V.S., Sawada, S., and Okada, H., 1977, 6-amino hexanoic acid cyclic dimer hydrolase. A new cyclic amide hydrolase produced by *Acromobacter guttatus* KI 72, Eur. J. Biochem, 80:489-495.

Kinoshita, S., Terada, T., Taniguchi, T., Takene, Y., Masuda, S., Matsunaga, N., and Okada, H., 1981, Purification and characterization of 6-aminohexonic-acid-oligomer hydrolase of *Flavobacterium* sp. KI72, Europ. J. Biochem., 116:547-551.

Mostov, K.E., Friedlander, M., and Blobel, G., 1984, The receptor for transepithelial transport of IgA and IgM contains multiple immunoglobulin-like domains, Nature, 308:37-43.

Muskavitch, M.A.T., and Hogness, D.S., 1982, An expandable gene that encodes a *Drosophila* glue protein is not expressed in variants lacking remote upstream sequences, Cell, 29:1041-1051.

Nagase, S., Simamune, K., and Shumiya, S., 1979, Albumin-deficient rat mutant, Science, 205:590-591.

Nishida, Y., Kataoka, T., Ishida, N., Nakai, S., Kishimoto, T., Bottcher, I., and Honjo, T., 1981, Cloning of mouse immunoglobulin ε gene and its location within the heavy chain gene cluster, Proc. Natl. Acad. Sci. USA, 78:1581-1585.

Ohno, S., 1970, "Evolution by Gene Duplication," Springer-Verlag, Heidelberg-Berlin-New York.

Ohno, S., 1972, So much "junk" DNA in our genome, in: "Evolution of Genetic Systems," H.H. Smith, ed., Brookhaven Symp. No. 26, Gordon and Breach, Inc., New York-London-Paris, pp. 366-370.

Ohno, S., 1981, Original domain for the serum albumin family arose from repeated sequences, Proc. Natl. Acad. Sci. USA, 78:7657-7661.

Ohno, S., 1984a, Birth of a unique enzyme from an alternative reading frame of the preexisted internally repetitious coding sequence, Proc. Natl. Acad. Sci. USA, 81:2421-2425.

Ohno, S., 1984b, Segmental homology and internal repetitiousness identified in putative nucleic acid polymerase and human hepatitis B surface antigen of human hepatitis B virus, Proc. Natl. Acad. Sci. USA, 81:3781-3785.

Ohno, S., Matsunaga, T., and Wallace, R.B., 1982, Identification of the 48-base-long primoridal building block sequence of mouse immunoglobulin variable region genes, Proc. Natl. Acad. Sci. USA, 79:1999-2002.

Ohno, S., Matsunaga, T., Epplen, J.T., Itakura, K., and Wallace, R.B., 1982, Identification of the 45-base-long primordial building block of the entire class I major histocompatibility complex antigen gene, Proc. Natl. Acad. Sci. USA, 79:6342-6346.

Ohno, S., and Epplen, J., 1983, The primitive code and repeats of base oligomers as the primordial protein-encoding sequence, Proc. Natl. Acad. Sci. USA, 80:3391-3395.

Ohno, S., and Yazaki, A., 1983, Simple construction of human *c-myc* gene implicated in B-cell neoplasma and its relationship with avian *v-myc* and human lymphokins, Scand. J. Immunol., 18:373-388.

Ohno, S., Matsunaga, T., and Lee, A.D., 1984, The invariably present Tryptophane loop as the core of all divergent antigen-binding pockets, Scand. J. Immunol., 20:377-388.

Okada, H., Negoro, S., Kumura, H., and Nakamura, S., 1983, Evolutionary adaptation of plasmid-encoded enzymes for degrading nylon oligomers, Nature, 306:203-206.

Saito, H., Kranz, D.M., Takagaki, Y., Hayday, A., Eisen, H.N., and Tonegawa, S., 1984, Complete primary structure of a heterodimeric T-cell receptor deduced from cDNA sequences, Nature, 309:757-762.

Saul, F.A., Amzel, L.M., and Poljak, R.J., 1978, Preliminary refinement and structural analysis of the Fab fragments from human immunoglobulin New at 2.0 A resolution, J. Biol. Chem., 253:585-597.

Shimizu, A., Takahashi, N., Yamawaki-Kataoka, Y., Nishida, Y., Kataoka, T., and Honjo, T., 1981, Ordering of mouse immunoglobulin heavy chain genes by molecular cloning, Nature, 289:149-153.

Siu, G., Clark, S.P., Yoshikai, Y., Malissen, M., Yanagi, Y., Strauss, E., Mak, T., and Hood, L., 1984, The human T-cell antigen receptor is encoded by variable diversity and joining gene segments that rearrange to generate a complete V gene, Cell, 37:393-401.

Sumegi, J., Wieslander, L., and Daneholt, B., 1982, A hierachic arrangement of the repetitive sequences in the Balbiani ring 2 gene of *Chironomus tentans*, Cell, 30:579-587.

Tanabe, Y., 1980, Phylogenetic relationships of dog breeds especially of Japanese native dog breeds determined by the blood protein polymorphisms, Reports of Japan. Soc. Res. on Native Stocks., 9:169-227.

Travers, P., Blundell, T.L., Sternberg. M.J.E., and Bodmer, W.F., 1984, Structural and evolutionary analysis of HLA-D-region products, Nature, 310:235-238.

Twining, S.S., and Attassi, M.Z., 1978, Antibody-combining sites can be mimicked synthetically: surface-simulation synthesis of the immunoglobulin New combining site of the γ-hydroxyl derivative of vitamin κ1, J. Biol. Chem., 253:5259-5262.

Yamada, Y., Avvedimento, V.E., Murdryj, M., Ohkubo, H., Vogeli, G., Irani, M., Pastan, I., and Crombrugge, B. de, 1980, The collagen gene: evidence for its evolutionary assembly by amplification of DNA segment containing an exon of 54 BP, Cell, 22:287-292.

Yazaki, A., and Ohno, S., 1983, The recurrence of 49 base decamers, nonomers and octamers within mouse Ig $C_{\mu H}$ genes and its primordial building block, Proc. Natl. Acad. Sci. USA, 80:2338-2340.

Ycas, M., 1972, De novo origin of periodic proteins, J. Mol. Evol., 2:17-27.

Zinkernagel, R.M., and Doherty, P.D., 1974, Immunological surveillance against altered self components by sensitized T-lymphocytes in lymphocytic choriomeningitis, Nature, 251:547-549.

ORGANIZATION OF MUTANT GENES IN MOUSE t-HAPLOTYPES

Dorothea Bennett

Laboratory of Developmental Genetics
Sloan-Kettering Institute for Cancer Research
New York, N.Y.

Embryonic lethal mutations at the "T-locus" in the mouse (Mus musculus) were first identified more that 50 years ago (Dobrovolskaia-Zawadskaia, 1927) and quickly recognized by L.C. Dunn and his colleagues as providing valuable material for studying the genetics of development (Chesley and Dunn, 1936). Studies of that locus have occupied generations of scientists since that time, and provided significant insights into both development and the organization of genes that regulate it. This paper will attempt to trace the evolution of our understanding of the structure of the "T-locus" and to speculate on what it may teach us in the future when its secrets finally succumb to the techniques of molecular biology.

The early genetics of the T-locus

The first mutation discovered was the dominant Brachyury (T) which produced (for a mouse) a major morphological defect in heterozygotes, namely a short tail, and was lethal to homozygotes. A straightforward situation rapidly became more complicated when it was found that another mutation, recessive to the wild type allele of T, interacted with T to produce a new phenotype, taillessness. This newly discovered mutation, designated t^0, behaved as a genetic allele to T since no recombination was observed between the two mutations, but was also lethal in homozygotes. Shortly later a second recessive mutation, t^1, was discovered; it interacted with T in the same way as t^0, but was homozygously lethal at a different stage, and, surprisingly, in fact yielded viable heterozygotes with t^0 (Dunn, 1937). These observations raised questions that were puzzling in the mid 1930's and that have only partly been solved today. The central question is so well put by Chesley and Dunn (1936) that I will quote it: ".....then two alleles, each of which is lethal when homozygous, are not lethal when combined with each other ... a situation which is sufficiently unique to be regarded with some suspicion as a final explanation." It was not in fact the final explanation, but it has taken 50 more years to reach an understanding of the real situation, and our understanding is still only a partial one. During that time the questions escalated as the number of different lethal genes identified grew to more than 16 (Klein et al., 1984), other genes affecting the differentiation and function of sperm were found to be associated with the lethal mutations (Bennett and Dunn, 1971), and it was found that these sets of genes occurred as high frequency natural polymorphisms in wild mouse populations (Bennett, 1978).

It was also found that the apparent allelism of recessive "*T*-locus" mutations to one another and to *T* reflects the fact that the recessive mutations occur in variants of chromosome 17 (now called *t*-haplotypes) in which a long region (about 1% of the mouse genome) undergoes recombination with wild type chromosomes at only about 1/100 of the normal frequency. Thus this whole segment of chromosome generally behaves as an "allele" of the mutation *T* which typically occurs in wild type chromosomes. The little we have known until recently about the structure of *t*-haplotypes has been derived from studies of the reciprocal products of the rare crossovers that do occur between them and normal chromosomes (Lyon and Meredith, 1964).

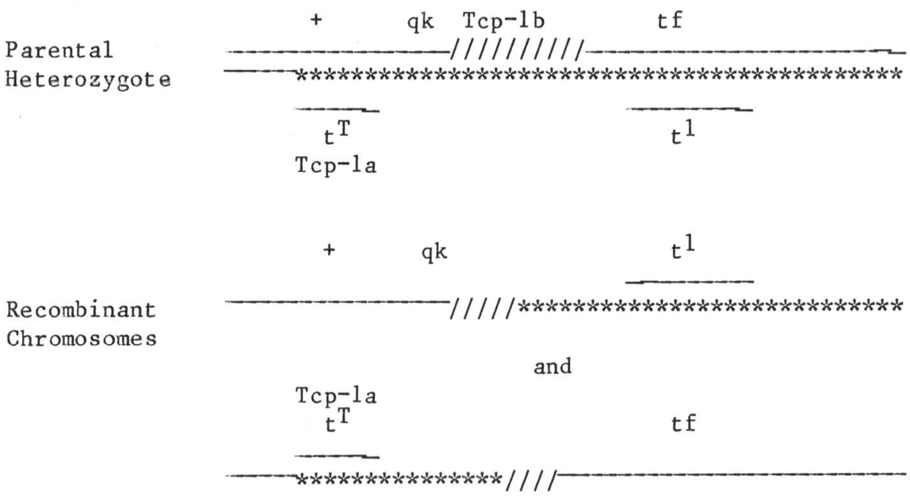

Fig. 1. Diagram of a *t*/+ heterozygote, not to scale, where —— indicates wild type chromosomes, ***** indicates the region of recombination suppression typical of *t*-haplotypes and

///// denotes the region bracketed by markers on wild type chromosomes where rare events of recombination have been detected.

When recombinant chromosomes from appropriately marked heterozygotes are analyzed (Fig. 1), it is evident that crossing over seems to be limited to a "hot spot" between the loci of qk and tf, and around the locus of Tcp-1b (Silver et al., 1980). Therefore two reciprocal products are obtained. One retains the lethal recessive mutation, but no longer interacts with T to produce taillessness, and the other continues to interact with T but no longer is homozygous lethal. In both types recombination suppression is retained over the length of the segment obtained from the parental t-haplotype. Thus, although these observations provided the rather surprising information that the ultimate phenotype produced by t-haplotypes depends on two separable mutations, the "tail interaction" factor t^T and a lethal gene, they did not approach the question of whether the lethal mutations were allelic or not.

The genetic and functional relationship among t-lethal mutations

The question of the relationship of the embryonic lethal mutations was finally resolved by Artzt and her colleagues who measured recombination between complete lethal t-haplotypes and an appropriately marked spontaneous recombinant chromosome carrying the distal lethal-containing segment of a different t-haplotype. These experiments showed that recombination occurred freely between regions where segments of t-haplotype chromosomes were paired, and thus made it possible to apply methods of classical genetics to an analysis of the "T-locus" (Silver and Artzt, 1981).

An extensive series of mapping studies has now been carried out, and showed that t-lethal genes with different embryonic effects mapped to different positions as much as 15 centimorgans apart and were thus clearly non-allelic (Artzt et al., 1982; Artzt, 1984).

These data thus solved the puzzle, stated so succinctly by Chesley and Dunn, by showing that their premise of allelism was wrong, and thus that the complementation they observed was actually between non-allelic genes. But in the meantime another aspect of that puzzle had become evident, namely that although the existence of complementation between lethal t-genes defined them as different, it was also tantalizingly obvious that complementation was usually far from complete, in the sense that the viability of complementing heterozygotes was almost always lower than normal, and ranged from 10 to 75% (Bennett, 1975). This clearly represented a paradoxical situation since, although complementation could be used as a criterion to define two recessive lethals as different, the fact that complementation was not complete seemed to imply some similarity between the two genes. This question has also become approachable now that recombination between two different _t_-haplotypes can be used to shuffle the genes they carry, and thus generate chromosomes that carry no _t_-lethals or more than one. In effect, we have been able to do "_cis_-_trans_" tests to examine whether defective complementation between two non-allelic recessive _t_-mutations does indeed result from some functional relatedness. As Fig. 2 shows, in some cases the viability of heterozygotes carrying two lethal genes in _cis_ is entirely normal, while heterozygotes carrying virtually identical components of _t_-haplotype chromosome, but with the lethal genes arranged in trans, are relatively inviable. In other cases, the viability of heterozygotes for two _t_-haplotypes remains defective regardless of the arrangement of the lethal mutations. This is not unexpected since it presumably results from homosygosity for parasitic deleterious genes that would be expected to accumulate on these lethal-bearing chromosomes during the course of evolution (Shin et al., 1983).

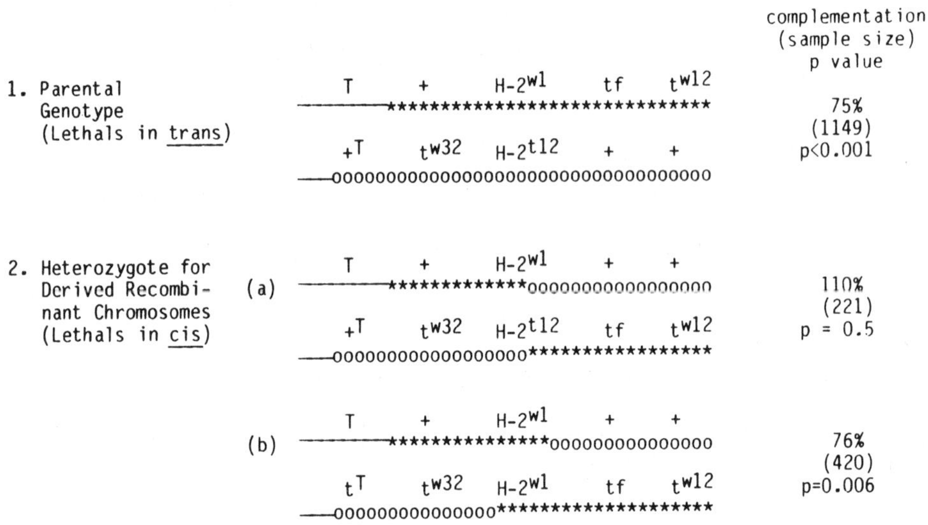

Fig. 2. Diagrammatic representation of cis-trans tests between two lethal genes. 1. The _trans_ configuration in a heterozygote that was used to generate the recombinant chromosomes shown in 2. 2a represents a heterozygote with a chromosomal content virtually identical to the genotype above, but with the lethal genes in _cis_. 2b shows a second heterozygote in which the t-lethals are also in cis, but which is homozygous for a region of chromosome around the H-2 complex.

Nevertheless, the fact that some cis-configurations (2a in Fig. 2) are fully viable leads to the surprising conclusion that on the criteria of the *cis-trans* test the *t*-complex seems to represent a functional unit that extends over more than 15 cM. The genetic and functional implications of this heretical observation are not at all clear, but may mean that the genetic organization of developmentally regulated genes is more complex than has been thought until now. Certainly the explanation must lie at the level of DNA, since cis- and trans-heterozygotes necessarily have identical gene dosage, and thus models based on protein interactions are not appropriate. There are no reported precedents for the very large physical separation between the genes involved in the putative functional unit identified by *t*-haplotypes. Nevertheless, it can be speculated that mechanisms based on somatic rearrangements of DNA, as in immunoglobulin genes, (Davis et al., 1980) or the use of genetic "cassettes" in expression loci, as in the mating type switch in yeast (Herskowitz, 1984) may be involved.

The relationship of t-genes to the H-2 complex

In the course of mapping *t*-lethal genes relative to one another we used as reference points three standard markers of chromosome 17: the mutations *T* and t^T, the hair growth variant *tufted* (*tf*) and the *H-2* complex. In wild type chromosomes these markers occur from the centromere in the order given above. Unexpectedly in double heterozygotes for *t*-haplotypes we found that the arrangement of these markers was different, with the order being t^T, *H-2*, and *tf* (Artzt et al., 1982). Thus *t*-haplotypes contain a major inversion, common to all of them, which is very likely responsible at least in part for the recombination suppression that exists between them and wild type chromosomes. A more detailed analysis of the structure of the MHC in *t*-haplotypes using a combination of serological reagents and cloned DNA probes to *TL*, *D*, $E\alpha$, $E\beta$, $A\beta$, and *K* showed that the entire MHC is involved in the inversion, with all of its components present in proper relative (albeit inverted) order. Even more interesting results emerged when the individual elements of the *H-2* complex were mapped relative to specific *t*-lethal genes, and we found that several of them map within the *H-2* complex itself (Shin et al., 1984). For example, two members of the same complementation group, t^{12} and t^{w32}, map between *TL* and *D* (i.e. in the *Qa* region) and the position of t^{w5} is so close to *K* that the two genes have not been separated in several hundred mice analyzed. These last results are particularly intriguing since several recent reports have suggested that genes in or linked to the *H-2* complex may be expressed during embryogenesis. For example, *H-2* linked genes are known to affect the rate of development of preimplantation mouse embryos (Goldbard et al., 1982). Furthermore, Brickell et al. (1983) have identified RNA transcripts of a class I sequence in teratocarcinoma cells, and Murphy et al. (1983) showed that a repetitive sequence associated with it is transcribed in embryos. Finally a Qa antigen has been described on the surface of preimplantation embryos (Gollnick, personal communication). Since there are many class I genes whose products and functions have not yet been identified, it can be speculated that some of them may represent _t_-genes that function during early development to regulate cell interaction and differentiation in ways analogous to the regulation of lymphocyte differentiation by "adult _H-2_" antigens.

The genetic and functional basis of transmission ratio distortion

Progress has also been made in understanding one of the most peculiar properties of _t_-haplotypes, namely transmission ratio distortion in male _t_/+ heterozygotes, and sterility in male complementing double t^x/t^y heterozygotes. Very briefly, males carrying a single complete _t_-haplotype have long been known to transmit their mutant chromosome to the majority of their offspring, at a frequency that usually approaches 100%. It is no doubt this advantage that allows these lethal chromosomes to persist in wild populations at polymorphic levels. Paradoxically, males that carry two different complementing complete _t_-haplotypes are completely sterile. We have recently resolved this apparent paradox by analyzing the breeding behavior of "tetraparental" chimeric males containing _t_/+ and +/+ genotypes which were distinguishable because the _t_/+ component was pigmented and the +/+ genotype was albino. The chimeric males were analyzed by crosses to albino females and produced progeny that were approximately half pigmented and half albino, thus showing that neither of the genotypes in the chimeras had a competitive advantage with respect to sperm production. Nevertheless almost all the pigmented progeny inherited the _t_-bearing chromosome of their father. These results show very clearly that the superiority of _t_-bearing sperm is achieved at the expense of their meiotic partners, and that they do not have any intrinsic advantage over other sperm that have developed independently of them in the same testis (Seitz and Bennett, 1985). Presumably therefore this phenomenon has its inception either at meiosis or during subsequent development when spermatids are linked by intercellular bridges. Likewise, in t^x/t^y males sterility probably results from mutual poisoning by the two _t_-chromosomes, and it is therefore likely that the same genes are responsible for the two sets of phenomena.

The genetic basis of transmission ratio distortion has been particularly frustrating to analyze, although basically it seems a relatively simply system. The available facts can be summarized briefly as follows: All naturally occurring wild lethal _t_-haplotypes are capable of being transmitted at frequencies close to 100%. This distorted transmission is clearly due to the synergistic activity of at least two different mutant genes within the region of recombination suppression, because neither of the two reciprocal products of the rare spontaneous recombination between _t_-haplotypes and wild type chromosomes exhibits the original property of very high transmission. Recombinants that retain the proximal end of _t_-haplotypes are transmitted either in mendelian proportions or at frequencies much lower than normal, i.e. about 15%, while those containing the distal end are generally higher than normal but quite variable. The variability inherent in the system has led some investigators, for example Lyon (1984) and Silver (1981), to propose that the interaction of at least three, and possibly more genes are responsible for ratio distortion, while others, like ourselves (Bennett et al., 1983) and Styrna and Klein (1981) have seen no reason to impute more than two. The resolution of this question will have to wait for further data, but we know already that the controversy has been generated partly by strong influences of genetic background on the transmission ratio shown by any given _t_-haplotype. The actual frequency of transmission of a haplotype like $\underline{t^0}$, for example, can vary between 30% and 95% under the influence of genes on the homologous chromosome (Bennett et al., 1983), and we have also just recently shown that an unlinked modifier gene can dramatically modify the transmission frequency of $\underline{t^{12}}$ (Gummere, Bennett and McCormick, unpublished results). Thus much of the uncertainty concerning genetic control of _t_-haplotype transmission distortion must arise from the use of genetically disparate stocks by different investigators. In any case, our new ability to dissect _t_-haplotypes by classical genetic recombinational analysis has already given the interesting information that the genes governing transmission ratio and sterility are different from those that produce embryonic death. As mentioned above, recombinant chromosomes that contain no lethal genes can be obtained by crossing over between two different lethal _t_-haplotypes. These chromosomes continue to have the high transmission ratio typical of the complete _t_-haplotype, and males homozygous for them are sterile (unpublished data). Thus the genes responsible for lethality must be separable from those involved in transmission ratio and sterility.

The origin and genealogy of t-haplotypes

_t_-Haplotypes exist as natural polymorphisms in wild mice, distributed world wide, and with a startlingly high gene frequency of over 10%. Yet the evidence suggests that these chromosomes have a common origin, and for the most part have maintained the DNA structure of their ancestral chromosome. The first reason for thinking this was that all _t_-haplotypes show strong linkage disequilibrium with a limited number of rather similar _H-2_ types (Hammerberg and Klein, 1975). Later, the genetic organization of _t_-haplotypes was analyzed at the level of restriction fragment patterns recognized by cloned DNA probes to class I genes, and they were found to be remarkably similar, almost identical, to one another even when they carried different H-2 complexes (Shin et al., 1982; Silver, 1982). Other types of analysis have also led to the conclusion that as a group t-haplotype chromosomes are relatively non-polymorphic both at the level of DNA and protein. For example, Rohme et al. (1984) isolated 9 random clones from wild type chromosome 17 DNA, and showed that all had t-specific restriction fragments different from wild type, but identical in all t-haplotypes studied. At the protein level, in an attempt to define random polymorphisms between t- and wild type chromosomes, 8 testicular cell proteins that differed between the two genotypes were identified on two-dimensional gels; however, no variation among different t-haplotypes was found (Silver et al., 1983). Likewise, the protein products of well studied chromosome markers such as glyoxylase 1, TL, and Qa are not polymorphic among t-haplotypes (Artzt et al., 1982). All of this evidence suggests strongly that a unique ancestral chromosome gave rise to the extant t-haplotypes. Furthermore, since these chromosomes are thought to have been in existence for about 10^6 years (Sage, 1981), it seems likely that their homologies have been preserved by their inability to undergo free recombination under natural conditions. A reasonable deduction from these observations is that in the course of evolutionary time, polymorphisms of DNA structure are generated primarily by recombinational errors rather than by other mutational events.

Very interestingly, an entirely different picture of the genetic structure of t-haplotypes is obtained when they are analyzed with respect to their I region genes. We have now studied 17 different t-haplotypes with probes to three class II genes, Eα, Eβ, and Aβ, and three restriction enzynmes by Southern blotting of genomic DNA. These 17 haplotypes fell into 4 groups with respect to Eα, 7 for Eβ, and 11 for Aβ. Thus there is a very high degree of sequence diversity around all these genes, which is in striking contrast to the general homogeneity of all other loci examined. This has important theoretical implications, since it must mean that Ia genes and especially Aβ have some mechanism for generating diversity that is not common to other regions of the chromosome (Artzt et al., 1985).

For the future

It is obvious that the last 50 years have produced a wealth of data on genetics of the region of mouse now called, for obvious reasons, the T/t complex rather than the T-locus. It is equally obvious that the essential question of how t-genes are regulated and function to govern early steps in embryogenesis have not been answered. The next few years should see the cloning of these genes and a final explanation of their genetics.

REFERENCES

Artzt, K., McCormick, P., and Bennett, D., 1982, Gene mapping within the T/t complex of the mouse. I. t-Lethal genes are nonallelic, Cell, 28:463.

Artzt, K., Shin, H-S., and Bennett, D., 1982, Gene mapping within the T/t complex of the mouse. II. Anomalous position of the H-2 complex in t-haplotypes, Cell, 29:969.

Artzt, K., Shin, H-S., Bennett, D., and DiMeo-Talento, A., 1985, Analysis of MHC haplotypes of t-chromosomes reveals the majority of diversity is generated by recombination, J. Exp. Med., in press.

Bennett, D., 1975, The T-locus of the mouse: a review, Cell, 6:441.

Bennett, D., 1978, Population genetics of T/t complex mutations, in: "Origins of Inbred Mice," H.C. Morse, ed., Academic Press, New York, pp. 615-632.

Bennett, D., and Dunn, L.C., 1971, Transmission ratio distorting genes on chromosome IX and their interactions, in: Proc. Symp. "Immunogenetics of the H-2 System," Liblice-Prague, pp. 90-187.

Bennett, D., Alton, A.K., and Artzt, K., 1983, Genetic analysis of transmission ratio distortion by t-haplotypes in the mouse, Genet. Res., Camb., 41:29.

Brickell, P.M., Latchman, D.S., Murphy, D., Willison, K., and Rigby, P.W.J., 1983, Activation of a Qa/Tla class I major histocompatibility antigen gene is a general feature of oncogensis in the mouse, Nature, 306:756.

Chesley, P., and Dunn, L.C., 1936, The inheritance of taillessness (anury) in the house mouse, Genetics, 21:525.

Davis, M.M., Kim, S.K., and Hood, L., 1980, Immunoglobulin class switching: developmentally regulated DNA rearrangements during differentiation, Cell, 22:1.

Dobrovolskaia-Zawadskaia, N., 1927, Sur la mortification spontanee de la queue chez la souris nouveau-nee et sur l'existence d'un caractere (facteur hereditaire) non-viable, C.R. Soc. de Biol., 97:114.

Dunn, L.C., 1937, A third lethal in the T (Brachy) series in the house mouse, Proc. Natl. Acad. Sci., USA, 23:474.

Goldbard, S.B., Verbanac, K.M., and Warner, C.M., 1982, Role of the H-2 complex in preimplantation mouse embryo development, Biol. Reprod., 26:591.

Hammerberg, C., and Klein, J., 1975, Evidence for postmeiotic effect of t factors causing segregation distortion in mouse, Nature, 253:137.

Herskowitz, I., 1983, Cellular differentiation, cell lineages, and transposable genetic cassettes in yeast, in: "Current Topics in Developmental Biology," 18, A.A. Moscona and A. Monroy, eds., Academic Press, New York, pp. 1-14.

Klein, J., Sipos, P., and Figueroa, F., 1984, Polymorphism of t-complex genes in European wild mice, Genet. Res., Camb., 44:39.

Lyon, M.F., 1984, Transmission ratio distortion in mouse t-haplotypes is due to multiple distorter genes acting on a responder locus, Cell, 37:621.

Lyon, M.F., and Meredith, R., 1964a, Investigations of the nature of t-alleles in the mouse. I. Genetic analysis of a series of mutants derived from a lethal allele, Heredity, 19:301.

Lyon, M.F., and Meredith, R., 1964b, Investigations of the nature of t-alleles in the mouse. II. Genetic analysis of an unusual mutant allele and its derivatives, Heredity, 19:313.

Lyon, M.F., and Meredith, R., 1964c, Investigations of the nature of t-alleles in the mouse. III. Short tests of some further mutant alleles, Heredity, 19:327.

Murphy, D., Brickell, P.M., Latchman, D.S., Willison, K., and Rigby, P.W.J., 1983, Transcripts regulated during normal embryonic development and oncogenic transformation share a repetitive element, Cell, 35:856.

Rohme, D., Fox, H., Herrmann, B., Frischauf, A-M., Edstome, J-E., Mains, P., Silver, L.M., and Lehrach, H., 1984, Molecular clones of the mouse t complex derived from microdissected metaphase chromosomes, Cell, 36:783.

Sage, R.D., 1981, Wild mice, in: "The Mouse in Biomedical Research," vol. 1, H.L. Foster, J.D. Small, and J.G. Fox, eds., Academic Press, New York, pp. 39-90.

Seitz, A., and Bennett, D., 1985, Transmission distortion of t-haplotypes is due to interactions between meiotic partners, not intrinsic superiority of t-bearing sperm, Nature, in press.

Shin, H-S., Bennett, D., and Artzt, K., 1984, Gene mapping within the T/t complex of the mouse. IV. The inverted MHC is intermingled with several t-lethal genes, Cell, 39: 573.

Shin, H-S., McCormick, P., Artzt, K., and Bennett, D., 1983, Cis-trans test shows a functional relationship between non-allelic lethal mutations in the T/t complex, Cell, 33:925.

Shin, H-S., Stavnezer, J., Artzt, K., and Bennett, D., 1982, The genetic structure and origin of t-haplotypes of mice, analyzed with H-2 cDNA probes, Cell, 29:969.

Silver, L.M., 1981, A structural gene (Tcp-1) within the mouse t complex is separable from effects on tail length and lethality but may be associated with effects on spermatogenesis, Genet. Res., Camb., 38:115.

Silver, L.M., 1982, Genomic analysis of the H-2 complex region associated with mouse t haplotypes, Cell, 29:961.

Silver, L.M., and Artzt, K., 1981, Recombination suppression of mouse t-haplotypes is due to chromatin mismatching, Nature, 290:68.

Silver, L.M., White, M., and Artzt, K., 1980, Evidence for unequal crossing over within the mouse T/t complex, Proc. Natl. Acad. Sci., USA, 77:6077.

Silver, L.M., Uman, J., Danska, J., and Garrels, J.I., 1983, A diversified set of testicular cell proteins specified by genes within the mouse t complex, Cell, 35:35.

Styrna, J., and Klein, J., 1981, Evidence for two regions in the mouse t complex controlling transmission ratios, Genet. Res., Camb., 38:315.

GENETIC ANALYSIS OF MORPHOGENESIS

Antonio García-Bellido

Centro de Biologia Molecular, CSIC
Facultad de Ciencias
Universidad Autonoma de Madrid
Madrid-34

INTRODUCTION

The biological realm is characterized by both the diversity of forms and their generative constancy. Thus, morphogenesis in evolution and in ontogeny is the paradigm of biology. During the history of biology the causal explanation of morphology has been alternating between a preformistic and an epigenetic notion of development and between internal coaptation and external adaptation as final causes in evolution.

These opposite notions were largely dependent on the nature of the analysis, the terms in which morphology had to be understood. I will discuss in this paper how modern developmental genetics is approaching that paradigm. For comparison I will rapidly consider more classical approaches.

Comparative anatomy tried first to classify morphologies in terms of homologies and analogies. To that end it had to decompose organisms into their anatomical parts. These studies incorporated fossil data and embryonic descriptions. The analysis uncovered the existence of discontinuous morphologies (modern phyla, orders, etc.) and the existence of "grundtypen" and derived forms within those groups. Embryonic stages would recapitulate morphologies progressing from the general to the particular (V. K. Baer) or, later, with the notion of organic evolution, from the primitive ancestor to the advanced form (E. Haeckel). Development thus becomes then "causally" related to evolution. This occurs by the addition of new characters to more simple ancestor forms. The particular morphologies of radiating forms are secondary caenogenetic adaptations of palingenetic or archetypic building principles.

The causal analysis of development (W. His; W. Roux) reduces even further the anatomical decomposition to the tissue and cellular level and looks for mechanical and physiological interactions between parts. Development proceeds in a predetermined way (mosaic development) or is susceptible to experimental modification (regulative development). The first is a "black box", the second can be further analysed (H. Driesch). This analysis uncovered inducers, diffusible effectors, later molecules, the provoque specific responses. Development thus appears as a chain of processes in which any stage not only precedes but determines the subsequent ones. It is a cascade process that leads to expanding diversity. In this view the genome responds to the epigenetic conditions, internal (developmental) and external (environmental) constraints, mobilizing all the required genetic variation and physiological interactions in order to obtain maximal "adaptation". It would provide merely for the peculiarities of the response. And evolution would result from the continuous modulation of such caenogenetic responses.

Chemistry and decomposed organisms into molecules of different classes, types and abundances. Some of them, the enzymes, perform specific metabolic functions. They appear at different times and in different cell types and, along with structural proteins, could represent the molecular counterpart of specificity during development. With the advent of molecular genetics it was finally shown that genes code for proteins and genetic variation is expressed in different proteins that change along evolution. With these analyses development and evolution have found a correlated representation in the structural variations at the molecular level.

The reducionist decomposition of forms to molecules has carried with it the notion that morphogenesis is a result of a complex net of gene interactions leading in cascade reaction to a parallel increase in morphological complexity. In fact physiological genetics has found that mutant specific syndromes result from pleiotropic effects far removed from the primary perturbation, e.g. in metabolic enzymes or structural proteins. Morphological evolution would then result from the adaptive fixation of allelic variations with pleiotropic phenotypes. Archetypes and morphogenetic building principles are dismissed as unreal abstractions. A mutationally plastic genome would provide for a continuous modulation of caenogenetic responses during evolution.

Recent studies in developmental genetics in *Drosophila melanogaster* begin to provide a different causal view of morphogenesis and morphological variation. Developmental genetics aims to identify gene functions in terms of their specific interactions determining operations that generate forms. In this view allelic variations would be responsible merely for differential morphologies. This analysis has already uncovered the existence of specific morphogenetic genes responsible for systemic morphological processes. As opposite to the morphological decomposition, the genetic one envisages morphology as the result of specific genetic operations which in different combinations, in time and space, generate forms. In this view, development and evolution appear not as a cumulative process, but as a combinational one, and form variations as different combinations of a finite number of genetic operations. We will now discuss some examples of such genetic operations.

Genetic organization of morphogenetic systems

Molecular systems

Genetic and molecular analyses have long ago discovered in microorganisms the existence of subsets of genes coding for proteins involved in specific operations. The first known examples are those of operons in bacteria, where a common regulatory gene modulates the expression of several genes of a metabolic chain. The controlled genes can be contiguous or dispersed in the genome. These gene interactions occur by molecular recognition and they represent therefore independent units of function. Similar functional systems are becoming identified and molecularly described in genetically accessible organisms. Classical examples are the ribosomes, the replicator complexes in bacteria, the DNA repair enzymes, the immunoglobulin genes, etc. They correspond to subsets of genes coding for products that perform specific operations.

The most relevant examples in this context is the genetic system that codes for capside proteins in viruses. The analysis of T4 phage morphogenesis (see Wood, 1973) has shown that the capside proteins are coded by genes transcribed in the "late" phase of phage transcription, possibly under the control of a specific RNA polymerase complex. The protein product of these genes, intact or secondarily processed, assemble in a chain of molecular recognition processes. The final capside is a multimeric complex of about 30 different components reiterated in different combiations in the anatomical parts of head, tail and tail fibers. The constraints imposed on this structure by molecular recognition and matching are so large that most mutations in the coding genes lead to inviable forms. In fact different phages have discontinuous morphologies, corresponding to as many compatible combinatorial solutions.

We already find in this system many generative rules that we will encounter in more complex systems. These are specialized gene sets interacting by molecular recognition, selfassembly, reiteration of elements, and combinatorial complexity. We will find, however, a major difference. Whereas the capside structure is isomorphic to the DNA linear structure that codes for its protein components, in the following systems morphological complexity results from a series of transformations: i.e. it cannot be predicted knowing its individual components.

Segmental specification in Drosophila

In multicellular organisms the genetic information is transmitted by cell division and its expression mainly restricted to cells. In the development of many organisms certain developmental processes are strictly mosaic: i.e. the final fate of the progeny of embryonic cells is determined early in development. These processes appear as "black boxes" in physiological experiments. The fate of these cells can, however, be changed by mutation and so genetic analysis can be used to study the genetic nature of this determination.

The possibility of generating what are called "genetic mosaics", developmental systems made out of cells of two genotypes that can be produced during development, allowed Stern (1954) to discover that mutations with systemic phenotypes were expressed autonomously in cells. One such type of mutation is called homoeotic and leads to the substituion of one organ by another, belonging to the same serial homology (W. Bateson). We now interpret these organs as homologous because they are generated by comomon genetics operations. This is the case of the insect segments and the dorsal or vental series of appendages. Clonal analysis of the development of these appendages has found that they are formed by the dichotomic segregation of alternative cell lineages (compartments) giving rise to specific parts of the final pattern. Since these clonal segregations are topologically homologous in different appendages, different segments and compartments within segments are considered generatively homologous structures (García-Bellido et al., 1979).

How are these structures encoded in the genome? Certain null mutants of homoeotic genes cause the transformation of derivatives of one segment into those typical of another one. These transformations follow a topological rule: cephalic segments, anterior to the thorax, are transformed into more posterior segments by mutants in the gene complex called Antennapedia (ANT-C) and thoracic and abdominal segments are transformed into more anterior segments by mutants in the bithorax-complex (BX-C) (Kaufman and Wakimoto, 1982; Lewis, 1978). The maximal transformation caused by the absence of either the ANT-C or the BX-C is to a segment with pro-mesothoracic characteristics. Embryos deficient for both complexes show a repetition of segments with these characteristics - the segment corresponding to the "developmental sink" of segmental specification (García-Bellido, 1975, 1977). Segments in insects are then serial structures which become differentially specified by combiations of genes whose function is presumably to define developmental alternatives to an archetypic segment. This segment was reiterated in a primitive ancestor - the annelid of the phylum to which the insects belong - and represented in the advanced forms by the pro-mesothoracic segments.

Interestingly, the anlage of prothoracic segment corresponds to the "differentiation center" of the blastoderm in all the insects, as detected in physiological experiments by Seidel (see Counce, 1973). This region is the first to initiate segmentation in all insects,, independently of the position and extent of the embryonic germ band in the egg. Segments appear by caudal proliferation (Anamery) in the trochophora larva of annelids and so do abdominal segments in primitive insects, but form in rapid sequence in *Drosophila* (Sander, 1981). Moreover, whereas the number of segments is undetermined in annelids, it is fixed in insects. Possibly the fixed number of segments in insects is a consequence of the finite number of gene combinations that define individual segments.

How then is the specific combination of ANT-C and BX-C genes set in the different regions of the egg? The phenotype of mutants in these complexes is cell autonomous in genetic mosaics produced after the time of blastoderm formation. This doubly indicates that: 1) the segmental specification is the result of the combination of active genes in the individual cells and 2) the function of the normal gene is required from the blastoderm stage on, throughout the cell proliferation period, to specify cell behaviour properties which result in the morphology of each of the segmental derivatives. Derepression mutations in the BX-C genes caused caudad transformations, the opposite to those found in lack-of-function mutants. Thus, the activity of the normal alleles must be restricted to those segments where lack-of-function mutations show the mutant phenotype. This genetic inference has been now corroborated in histological sections with DNA to RNA *in situ* hybridization (Akam, 1983; Levine et al., 1983). The differential distribution of transcripts indicates transcriptional regulation of the corresponding genes.

The conditions that cause the activation of the genes of these complexes in particular regions of the blastoderm seem to be defined by the products of trans-regulatory genes, called "activators" (García-Bellido, 1975). Studies of double mutants, changes in relative gene-dosage between genes and clonal analysis (García-Bellido and Capdevila, 1978; Lewis, 1978; García-Bellido, 1981a; Capdevila and García-Bellido, 1981; Struhl, 1983; Duncan and

Lewis, 1982; Capdevila and Botas, unpublished data) are consistent with the following model of regulation. Variations in the concentration of effectors, coded by these activator genes, a repressor (*Polycomb*, *Pc*, (and others) and an inducer (antirepressor) (*Regulator of bithorax*, *Rg-bx*) deterine the differential spatial activation of these complexes. The postulated variations in concentration of these effectors along the anterior posterior axis of the embryo seems to result from their mutual titration in the blastoderm cells. Whereas the *Rg-bx* products are fundamentally released in the oocyte, the *Pc* products are zygotic. The pattern of distribution of the unbound effectors seems to be due to the initiation and prolonged synthesis of the *Pc* products in the differentiation center (presumptive pro-mesothorax) and cephalic segments and decreased production along more caudal, later appearing segments. The increased concentration of *Pc* products causes the corresponding depletion of the *Rg-bx* products, giving rise to the pattern of distribution along the anterior posterior axis of the blastoderm seen in Fig. 1. Mutations in the Pc locus cause a transformation of all segments into an 8th abdominal segment - probably by derepression of the BX-C and that, in turn, causing the repression of the ANT-C in the cephalic segments.

If this interpretation of the spatial regulation of the ANT-C and BX-C is correct we can envisage the evolutionary origin of the BX-C as secondary to that of the ANT-C. If we imagine primitive arthropods, like trilobites, to have the gnathocephalic segments and the terminalia (segments distal to the gonadal exit and anus) already specified by the ANT-C, the BX genes would have arisen to specify the undefined intercalary segments of thorax and abdomen. Those are the segments that grow from the prothorax by anamery in arthropods and primitive insects. BX genes may have used to that end the gradients of *Pc* products as the effector signal. Lewis (1978) has suggested that the genes of the BX-C may have derived by tandem duplication from a primitive single gene. In this model, the relative concentration of two interacting signals, *Pc* and *Rg-bx*, defining position along the blastoderm, modulates the activity of two series of genes, ANT-C and BX-C, which transform position into segmental specificity.

Segmental specificity is further defined by a combinatorial superposition of the ANT-C and BX-C with other genes of systemic functions. Thus, the gene *engrailed* specifies posterior versus anterior lineages (compartments) within all the segments. And mutants of the BX-C not only transform epidermal characteristics (cuticular pattern) but also neuromere specificity (segmental glanglia) and mesoderm specificity (muscles, mesodermal envelop in gonads) possibly in combination with other genes defining as many developmental alternatives. I have designated "selector" genes these genes which specify developmental alternatives under the control of activator genes which determine their on/off condition

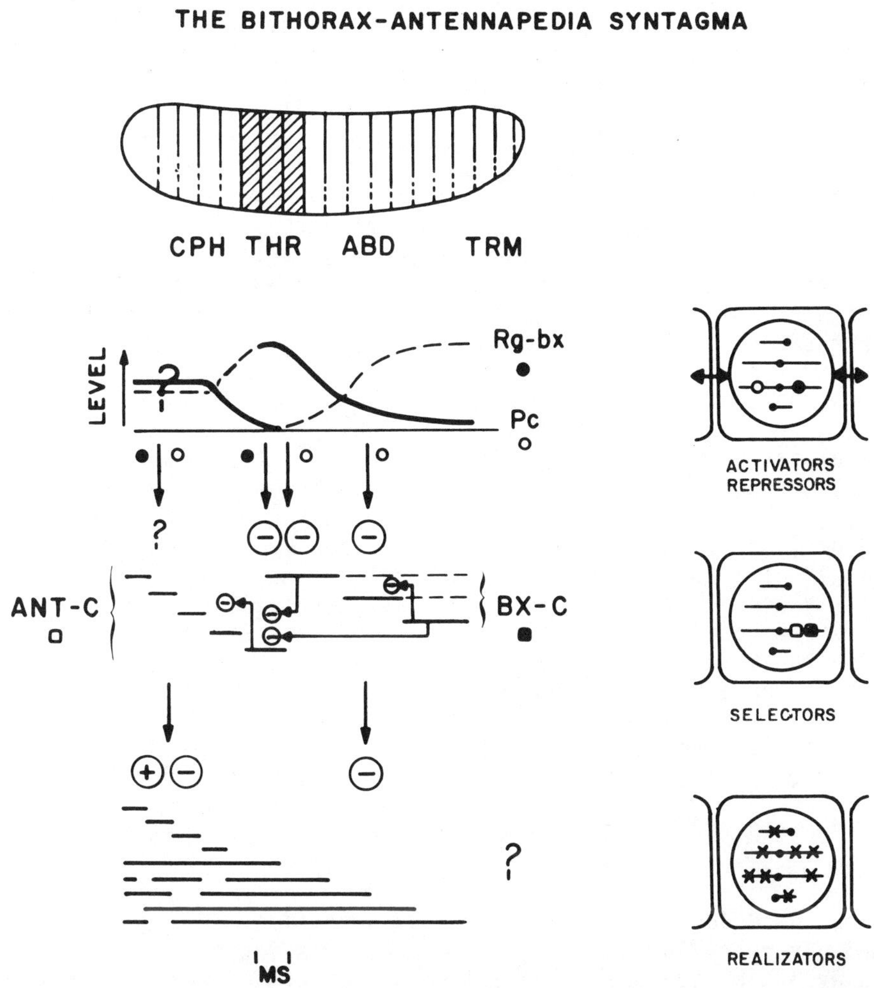

Fig. 1. Genetic interactions involved in the specification of segments. CPH: cephalic, THR: thoracic, ABD: abdominal; TRM: terminalia. + and -: activation and repression. Horizontal bars: segments in which the corresponding genes are active. MS: Mesothorax. Arrows connecting cells: cell interaction. Rg-bx: Regulator of bithorax, PC: Polycomb. ANT-C: Antennapedia gene complex; BX-C: bithorax gene complex.

(García-Bellido, 1975). Different developmental pathways seem to be, then, genetically specified by a "genetic address" that defines the combination of selector genes in the on or off condition (García-Bellido et al., 1979).

How then is the specific combination of genes expressed in the specific morphologies of segments? The combinatorial action of selector genes must be expressed via other genes, hierarchically downstream, that transform their signals into developmental operations leading to the observable morphology. I have called these genes "realizators" (García-Bellido et al., 1975). Within the epidermal pathways they would generate cuticular patterns, size and shape of organs and appendages. We do not know the nature of these genes, their primary functions, nor the operations they are involved in. It is however conceivable that they are involved in basic, ubiquotous, cell behaviour properties (cell proliferation, cell adhesion and recognition, etc.) which are independently modulated by these selector genes. We envisage their control to be by repression of the realizator genes depending on the BX-C, preventing part of the genes of the archetypic thoracic segment (pro-mesothorax) of being expressed in more caudal segments. In fact _in situ_ antibody labelling to bithorax protein has uncovered its localization in the nucleus of the cells of the segments specified by bithorax - an indication of its regulatory nature (Hogness, pers. comm.; White and Wilcox, 1984). The gnathocephalic appendages, on the other hand, could have derived from primitive appendages of a primitive segment secondarily modified by positively and negatively controlled realizators under the activity of the ANT-C genes. Recently, it was found that several genes of both the ANT-C and the BX-C share a DNA sequence -called homoeobox- that codes for a protein sequence rich in basic aminoacids (McGinnis et al., 1984). This peptide sequence could be instrumental in recognizing regulatory regions of realizator genes common to both complexes. This homoeobox is also present in the _engrailed_ locus (W. Gehring and T. Kornberg, pers. comm.) suggestion that this gene operates in the same set of realizator genes.

Thus genetic analysis of segmental specification suggests that spatial diversity in morphological characteristics has arisen during evolution by local modulations of subsets of realizator genes that show a maximal combinatorial expression in a primitive reiterative segmental pattern. It is interesting in this context that mutants or abnormal dosage relations between selector genes, can produce patterns which are normal in other _Drosophila_ species (propleural chaetaes, Visternite pattern, extra sex combs in several tarsal segments of the foreleg, etc. unpublished observations).

Genes analysis suggests the existence of built-in genetic operations that specify morphology in a combinatorial fashion. Each operation is defined by specific molecular interactions between genes constituting a "syntagma" (S), like the activators, selectors and realizators in the BX-S (García-Bellido, 1981a). Several independent syntagmas are at work in any particular cell defining the on/off condition of realizator genes, i.e. the active genome which specifies cell behaviour properties, during proliferation and terminal differentiation. The final visible forms are thus generated by the combinatorial superposition of independent genetic operations.

The existence of independent developmental pathways -differentially specified- allows for local caenogenetic modifications. Such are the modifications of the mouth parts and leg-bearing segments in different orders of arthropods, the adaptation of the first, second or third leg to perform different mechanical functions, the substitution of the metathoracic wing (*Diptera*) or the mesothoracic one (*Strepsiptera*) by a haltere, or the total absence of wings in *Aphaniptera* and in certain parasitic forms or casts of insects.

c) *The pattern of peripheral sensory elements*

We have seen in the case of segmentation that its genetic decomposition is not isomorphic with the final morphology. The same situation holds for the patterned distribution of sensory elements in the cuticle of the larva and imago of *Drosophila*.

Chaetae and sensillae are sensory organs (mechano- and chemoreceptors) distributed over the surface of the body in species-specific positions. They perceive movements and contacts between different parts of the body, as well as external signals. They have specific projections to the segmental ganglia of the central nervous system. Those organs are ectodermal derivatives of the epidermal lineage. Thus e.g. in the imaginal discs, during the cell proliferation period, a chaeta-mother cell becomes singularized and during metamorphosis gives rise by two differential divisions to two cuticular and two nervous elements. Classical studies on pattern formation have suggested that spatial differentiation results from two operations: one defining signals of position, other responding to them with differentiation. Genetic analysis of the achaetescute complex suggests another interpretation (Fig. 2).

Mutants in this gene complex (AS-C) affect both the differentiation and the position of chaetae. Flies deficient for the AS-C lack of chaetae and sensillae on the whole body. This phenotype is cell autonomous in genetic mosaics. Lack of function mutations reduce the density of microchaetae in the notum and of macrochaetae in the tergites. Excess-of-function mutations (*Hairy-wing*, *Hw*) determine the appearance of extrachaetae, in regions normally void of them or increase their density in regions with chaetes. These phenotypes are also produced by manipulating

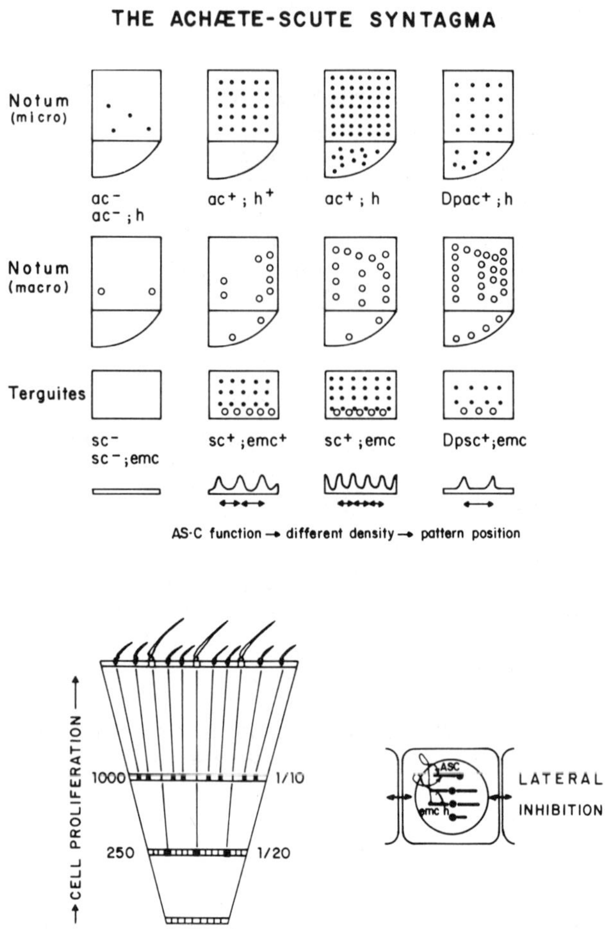

Fig. 2. Genetic interactions involved in the specification of spatial differentiation of chaetae (macro- and micro-) in notum and terquites. ac: achaete, sc: scute, Dp: duplication of the locus. h: hairy, emc: extramacrochaetes. AS-C: Achaete-scute gene complex. Upper part: increasing amount of AS-C function from left

the activity of trans-regulatory genes of the AS-C: *hairy* (*h*) and *extramacrochaete* (*emc*). Those mutants cause the appearance of extrachaetae in a new pattern, in the case of *hairy* very similar to that caused by *Hw*. This effect is exagerated in flies doubly mutant for these regulatory genes and carrying extradoses of AS-C wildtype genes, an indication of increasing derepression of the AS-C. This dose dependence has a limit, after which excess of function of the AS-C leads to less chaetae distributed in a lower density. The model accounting for this genetic behaviour postulates that the initiation of the sensory organ pathway (differentiation of chaete mother cells) in the epidermal cell population, occurs when AS-C escapes the *emc*$^+$ and *h*$^+$ repression (Fig. 2). After this event cells with an active AS-C will inhibit neighbouring cells (lateral inhibition) to enter the same pathway. It is interesting that chaete determination occurs, e.g. in the notum anlage, of wildtype flies, at a density of 1/20 cells, and in flies with maximal number of extrachaetae, at a density of 1/5 cells, corresponding to radial distances of 1-2 cells (García-Bellido, 1981b). Higher derepression of the AS-C leads to a dominance of lateral inhibition over activation. Thus, chaeta pattern seems to result from the transformation of quantity of gene activity of the AS-Syntagma (AS-S) into distances between differentiating cells and this, in turn, is expressed into the visible pattern (Moscoso del Prado and García-Bellido, 1984a, b).

This system is interesting in an evolutionary context on several grounds. Comparative anatomy of the chaeta pattern in Drosophilids uncovers variations that can be explained in terms of variable amounts of activity of the AS-S. There are variations in the patterns between different segments and segments regions of the adult *D. melanogaster*. In the chaeta pattern of this species there are two types of chaetae: macro- and microchaetae. Their mother cells are determined at different times in development (García-Bellido and Santamaria, 1978) (Fig. 2). Moreover the

---

to right and model of specification of position based on density modulation of AS-C function and lateral inhibition. Lower part, left: determination of macro- and microchaetae at different stages (cell number) of the development of a compartment (mesonotum) expressed as a fraction of the total number of cells. Right: cell autonomous and cell interactive components involved in the determination of chaetae. Arrows indicate repression.

macrochaetae are affected by mutants of the *scute* region of the complex whereas the microchaetae are affected by *achaete* mutants (their corresponding mutants map in different coding regions in the DNA of the complex, Carramolino et al., 1984). Finally, they are differentially regulated by *h* and *emc*. Thus, the global chaeta pattern can be decomposed in *D. melanogaster* into two genetic systems, possibly deriving from a common ancestor gene. In other Diptera, e.g. *Musca*, there are more classes of chaetes, differing in size and pattern distribution, showing as a rule a negative correlation between their size and their density. They may reflect differential times of determination and perhaps different coding regions of the AS-C. Especially interesting are the variations in the pattern of macrochaetae in the mesonotum. In different Drosophilidae we observe patterns similar to those of *scute* mutations and patterns similar to those caused by *emc* mutants the macrochaeta patterns consists of a regular grid of macrochaetae as it is found in species of the Genus *Leptocera*. It is conceivable that this is the archetypic pattern from which non-regular patterns are derived by local suppression of this regularity (García-Bellido, 1981b, 1983).

Chaetae in *Drosophila* are not distributed regularly throughout the body surface. Certain regions are void of them (e.g. thoracic and abdominal pleura, wings, etc.). Mutations in the AS-S lead to the appearance of extrachaetae in these regions. Interestingly, species in other *Diptera* show these extrachaetes in wildtype individuals. In the generative model outlined above, local variations in chaeta pattern can be explained by degrees of activity of the AS-C and by the cell proliferation peculiarities of the particular region. These wildtype patterns in other species can be due to variations of either or both parameters. Another modification of the regular stochastic pattern of chaetae distribution is its arrangement in rows. As shown by clonal analysis the microchaeta pattern of the mesonotum, the rows of chaetae in the legs and the chaetae in tergites seem to result from local, short range migration, of chaeta mother cells to occupy the definitive location in the pattern (García-Bellido and Merriam, 1971; Lawrence et al., 1979). These would correspond to as many genetic operations combinatorially superposed to generate pattern constancy and specificity. Thus, like in segmental specificity, evolutionary and developmental diversity in chaeta patterns seem to result from the local modification of homogeneity.

The genetic mechanisms involved in this evolutionary variation have been explored in experiments of artificial selection. Selection experiments to increase -or decrease- the number of chaetae in a given region of the adult pattern have been carried out for a long time in *D. melanogaster* (see Rendel, 1967). In the analysis of the response two major types of genetic variants have been found. "Major modifiers" causing discontinuous responses, are easily mappable and affect the overall pattern, possibly perturbing the general mechanism of chaeta differentiation. In the second group fall the "minor modifiers". They account for most of the continuous response, are additive in their effects, and have a weak contribution to the phenotype. They are therefore difficult to map genetically, supposed to be numerous and may correspond to functions not related to chaete differentiation. Remarkably, however their additive effects are restricted to the particular region where the selective pressure has been applied. Thus, in extrascutellus selection lines for extra chaetes in the scutellum these mainly appear in the scutellum and much less in other regions of the body (Rendel, 1963). Also locally restricted are the extrachaetae in flies of sternopleural selection lines. However, selection for extrachaetae in the sternite plates of a given abdominal segment gives a correlated response with extrachaetae in the sternites of all the segments.

We have carried out similar selection experiments in flies genetically determined (Hw, h or emc) to present extrachaetae in several regions of the prothorax and mesothorax. After selection, the increase in extrachaetae is restricted to the particular region exposed to selection or has a correlated response in the homologous region of other segments. Therefore, minor modifiers are not affecting the general activity of the AS-C, h or emc loci. Presumably they affect in an additive way the local conditions of cell prliferations, i.e. the developmental context, which leads the AS-C genes to become derepreseed. Possibly, natural selection has operated in evolution accumulating genetic variants which subliminally affect many developmental processes, but have maximal additive effects in particular regions. We do not know the nature of these genetic variants but they could cause changes in efficiency of transcription rather than qualitative changes in proteins (unpublished data). Such a mechanism for local modulation could account for coaptation between parts of the organism -like the mechanoreceptors to monitor movements of parts of the body relative to each other- or for integration of colour patterns in both wings of butterflies, or in mimetic forms and in general in the phenomenon called by Lamarck "mosaic selection".

Sexual differentiation

Sexual differentiation usually occurs in separated individuals (gonochorism). The morphological differences between sexes can encompass the overall morphology, be reduced to certain sexually dimorphic characters, or only affect the reproductory organs and gametes. In *Drosophila*, and possibly in all insects, sexual differences affect morphology, behaviour and physiology and even pervade all the morphological characters down to quantitative traits, such a cell size and number of cells in a given organ. Even these quantitative characters are cell autonomous, in gynandromorphs - mosaic individuals with male and female tissues.

Classical genetic analysis has shown that in *D. melanogaster* sex determination responds to a chromosomal determining mechanism: the X-chromosome/autosome ratio (Bridges, 1925). More recent studies (see review by Baker and Belote, 1983) have analyzed the subsequent steps leading to somatic and germ line sex differentiation. The X/autosome ratio is somehow assessed by genes, one of which located in the X/chromosome (Sxl) controls the alternative male-female in somatic differentiation (Cline, 1983). when *on* it determines the female pathway, or failing to activate them *in* the male one. The signals of these genes finally reach a complex gene, *double sex*, (*dsx*) which codes for two different functions, either a male- or a female-determining gene product. Presumably the products of this complex gene in turn control two sets, male and female, of sex-realizator genes. The sex-differentiation (SX) group of genes constitute then a syntagma (SX-S) similar to that of BX, ANT or AS (Fig. 3). There are phenotypic differences, at the cellular level, between the different elements of the SX-S. Cells are either male or female in triploid intersex flies as they are in gynandromorphs or *tra*-genetic mosaics. But they have both male and female characteristics in *dsx* null mutants.

In the SX-S *double-sex* formally corresponds to a "selector" in other syntagmas. However, contrary to other selector genes, there are two signals emanating from *dsx* determining alternative pathways, and in their absence (null alleles) the two exclusive activities, male and female, appear simultaneously in the cell. The function of this gene possibly is to repress female realizator genes genes in the male-on condition, and male-realizator genes in the female-on condition. some of the products of the realizator genes are sex-specific: yolk proteins and the chorion proteins in females and a sex peptide in males. Using temperature sensitive alleles of a *tra* locus it has been shown that the synthesis of yolk proteins in the fat bodies of *xx*; *tra*^{ts} stops, following treatment at the restructive temperature. *tra*^{ts} seems, then, to act via *dsx*, this repressing male realizator genes and allowing female ones to be expressed.

Sex differentiation is one of the oldest evolutionary discoveries. In *Drosophila* sex is determined by a chromosomal mechanism, but in other organisms there are no differentiated sex chromosomes, and in other organisms sexual determination may even depend on changes in environmental external conditions. It is, however, possible that all maintain the selector-realizator elements of the syntagma. In this sense the mating-type gene of yeast would be the operational equivalent to *double-sex* in *Drosophila*.

THE SEX-DIFFERENTIATION SYNTAGMA

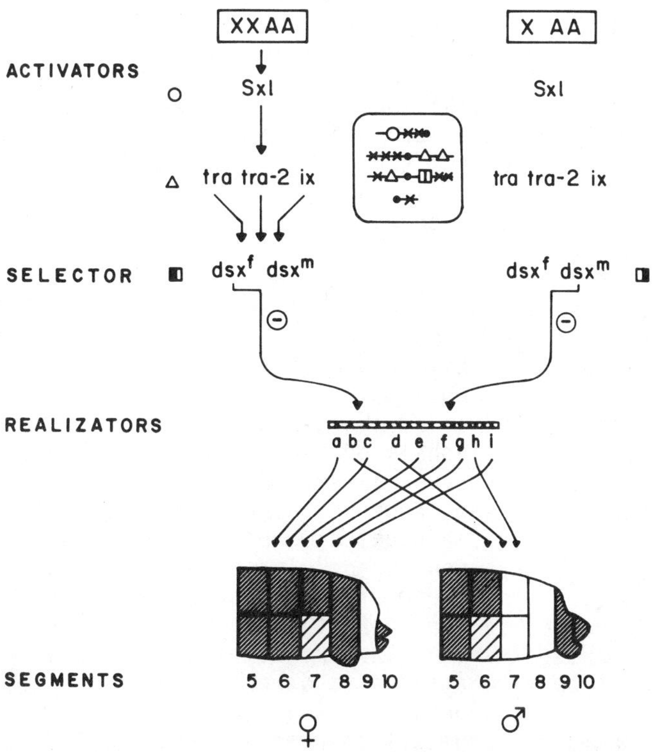

Fig. 3. Genetics interactions involved in sex determination and differentiation of somatic cells (as examplified in the last abdominal segments). A: Autosomes, X: X-chromosomes. Sxl: sex-lethal, tra: transformer (in 3rd chromosome); tra-2: in 2nd chromosome; ix: intersex, dsx: double-sex with the m and f functinal states. Arrows from a,b,...i: active realizator genes that escape, differentially, repression by either dsx^f or dsx^m.

The secondary sexual characters expressed in the morphologically dimorphic regions result from combinatorial superposition of the SX-S functions and those of other syntagmas controlling other developmental pathways. In *D. melanogaster* sexual dimorphism is very pronounced in the last abdominal segments (Fig. 3) and forelegs. In wildtype flies, abdominal segment A10 (+11?) develops the analia in both sexes; A9 does not produce adult structures in females, but form the genitalia in males; A8 is not represented in adult males but develops the female genitalia and a reduced tergite 8 in females; the anlage (histoblasts) of the A7 tergite and of the A6 sternite, clearly present in both sexes, do not develop certain adult structures in males (Nöthiger et al., 1982). These sexually differential pathways are expressed autonomously in cells in genetic mosaics for male and female determining genotypes (Weischaus and Nöthiger, 1982). Certain mutations in the BX-C or in their activator genes can awake the A6 and A7 anlage of males, transforming them into anlagen of more anterior segments that differentiate adult structures, and transform the anlage of A8 in the same direction. Similarly sexual dimorphism in forelegs can be affected by sex-transforming genotypes, and mutant combinations in the BX-S and ANT-S can extend it to the second and third thoracic legs.

Comparative studies of the sexually dimorphic characters between different *Drosophila* species show a remarkable variation in shapes and chaetal patterns in several regions of the male fly. Flies of the female sex, however, appear very constant (see García-Bellido, 1983). Sexual dimorphic characters can largely vary in extent (shape, number of elements and wealth of pattern in different segments) between different species and orders of insects. However, it is interesting that it usually is the male sex that departs more from the archetype of the taxonomic group. That observation also applies to other animal groups like vertebrates. It is conceivable that this wealth of morphological variation in males results from repression of realizator genes at work in many morphogenetic pathways of the species - as expressed in the female sex. The set of sex realizator genes would then consist of those which are exclusive to either sex (gametogenesis, sex specific physiological gene products) and a fraction of the morphogenetic genes which are mainly modulated in the male sex.

Genetic considerations in evolutionary morphology

To the anatomical description, morphogenesis appears as a divergeny and increasing complexity of forms. To the physiological analysis, morphogenesis is a consequence of interactions between parts and of cascade events in which one stage precedes and determines the next. Development in these views is a linear (branching) cumulative process. From what we have seen, to a developmental genetic analysis morphogenesis results from built-in genetic operations which in different combinations, specify inventories of cell behaviour. Development thus appears as a combinatorial process. These different views give contrasting interpretations of morphological variation in evolution (Fig. 4).

Physiological genetics has been confronted with mutant pleiotropic phenotypes as a consequence of cascade events, following primary unspecific perturbations. Classical examples are the syndromes of cartilage deformation in sickle-cell anemia and strabism in siames-cats as result of aminoacid substitution in flobin and in tyrosinase respectively. Since their phenotypes are mutant specific the notion was derived that morphological diversity results from pleiotropic effects of allelic variations. The parallele notion in population genetic studies is that most allelic substitutions affect fitness. Similarly, in artificial selection experiments, morphological characters seem to be easily modifiable by the cumulative effect of alleles in multiple loci. In this view morphological evolution would result from changing external constraints acting upon an unspecific and mutationally plastic genome. And organisms would be in a maximal adaptation, visualized morphologically in the constancy of forms of the species. The genetic analysis of the morphogenetic systems discussed in this paper invites exploration of a different notion. In it, the are the genetic constraints of the interacting genome which limit genetic variation and determine the new forms to be tested by natural selection.

An important finding of developmental genetics is that most morphogenetic mutations are cell autonomous and consequently the corresponding wildtype genes must be involved in generating the particular cell behaviour affected by the mutation. The spatial and temporal coordination of morphogenetic processes results from local cell interactions (e.g. morphogenetic fields, embryonic inductions, lateral inhibition) or interaction between separated cells by means of diffusible hormones or growth factors. However, the specificity of the response resides in the competent cell or tissue, and this is again cell autonomous. Pleiotropy then has to be analyzed at the cellular level. In fact, mutations in the morphogenetic systems here studied are pleiotropic. We have seen that their pleiotropy results from their specific combinatorial effects on the pathway in which their wildtype genes intervene, and

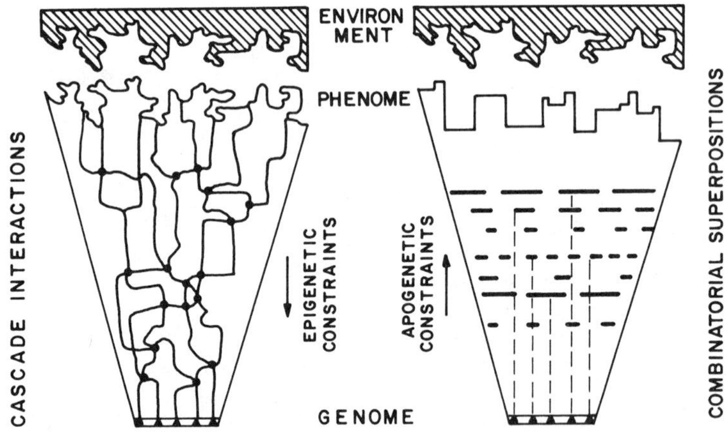

Fig. 4. Contrasting models -epigenetic and apogenetic- of genetic control of morphogenesis. In the epigenetic model phenotypes are at maximal adaptation with the environment of the species, through cascade modulation of gene interactions. In the apogenetic model different gene sets (syntagmas) define genetic operations, which, in a combinatorial fashion, produce constant phenomes (forms) with sufficient adaptation to the environment.

stage of these combinatorial pathways is expressed autonomously in cells.

We have seen that some morphogenetic processes can be genetically decomposed in independent genetic operations affecting particular aspects of cell behaviour. Initiation and specification of cell lineages and compartments, different modes of cell proliferation, cell recognition, interaction and migration in spatial differentiation, and different types of terminal differentiation, are examples of these operations. In some cases we know the genetic elements ivolved, and their functional interactions. They form a syntagma, i.e. a set of genes, functionally related, to perform a genetic operation (e.g. the early and late genes of viruses, those controlling segmental specification and the sex differentiation genes). Many syntagmas are operating in the same cell, and its specific behaviour corresponds to the combination of those active genes of the synthagmas operating in it I will call this combination of active genes the "apogenome" of the cell. Different lineages, different tissues and organs can then be described by their apogenomes, and the species by an algorithm of apogenomes. Thus, from a developmental genetic point of view morphological evolution results from variations in apogenomes.

Syntagmas are built-in entities of genes related functionally by molecular recognition. This imposses constraints to variation. The inclusion of new genes in a syntagma must be rare. Syntagmas are also largely independent from each other. Thus, their number must have been kept small, and only their combinations increase leading to new and different apogenomes along evolution. The observed morphological invariance in organisms of the same phylum or major taxonomic groups is then possibly based on the genetic inertia of syntagmas and their combinations in apogenomes. If the conservation of syntagmas is based on the interaction between their elements, allelic substitutions that retain the function, i.e. these interactions, may be incorporated. Specially those which affect efficiency of transcription (e.g. inserts in non-coding regions) will be operational in generating new apogenomes. They will appear by spatial and temporal shifting of the same syntagmas to or from different cells.

New forms seem to derive from more primitive ones by local variations of simpler and reiterated apogenomes, generating diversity from homogeneity. In fact, strong mutations in activator or selector genes cause more primitive ("ancestral" or "atavic") patterns or forms. Thus, the notion of specific genes for early versus late morphogenetic processes, or for general versus specific developmental properties, possibly does not have a genetic representation. Conceivably, the same morphogenetic operations may participate in early and late stages and in most tissues and organisms. Possibly, their genetic bases have been conserved throughout evolution in different phyla.

The explanation of morphological diversity in terms of the combinatorial superposition of independent genetic operations at work in different cell lineages allows us to understand how selection can modulate particular subsystems (e.g. segments in insects) without cascade effects in the whole organism. This means that different apogenomes can be tested and changed in different parts and stages of development. That, in turn, makes it easy to understand how similar apogenomes could generate forms as distinct as those found in organisms with metagenesis, metamorphosis or cast polymorphism. In the case of metamorphosis and cast polymorphisms hormonal conditions cause alternating forms, but it is difficult to imagine that they are generated by exclusive syntagmas. Holometabolous insects, for example, can hardly derive from hemimetabolous forms by the addition of new sets of genes. Caenogenetic adaptations can also occur by temporally shifting (heterocrony) those independent genetic operations. Thus, neotenic or progenetic species are explained by relative shifts of genetic operations leadig to sexual maturation relative to somatic development. This could be also the underlaying mechanism of the rule of increasing precocity of developmental processes in derived forms of the same phylum. These extreme neotenic and progenetic forms probably retain most of the syntagmas of their ancestors at work in early stages of development. It is from these that new apogenomes may derive to generate the new forms of major taxonomic groups.

This generative notion of morphology helps to approach classic taxonomical paradoxes. Cladistic grouping of related species produce conflictive seriations depending on the "key characters" one uses for classification: related species containing overlapping within non-overlapping characters. Similarly, phylogenetic trees have to postulate missing forms (X-trees) to produce branching deviations of species (see Henning, 1966). However if mutational changes affect combinatorial apogenomes, morphological variations between species could be large, non-serial and discontinuous. They would also be potentially reversible in phenotypic terms: i.e. entire patterns disappear and reappear in the same phylogeny. This seems to occur in the chaete patterns of the species of the genus

Drosophila (García-Bellido, 1983). Thus, it is conceivable that genetic isolation at speciation, may establish at once complex morphological changes. They would correspond to slightly different apogenomes rather than to macromutations. The viability of these new apogenomes is largely ensured because the intervining operations have been independently tested in their ancestors. Stabilizing selection would later fix them. These notions force upon us two questions: how much morphological noise is compatible with fitness- and how far morphological adaptation is not merely our interpretation of the objective morphological constancy of a given species.

ACKNOWLEDGEMENTS

I thank Drs. M. P. Capdevila, J. Botas, L. García-Alonso, B. L. Sheldon and I. Franklin for permission to quote unpublished results of our work. Comments to the manuscript, especially by Dr. R. Nöthiger on the sex syntagma, are most appreciated. The unpublished work was supported by the Spanish CAICYT and FIS and CSIRO, Australia.

REFERENCES

Akam, M. E., 1983, The location of Ultrabithorax transcripts in Drosophila tissue section, EMBO J., 2:2075-2084.

Baker, B. S., and Belote, J. M., 1983, Sex determination and dosage compensation in Drosophila melanogaster, Ann. Rev. Genet., 17:345-393.

Bridges, C. B., 1925, Sex in relation to chromosomes and genes, Am. Natur., 59:127-137.

Capevila, M. P., and García-Bellido, A., 1981, Genes involved in the activation of the bithorax complex of Drosophila, Wilhelm Roux's Archiv., 190:339-350.

Carramolino, L., Ruiz-Gomex, M., Boronoat, A., and Modolell, J., 1984, Molecular genetics of the achaete-scute gene complex of Drosophila melangoaster, Cell, in press.

Cline, T. W., 1983, The interactions between daughterless and sex-lethal in triploids: a lethal sex transforming maternal effect linking sex determination and dosage compensation in Drosophila melanogaster, Devlop. Biol., 95:260-274.

Counce, S. J., 1973, The casual analysis of insect embryogenesis, in: "Developmental systems: insect," vol. 2, S. J. Counce and C. H. Waddington, eds., Acad. Press, New York, pp. 1-156.

Duncan, I., and Lewis, E. B., 1982, Genetic control of body segment differentiation in Drosophila, in: "Developmental order: its origin and regulation," S. Subtenly, ed., Liss, New York, pp. 533-554.

García-Bellido, A., 1975, Genetic control of wing disc development in Drosophila, in: "Cell Patterning," Ciba Foundation Symposium 29, Elsevier, Amsterdam, pp. 161-182.

García-Bellido, A., 1977, Homoeotic and atavic mutations in insects, Amer. Zool., 17:613-629.

García-Bellido, A., 1981a, The bithorax syntagma, in: "Advances in Genetics, Development, and Evolution of Drosophila," VII European Drosophila Research Conference, S. Lakovaara, ed., Plenum Press, pp. 135-148.

García-Bellido, A., 1981b, From the gene to the pattern: chaeta differentation, in: "Cellular controls in Differentiation," C. W. Lloyd and D. A. Rees, eds., Acad. Press, pp. 281-304.

García-Bellido, A., 1983, Comparative anatomy of cuticular pattern in the genus Drosophila, in: "Development and Evolution," B. C. Goodwin, N. Holder and C. C. Wylie, eds., Cambridge University Press, pp. 227-255.

García-Bellido, A. and Capdevila, M. P., 1978, The initiation and maintenance of gene activity in a developmental pathway of Drosophila, in: "The clonal basis of development," vol. 7, S. Subtelny and I. M. Susses, eds., Academic Press, New York, pp. 3-21.

García-Bellido, A., Lawrence, P. A., and Morata, G., 1979, Compartments in animal development, Scientific American, 241:102-110.

García-Bellido, A., and Merriam, J. R., 1971, Clonal parameters of tergite development in Drosophila, Devel. Biol., 26:264-276.

García-Bellido, A., and Santamaria, B., 1978, Developmental analysis of the achaete-scute system of Drosophila melanogaster, Genetics, 88:469-486.

Hennig, W., 1966, "Phylogenetic Systematics," Univ. Illinois Press, Urbana, Ill.

Kaufman, T. C., and Wakimoto, B. T., 1982, Genes that control high level developmental switches, in: "Evolution and Development," J. T. Bonner, ed., Spring Vlg. Berlin, pp. 189-205.

Lawrence, P. A., Struhl, G., and Morata, G., 1979, Bristle patterns and compartment boundaries in the tarsi of Drosophila, J. Embryol. Exp. Morph., 51:195-208.

Levine, M., Hafen, E., Garber, R. L., and Gehring, W. J., 1983, Spatial disbtribution of Antennapedia transcripts during Drosophila development, EMBO J., 2:2037-2046.

Lewis, E. B., 1978, A gene complex controlling segmentation in Drosophila, Nature, 276:565-570.

McGinnis, W., Garber, R. L., Wirz, J., Kuroiwa, A., and Gehring, W. J., 1984, A homologous protein coding sequence in Drosophila homoeotic genes and its conservation in other metazoans, Cell, 37:403-408.

Moscoso del Prado, J., and García-Bellido, A., 1984a, Genetic regulation of the Achaete-scute complex of Drosophila melanogaster, Roux's Arch. Dev. Biol., 193:242-245.

Moscoso del Prado, J., and García-Bellido, A., 1984b, Cell interactions in the generation of chaetae pattern in *Drosophila*, *Roux's arch. Dev. Biol.*, 193:246-251.

Nöthiger, R., Dübendorfer, A., and Epper, F., 1977, Synandromorphs reveal two separate primordia for male and female genitalia in *Drosophila melanogaster*, *W.R. Arch. Entw. Mech. Org.*, 181:367-373.

Rendel, J. M., 1963, Correlation between the number of scutellar and abdominal bristles in *Drosophila melanogaster*, *Genetics*, 48:391-408.

Rendel, J. M., 1967, "Canalization and gene control," Acad. Press, London.

Sander, R., 1981, Pattern generation and pattern conservation in insect ontogenesis: problems, data and models, *in*: "Progress in Developmental Biology,", H. W. Sauer, ed., *Fortschr. Zool.*, 26:101-119.

Scott, M. P., and Weiner, A. J., 1984, Structural relatinships among genes that control development: sequence homology between Antennapedia, Ultrabithorax and fushi tarasu loci of *Drosophila*, *Proc. Natl. Acad. Sci. USA*, in press.

Stern, C., 1954, Two or three bristles, *Am. Sci.*, 42:213-247.

Struhl, G., 1983, Role of the esc^{+} gene product in ensuring the selective expression of segment-specific homeotic genes in *Drosophila*, *J. Embryol. Exp. Morph.*, 76:297-331.

White, R.A.H., and Wilcox, M., 1981, Protein productsw of the bithorax complex in *Drosophila*, *Cell*, in press.

Wieschaus, E., and Nöthiger, R., 1982, The role of transformer gnees in the development of genitalia and analia of *Drosophila melanogaster*, *Devel. Biol.*, 90:320-334.

Wood, W. B., 1973, Genetic control of bacteriophage T4 morphogenesis, *in*: "Genetic mechanisms of development," F.H. Ruddle, ed., Acad. Press, New York, pp. 29-46.

CELL LINEAGE AND CELL INTERACTIONS IN THE DETERMINATION OF DEVELOPMENTAL CELL FATES

Marty Shankland and Gunther S. Stent

Department of Molecular Biology
University of California
Berkeley, California 94720

INTRODUCTION

A central problem of developmental biology is to understand the sequence of cellular and molecular events which transforms a fertilized egg into a mature organism composed of a precise spatial array of distinct cell types. A significant consideration in addressing this problem is the fact that each of the many cells in the mature organism is descended from the egg via a lengthy sequence of progenitor cell divisions. In order to reduce the number of variables associated with the problem, many embryologists have concentrated on invertebrates such as nematodes, annelids, molluscs and tunicates in which the sequence of embryonic cell divisions is highly stereotyped. Because of this regularity it is possible to portray the development of the entire organism by the genealogical relationship between the composite cells, and to identify individual cells on the basis of their lineage (Zur Strassen, 1896; Conklin, 1905; Schleip, 1936). A determinate cell lineage offers clear advantages for studying developmental mechanisms at a cellular level: if the developmental state of any given cell is the product of commitments made by its ancestors, then it will be more straightforward to characterize the commitment process if the cells of the mature organism arise from the egg by a known line of descent.

This argument assumes that the developmental fate of a given cell is at least partially determined by ancestry, and that its lineal inheritance, i.e. properties inherited from its mother cell, can restrict the cell's potential response to different embryonic environments. This idea has received support both from experimental analyses of the relationship between cell lineage and developmental potential (Sulston and Horvitz, 1977; Blair, 1983), and from the discovery of heritable cytoplasmic components which do play a role in the determination of cell fates (Illmensee and Mahowald, 1974; Whittaker, 1980).

The state of a cell's developmental commitment is not always determined at the time of its birth, but can also arise secondarily from interaction with its environment. Embryos which exhibit a determinate pattern of cell lineage are especially useful for unraveling the relative contribution of lineal inheritance and cell interaction in particular commitment events (Sulston and White, 1980; Kimble, 1981; Weisblat and Blair, 1984), because individual cells can be identified on the basis of their ancestry, and the regularity of development permits detailed characterization both of the physical relationship between different cells and of their normal descendant fates. Our purpose here is to review a series of recent studies which examine the significance of cell lineage and cell interactions in the differential commitment of the O and P cell lines of the glossiphoniid leech embryo. The results offer insight into the means by which cell interactions modify inherited factors at the cellular level, and help establish a framework in which to consider the relationship between developmental mechanics and phylogenetic adaptation.

CELL LINEAGE ANALYSIS OF LEECH EMBRYOGENESIS

The eggs of leeches and other annelids develop through a stereotyped sequence of embryonic cleavages (for a review, see Anderson, 1973). The most complete descriptions of embryonic cell lineage have been obtained from leeches of the family Glossiphoniidae, whose eggs are larger than those of most other annelids and develop directly into the adult morphology without passing through an intervening larval stage. The early cell divisions have been characterized by direct microscopic observation of intact embryos (Whitman, 1878; Schleip, 1936; Fernandez, 1980), but the later stages of embryogenesis did not prove amenable to detailed cellular analysis until the advent of intracellular lineage tracers (Weisblat et al., 1978; 1980a). By injecting individual embryonic blastomeres with compounds such as horseradish peroxidase (HRP) or fluorescent dyes conjugated to carrier molecules, it has been possible to selectively label specific embryonic cell lines and to trace their developmental fate all the way through to the terminal differentiation of their postmitotic descendants (Weisblat et al., 1980b, 1984; Shankland and Weisblat, 1984; Kramer and Weisblat, 1985; Gimlich and Braun, 1985). Lineage tracers have also proven useful in extending cell lineage descriptions past the stage at which the individual cells can no longer be readily visualized in the living embryo (Zackson, 1984).

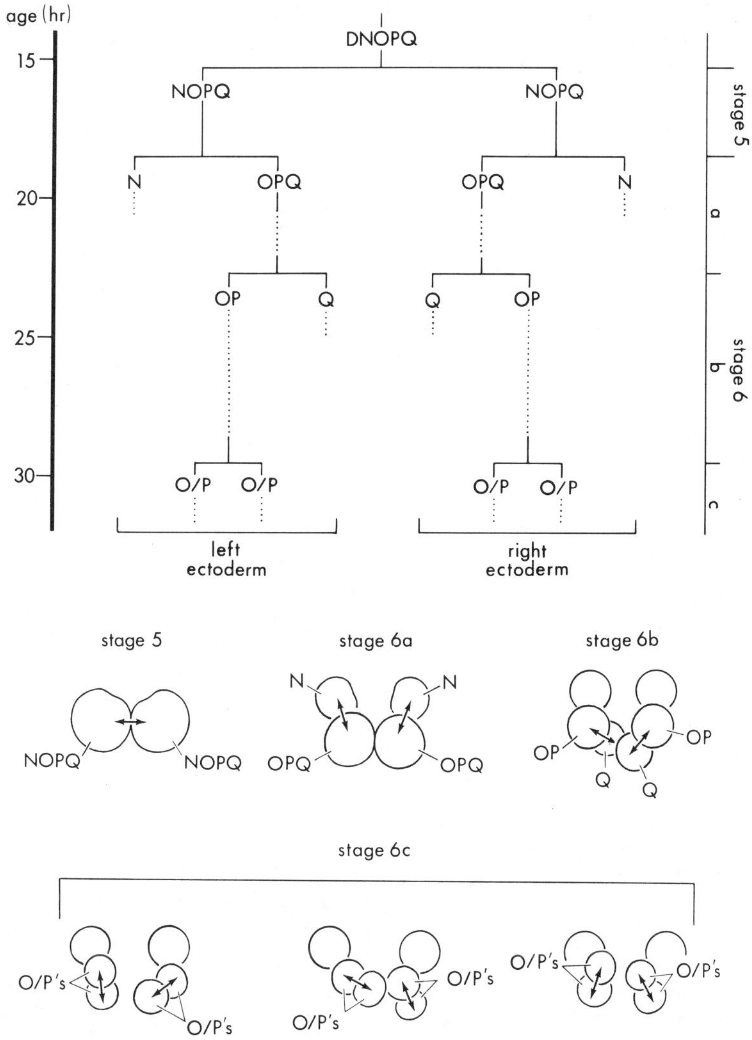

Fig. 1. The definitive ectoderm arises from cell DNOPQ via a stereotyped lineage. The cell genealogy at the top depicts the cleavage of cell DNOPQ to form the four ectodermal cell lines. Dotted lines represent periods of time during which the large blastomeres bud off micromeres or blast cells. The camera lucida tracings at the bottom show the orientation of these cleavages in embryos of the species Helobdella triserialis. Most cleavages have a fixed orientation, and are symmetrical on the right and left sides of the embryo. However, the cleavage of cell OP to form the sister O/P teloblasts varies considerably between individuals, and often between the right and left sides of the same egg, thus leading to ambiguity in the identification of O/P sisters.

Despite these technical advances, cell lineage analysis of leech embryogenesis has been limited by the cellular complexity of the mature organism. The nematode *Caenorhabditis elegans*--whose hermaphroditic adult contains 959 identified somatic cell nuclei--is the only animal whose developmental cell lineage has been described in complete detail (Sulston et al., 1983), and by way of comparison the adult leech contains approximately two orders of magnitude more cells. Scientific analysis of the leech's cell lineage can, however, be simplified somewhat by noting that the leech body is comprised of 32 bilaterally symmetrical segments (plus a nonsegmental prostomium) which arise from homologous sublineages within the overall pattern of embryonic cell division. These homologous sublineages appear to be identical through the first few cell divisions (Zackson, 1984), indicating that segmental diversification does not become manifest until some later stage. In the leech species *Helobdella triserialis*, the development of the O cell line has been traced through 16 distinct cell divisions (not including iterated divisions) from the uncleaved egg, and there would seem to be only 3-4 additional, as yet undescribed rounds of division prior to the terminal differentiation of postmitotic descendants in this cell line (Shankland and Weisblat, 1984). Thus, it may soon be possible to trace some branches of the leech's developmental cell lineage division-by-division all the way from the uncleaved egg to identified postmitotic cells.

ORIGIN OF THE ECTODERMAL CELL LINES

During leech embryogenesis the first two rounds of cleavage lead to the formation of four large blastomeres designated as A, B, C and D. Cell D later divides to yield cells DM and DNOPQ (terminology of Stent et al., 1982) which give rise respectively to the segmental mesoderm and segmental ectoderm of the mature leech. Cell DM cleaves symmetrically to produce the right and left M cells, which serve as hemilateral mesodermal precursors. Cell DNOPQ also cleaves symmetrically, producing the right and left NOPQ cells, and then each NOPQ cell undergoes a series of three additional cleavages to establish four separate ectodermal cell lines. At the first of these cleavages cell NOPQ cleaves asymmetrically to produce a larger daughter cell OPQ and a smaller daughter cell N. Cell OPQ then cleaves asymmetrically to produce a larger cell OP and a smaller cell Q. Finally cell OP cleaves symmetrically to produce two sister O/P cells (Fig. 1).

Cells M, N, O/P and Q are referred to as *teloblasts*, and soon after their birth undertake a stem cell mode of division, i.e. an iterated series of highly asymmetric cleavages in which the larger daughter retains the characteristics of the mother cell, and hence is referred to by the same name. The teloblasts' smaller daughters are called *primary blast cells*, and depart from the teloblast in

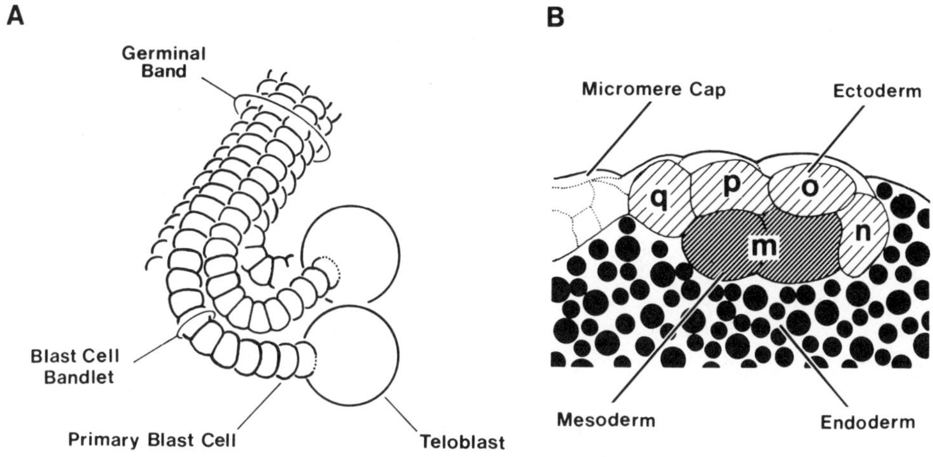

Fig. 2. A. Each teloblast undergoes a stem cell mode of division and gives rise to a bandlet consisting of several dozen primary blast cells. Bandlets arising from the ipsilateral teloblasts come together in parallel to form a unified germinal band. This merger begins the oldest blast cells, i.e. those lying furthest from the teloblast, and the band lengthens by the progressive addition of younger cells more posteriorly. B. Transverse section showing the organization of bandlets within the germinal band. The n, o/p and q bandlets form the ectodermal cell layer, with the two o/p bandlets being designated as o and p bandlets on the basis of position. The ectoderm is supported by the mesodermal m bandlet, which is two cells wide by the time it enters the germinal band. The band itself lies over the yolky endodermal macromeres, and is surmounted by a provisional integument extending outward from the micromere cap.

discrete columns or bandlets. Both the bandlets and their constituent blast cells are designated by lower case letters corresponding to their teloblasts of origin--m, n, o/p or q.

Beginning with the firstborn blast cells the four ectodermal bandlets and the one mesodermal bandlet from the same side of the embryo come together in parallel to form a unified germinal band (Fig. 2A). The two o/p bandlets lie between the n and q bandlets in the more superficial, ectodermal cell layer, and are henceforth designated as o and p bandlets on the basis of position. The o bandlet lies adjacent to the n bandlet on the side of the germinal band away from the micromere cap, and the p bandlet lies adjacent to the q bandlet on the side nearer the micromere cap (Fig. 2B). After entering the germinal band the blast cells divide repeatedly, and it is as a result of this proliferation that the two germinal bands will ultimately give rise to the epidermis, musculature, nervous system and various internal organs on the right and left sides of the mature leech (Weisblat et al., 1978).

EQUIVALENCE AND NONEQUIVALENCE OF LINEALLY IDENTIFIED CELLS

Before discussing the way in which a cell's ancestry contributes to its developmental fate, we must first consider the degree to which ancestry can be used to identify cells in a lineage such as that depicted in Fig. 1. This particular lineage is comprised of an orderly sequence of cell divisions shared by all glossiphoniid leech species. However, the fact that there is a determinate sequence of cell divisions does not necessarily entail that every cell in the lineage can be uniquely identified from one individual to the next. Cells not derived from the same mother cell can be distinguished by their ancestry, but sister cells must be distinguished either on the basis of physical characteristics, such as size or cytoplasmic composition, or by their differing positions with respect to other structures in the egg.

In many cases these criteria are sufficient for reliable identification. For example, the two NOPQ cells undergo mirror symmetrical divisions to yield a larger, posteromedial daughter--cell OPQ--and a smaller, anterolateral daughter--the N teloblast. In other cases sisters can be distinguished by only one of these two criteria. Thus, sister NOPQ cells are identified by position, but are physically indistinguishable and express essentially identical developmental fates on the right and left sides of the leech body (Weisblat et al., 1978).

There are some cases in which it is not valid to draw a distinction between the daughters of a division. For example, sister O/P teloblasts are equal in size, and in *H. triserialis* the orientation of their formative cleavage varies considerably, both between individuals and between the two sides of the same egg (Fig. 1). Thus, the position of O/P sisters in one embryo can not be meaningfully extended to other members of the same species. Sister O/P teloblasts are therefore deemed to be lineally equivalent, and have been assigned a common name accordingly.

The apparent equivalence of the O/P teloblasts is further supported by experimental studies which show that the two sister teloblasts have the same developmental potential, and hence do not bear any covert differences of developmental significance (Weisblat and Blair, 1984; Shankland, in press). The OP cleavage is of fixed orientation in another glossiphoniid species, *Theromyzon rude* (Fernandez, 1980). However, as in *H. triserialis*, there is no fixed correlation between the position of these two teloblasts and their normal developmental fate in this other species.

CELL LINEAGE AND COMMITMENT

Tracer injections reveal that each of the leech's lineally identified blastomeres has a precise developmental fate during the course of normal embryogenesis. These fates have been described both in terms of the spatiotemporal pattern of blast cell divisions (Weisblat et al., 1980b; Zackson, 1984), and also the cellular composition of a cell's descendant clone in the mature leech (Weisblat et al., 1978; 1980b; 1984). If a cell's lineage history does play a role in determining its developmental fate, then there should be some cell intrinsic property, not position, preventing it from taking on the fate normally associated with another cell. In the leech embryo this issue has been addressed by selectively labeling one cell line with a lineage tracer, and following its fate when other, neighboring cell lines have been ablated.

The determinative role of cell lineage is best appreciated by considering the degree of developmental autonomy manifested by the various blast cell bandlets. The m, n and q bandlets all arise from lineally distinct teloblasts, and experimental studies have shown that all three undergo essentially normal patterns of cell division and give rise to descendant clones of seemingly appropriate histotypic composition following the elimination of one or more of their neighboring bandlets (Blair, 1982; 1983; Zackson, 1984). Bandlet interactions do play a critical role in certain aspects of morphogenesis (Blair, 1982; Blair and Weisblat, 1982, 1984), but here it seems that bandlet interactions are influencing not the establishment but rather the expression of cell intrinsic identities. Thus, the experimental evidence is consonant with the observed lineal nonequivalence of the M, N and Q teloblasts, and suggests that their differing ancestries convey upon them some intrinsic differences in developmental potential which will insure that their progeny follow divergent developmental pathways.

Although the sister O/P teloblasts are initially indistinguishable, during the later development these two cell lines take on divergent O and P fates which correlate with the respectively o or p position of their bandlets within the germinal band (Weisblat and Blair, 1984). The o and p bandlets show a pronounced degree of developmental autonomy with respect to the elimination of the m, n or q bandlets (Zackson, 1984), consistent with the idea that o/p blast cells possess an intrinsic identity which distinguishes them from the other, lineally nonequivalent cell lines. However, the O or P fate of a given o/p blast cell can be changed by altering the spatial relationship of the two o/p bandlets with respect to one another (Weisblat and Blair, 1984; Zackson, 1984; Shankland and Weisblat, 1984; Shankland, in press), indicating

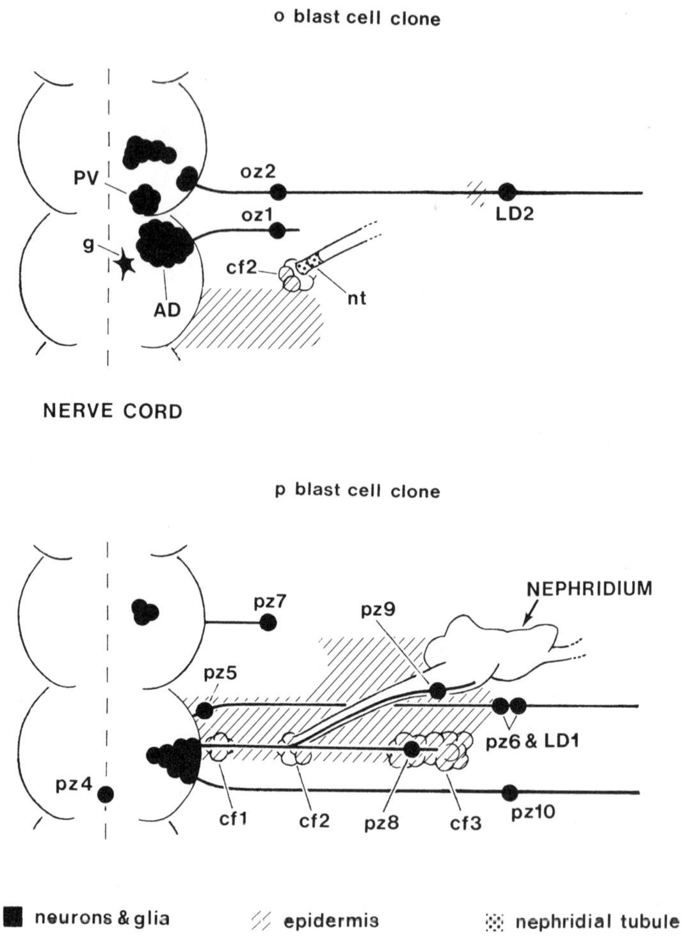

Fig. 3. Pattern elements descended from primary blast cells in the o and p bandlets. The nerve cord consists of a chain of segmental ganglia lying along the ventral midline (dashed line). Both o and p blast cells contribute descendants to two consecutive segmental ganglia, and to a corresponding portion of the body wall. Included in the o blast cell clone are anterodorsal (AD) and posteroventral (PV) clusters of central neurons, a central glia located near the ganglionic midline (g), three positionally identified peripheral neurons (oz1, oz2 and LD2), part of the nephridial tubule (nt) and part of an epidermal specialization called cell floret 2 (cf2). The p blast cell clone includes seven peripheral neurons (pz5-10 and LD1), all of cell florets 1 and 3 (cf1, cf3) and the remainder of cell floret 2. Some but not all p blast cell clones include an unpaired interneuron (pz4) located at the ganglionic midline.

that o/p blast cells are not committed to either the O or the P developmental pathway at the time of their birth. Thus, the lineally equivalent blast cells of the o and p bandlets do not manifest any intrinsic differences in developmental potential, and their subsequent commitment to the O or P pathway arises secondarily in a manner which depends upon interaction between the cells of the two bandlets.

COMMITMENT OF BLAST CELLS TO THE O AND P DEVELOPMENTAL PATHWAYS

As mentioned above, the progeny of the O/P teloblasts are designated as o blast cells or p blast cells depending upon which of two positions they come to occupy within the germinal band. The blast cells in these two bandlets do not begin to manifest noteworthy differences in developmental fate until after they have occupied these positions for a period of several hours. The first evidence of divergence is seen in the pattern of blast cell division (Zackson, 1984). The primary o and p blast cells undergo their first division approximately 10 hr after entering the germinal band, with the primary p blast cell dividing nearly symmetrically to yield an anterior daughter, cell p.a, and a posterior daughter, p.p, whose volume ratio is approximately 1.5:1, and the o blast cell dividing more asymmetrically to yield an anterior daughter, cell o.a, and a posterior daughter, o.p, whose volume ratio is approximately 8:1. Additional differences in the geometry and timing of later divisions have also been described (Zackson, 1984).

Lineage tracer injections show that the o and p bandlets give rise to distinct, segmentally iterated patterns of descendant tissue in the mature leech (Weisblat and Blair, 1984). Each primary blast cell gives rise to one metameric unit of the teloblast-derived pattern, and thus, with the exception of a very few segmental differences, all of the blast cells in a given o or p bandlet express the same developmental fate in different segments (Weisblat and Shankland, in preparation). However, the pattern of mature tissues derived from the two bandlets differs considerably, and a given blast cell clone can be distingushed as being O-type or P-type on the basis of a number of uniquely identified pattern elements, including individual neurons and neuron clusters, epidermal specializations known as cell florets, and cells of the nephridial tubule (Fig. 3).

Role of Blast Cell Interactions

The commitment of o/p blast cells to the O and P developmental pathways is influenced by interaction between the blast cells of these two bandlets. If the original p bandlet is eliminated by ablation of its parental teloblast (Weisblat and Blair, 1984; Zackson, 1984) or direct photolesion (Shankland and Weisblat, 1984), the blast cells in the adjacent o bandlet change fate, undergoing the division pattern and giving rise to descendant clones typical of p blast cells (Fig. 4). The converse experiment--elimination of the o bandlet--alters neither the division pattern nor the pattern of descendants produced by blast cells in the adjacent p bandlet, indicating that the interaction between the two bandlets is not reciprocal. These and other related findings have been explained by assuming (i) that the o/p blast cells are initially capable of following either the O or the P developmental pathway, and by further assuming (ii) that each individual o/p blast cell becomes committed to one or the other pathway in accordance with its o or p position within the germinal band. This positional specification model has received support from the observation that a p bandlet transposed into the o bandlet position displays a highly asymmetric first division and gives rise to unambiguous O pattern elements among its descendant tissues (Shankland, in press).

In order for the positional specification model to account for the lack of reciprocity in response to elimination of the o and p bandlets, there must be a developmental hierarchy in the assignment of cells to the O and P developmental pathways. Blast cells in a single, experimentally isolated o/p bandlet become committed to the P pathway regardless of their original fate, which suggests that during normal development the adjacent o and p bandlets interact in such a way that cells in the o bandlet are excluded from the P pathway--which they would have followed in isolation--into an alternative O pathway. The p blast cells are able to exert this influence over a distance of no more than 1-2 cell widths (Zackson, 1984; Shankland and Weisblat, 1984), and it may be that this influence depends upon a localized, possibly contact-mediated interaction between neighboring blast cells in the two o/p bandlets.

Time-course of Blast Cell Commitment

To discover when the o blast cell clone becomes autonomously committed to the O developmental pathway, the technique of dye-sensitized photolesioning (Shankland, 1984) was used to selectively ablate the entire p bandlet throughout the length of the germinal band. Interpretation of this experiment relies on the fact that each of the blast cells in the o bandlet has an essentially identical developmental fate, and that they follow the same developmental timetable offset by differences in birthdate (Zackson, 1984). The spatial order of blast cells within the o bandlet reflects the temporal order of birth, so that at any given time the older cells have progressed further into the germinal band. Thus, one can ascertain the relationship between the age of an o blast cell and its state of commitment with respect to the O pathway by comparing the response of o blast cells to p bandlet ablation along the bandlet's length. In these experiments the response of individual o blast cells was determined by labeling the parental teloblast with

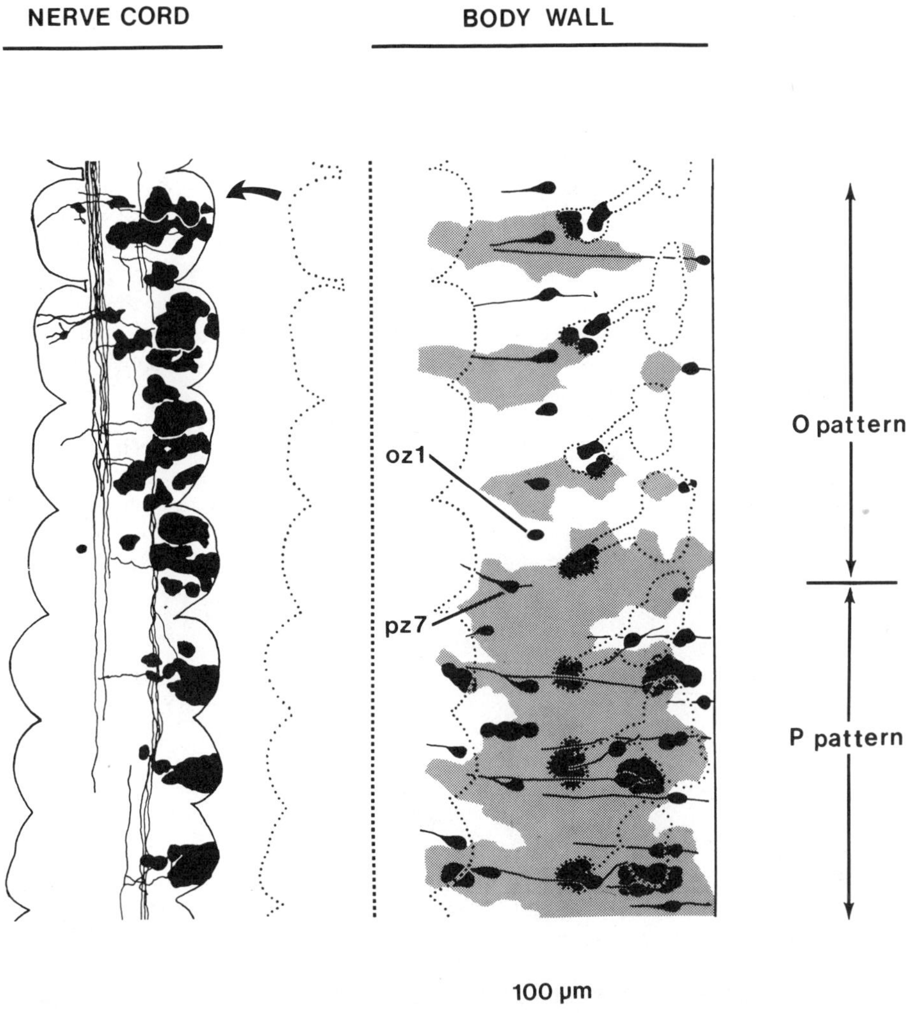

Fig. 4. Embryo in which the o bandlet was labeled with HRP, and the posterior half of the p bandlet eliminated by teloblast ablation. This camera lucida tracing shows the pattern of HRP-labelled descendants in a series of seven midbody segments, with anterior at the top. Those o blast cells which had p blast cell neighbors gave rise to a normal O descendant pattern, but more posterior o blast cells were deprived of their normal p blast cell neighbors and followed the P pathway instead, giving rise to an essentially normal P pattern of descendants. Shading represents labelled portions of the bodywall epidermis. Neurons are designated as in Fig. 3.

HRP, and scoring the clones descended from the individual blast cells for the presence or absence of five unambiguously identifiable O pattern elements and six unambiguously identifiable P pattern elements.

Photoablation of the p bandlet has revealed that the o blast cells become autonomously committed to the O developmental pathway, i.e. lose their dependence on the p bandlet, shortly after they become incorporated into the germinal band (Shankland and Weisblat, 1984). Blast cells which have just entered the band as part of the o bandlet will respond to p bandlet ablation by giving rise to all six unambiguous P pattern elements and none of the five unambiguous O pattern elements (Table 1). This finding demonstrates that the primary o blast cells are capable of following either the O or the P pathway when they first assume the o bandlet position. However, as the blast cell progresses further into the germinal band it ceases to require the continued presence of the p bandlet in order to follow the O pathway, and eventually reaches a state in which it will give rise to its normal complement of O pattern elements following photolesion of the neighboring p blast cells.

The commitment of an o blast cell to the O developmental pathway is not a single event, but rather a sequence of three discrete commitment events, each of which independently affects the O or P fate of a specific subset of the cells in its descendant clone (Table 1). If the p bandlet is present during the first commitment event, the o blast cell will be committed to the production of neuron oz2, while the fate of the other four unambiguous O pattern elements will remain contingent on the continued presence of the p bandlet. In addition, the o blast cell clone loses the capacity to produce one of the unambiguous P pattern elements, a neuron of the pz6/LD1 pair, concomitant with its commitment to produce neuron oz2.

The second and third commitment events convey additional restrictions upon the developmental potential of the o blast cell clone. If the p bandlet is present at the time of the second commitment event the o clone will be committed to produce the four remaining unambiguous O pattern elements, with a concomitant loss of three of the unambiguous P pattern elements, and if the p bandlet is present at the third commitment event the o clone loses the capacity to produce the two remaining unambiguous P pattern elements. The third event thus differs from the preceding two in that it adds no unambiguous O pattern elements to the descendant clone; however, we can not discount the possibility that this event does involve the commitment of some of the other, less readily identifiable O pattern elements which were not scored in this study.

These three experimentally separable commitment events would seem to reflect a sequence of three independent critical periods for

Table 1. Set of unambiguous O and P pattern elements produced by a primary o blast cell clone following ablation of the adjacent p bandlet.

	age of clone at time of ablation			
	0–20 hr	21–25 hr	26–31 hr	32+ hr
<u>O</u> <u>pattern</u> <u>elements</u>				
nephridial tubule	–	–	+	+
AD neuron cluster	–	–	+	+
PV neuron cluster	–	–	+	+
neuron oz1	–	–	+	+
neuron oz2	–	+	+	+
<u>P</u> <u>pattern</u> <u>elements</u>				
neuron pz7	+	+	+	–
cell floret 3	+	+	+	–
cell floret 1	+	+	–	–
neuron pz9	+	+	–	–
neuron pz6/LD1	+	+	–	–
neuron pz6/LD1	+	–	–	–
Commitment Events:	▲ 1st	▲ 2nd	▲ 3rd	

Each vertical column represents the typical set of unambiguous O and P pattern elements produced by a primary o blast cell following ablation of the neighboring p bandlet. Neurons pz6 and LD1 were not distinguished in these experiments, and it is not known which of these cells disappears in association with the first commitment event. Adapted from Shankland and Weisblat (1984).

the interaction of the o and p bandlets. During each critical period the presence or absence of p blast cell neighbors would determine whether a portion of the o blast cell clone becomes committed to the O or the P pathway respectively. Normally the p bandlet is present during all three critical periods, with the result that the o blast cell clone undergoes a complete and stepwise commitment to the O pathway. If the p bandlet is eliminated before an o blast cell has reached its first critical period, then the o blast cell clone would become committed to the P pathway in its entirety. If, on the other hand, an o blast cell has already undergone one or two partial commitments to the O pathway at the time of the ablation, then only a subset of its descendants would be able to follow the P pathway, and it would respond to the ablation by giving rise to a mixture of both O and P pattern elements in its descendant clone (Table 1).

Association of Commitment Events with Blast Cell Divisions

In order to understand the way in which blast cell interactions govern the fate of the o blast cell, we have recently focussed our attention on the cellular basis of the three commitment events described in the preceding section. These events occur in phase with the stereotyped sequence of divisions manifested by the o blast cell after it enters the germinal band (Shankland and Weisblat, 1984), and there are clear indications that at least the first commitment event is functionally related to a particular cell division.

As previously mentioned, the primary o blast cell first divides to yield a large anterior daughter, cell o.a, and a small posterior daughter, cell o.p. Cell o.a divides approximately 5 hr later to yield an anterior daughter, cell o.aa, and a posterior daughter, cell o.ap (Zackson, 1984). By injecting cell o.a and its daughters with fluorescent lineage tracers and following their labeled clones (Shankland, in preparation), it has been possible to characterize the segregation of descendant cell fates that is normally associated with its division. Cell o.a gives rise to all five of the unambiguous O pattern elements scored in the photoablation experiments, and these pattern elements are segregated between its two daughters in a stereotyped manner (Fig. 5A).

The segregation of these five O pattern elements at the division of cell o.a parallels the segregation of committed and uncommitted pattern elements at the first commitment event. This correlation could be explained if cell o.aa were committed to the production of its normal complement of O descendant pattern elements as a result of the first commitment event, while cell o.ap remains uncommitted, i.e. capable of giving rise to either O or P pattern elements, until some later time (Fig. 5B). The relative timing of commitment events and blast cell divisions has been estimated

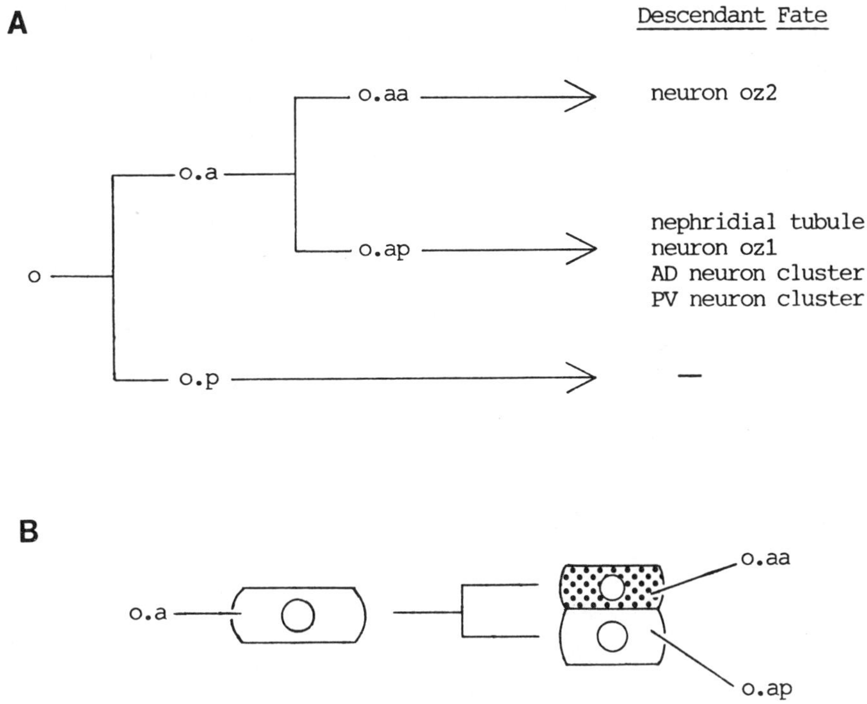

Fig. 5. A. Segregation of descendant cell fates during the first two divisions of the o blast cell, as ascertained by iontophoretically injecting identified blast cell daughters with a fluorescent lineage tracer and characterizing their descendant clones (Shankland, in preparation). During normal development cell o.a, the anterior daughter of the first division, gives rise to the five unambiguous O pattern elements. Cell o.a divides to produce an anterior daughter, cell o.aa, which gives rise to neuron oz2, and a posterior daughter, cell o.ap, which gives rise to the other four unambiguous O pattern elements. This pattern of segregation of descendant cell fates parallels the segregation of committed and uncommitted pattern elements at the first of the commitment events described in the text. B. Cellular basis of the first commitment event. Cell o.a can contribute descendants to either the O or P pathway and hence is uncommitted with respect to this choice. Under the influence of the adjacent p bandlet cell o.a divides to produce daughter cells with different states of commitment. The posterior daughter o.ap is likewise uncommitted with respect to the O and P pathways (unstippled), while the anterior daughter o.aa is committed to the O pathway (stippled), and will give rise to O descendant pattern elements regardless of any subsequent p bandlet ablation.

experimentally (Shankland and Weisblat, 1984), and it appears that the first commitment event occurs several hours prior to the o.a cell division. Taken together these findings suggest that during the mitotic cycle of cell o.a there is a critical period for interaction with the adjacent p bandlet, and if the p bandlet is present during this critical period then cell o.a will produce an anterior daughter which is committed to give rise to specific O pattern elements regardless of the continued presence of the p bandlet.

Our preliminary findings suggest that the second and third commitment events are associated with subsequent divisions of the uncommitted cell o.ap, and furthermore that the second commitment event, like the first, may involve a differential commitment of the two daughter cells of a division.

Integration of Intrinsic and Extrinsic Factors

By comparing the sequence of cellular events which constitute the O and P developmental pathways we can to a first approximation ascertain the nature of the p bandlet's influence on o blast cell development. The kinds of divergence observed are of course to a large degree reflective of the methods of description and experimental perturbation that have been used. Thus, we can not at this time say how many biochemical steps may intervene between the actual transmission and reception of an intercellular signal and observed effect it has on the course of o blast cell differentiation.

The interaction between the o and p bandlets clearly influences the outcome of at least the first two o blast cell divisions. If the first division occurs in the presence of the p bandlet, the primary o blast cell divides much more asymmetrically than it would if the p bandlet had been ablated. Thus, it would seem that the interaction between o and p blast cells induces a more posterior positioning of the mitotic spindle during the cytokinesis of the o blast cell.

The pattern of o and p blast cell divisions is largely unaffected by elimination of the other, surrounding bandlets (Zackson, 1984). It therefore seems likely that the p-type division pattern is dependent upon intrinsic factors inherited from the parental teloblast, and that the interaction between the two bandlets modifies the expression of these intrinsic factors in cells of the o bandlet. One could imagine that there are two alternative stable configurations for the cytoskeletal architecture of the primary o/p blast cell which lead respectively to nearly symmetrical (p-type) or highly asymmetrical (o-type) first divisions. In the absence of a second, adjacent o/p bandlet the primary o/p blast cell would express the former configuration, but interacting o/p bandlets would generate an intercellular signal that biases the cytoskeleton of the cells in the o bandlet towards the latter configuration.

The expression of an o-type first division has not been shown to play any role in committing the o blast cell clone to the production of O pattern elements. However, the experiments performed to date are not a valid test of the long term significance of the first o blast cell division because the smaller of the two daughters of this division, cell o.p, does not give rise to any of the unambiguous O pattern elements (Fig. 5A). Hence, even if cell o.p were committed to the O pathway, this commitment could not been detected in the group of pattern elements that were examined following p bandlet ablation.

The second o blast cell division is also clearly influenced by interactions with the p bandlet, but here the known effect is a change in a long term commitment to produce particular descendant pattern elements. This change in fate may also arise from an intercellular modulation of events that are occurring as a part of an intrinsically determined timetable of blast cell differentiation. That there is a cell intrinsic mechanism for differential daughter cell commitment seems likely, because the majority of the blastomere cleavages yield differentially committed daughter cells, and ablation experiments have indicated that such commitments are for the most part unaffected by the surrounding blastomeres. What we have shown here is that cell o.a divides to yield a pair of daughter cells which express different states of commitment with respect to the O and P pathways, and that in this case the pattern of commitment is dependent upon the influence of the neighboring p bandlet. It may be that cell o.a has intrinsic properties which insure that it will give rise to a committed anterior daughter and an uncommitted posterior daughter, and that it is the presence or absence of the intercellular signal which decides whether the anterior daughter will be committed to the O or the P pathway respectively.

ROLE OF CELL INTERACTIONS IN THE DIVERSIFICATION OF CELL FATES

The interaction between the o and p bandlets causes these two cell lines to undergo a secondary diversification so that initially equipotent blast cells produce very different complements of differentiated pattern elements, cf. Fig. 3. Despite the dramatic differences in fate engendered by the o/p blast cell interaction, some descendants do manifest a pronounced degree of phenotypic similarity. The most striking example is the O-derived dorsal and P-derived ventral pressure-sensitive mechanosensory neurons (Kramer and Weisblat, 1985) which exhibit similar sensory modality, membrane properties, and synaptic connections but innervate different regions of the skin (Nicholls and Baylor, 1968; Nicholls and Purves, 1970).

It may be that the p bandlet alters o blast cell development in a manner which has a differing degree of influence on the phenotype of various elements in the descendant clone.

Experimental analysis of developmental cell lineages in the nematode *Caenorhabditis elegans* (Sulston and White, 1980; Kimble, 1981; Sulston et al., 1983) and the grasshopper *Schistocerca americana* (Taghert et al., 1984) have uncovered other instances in which equipotent founder cells interact so as to bring about a secondary diversification of their descendant lineages. Such interactions are not limited to embryonic progenitors: pairs of postmitotic sister neurons in the grasshopper's embryonic nerve cord undergo divergent patterns of cytodifferentiation as the result of an interaction which occurs a few hours after their birth (Goodman et al., 1984). A comparable phenomenon may also occur in the nerve cord of the leech. The seven most anterior abdominal ganglia generate a pair of bilaterally symmetrical, N-derived serotonergic neurons, but in each ganglion one of these two neurons ceases to synthesize this neurotransmitter, possibly indicative of cell death, during the period of axonogenesis (Stuart et al., 1983). If either the right or left N teloblast is ablated, only one of these two neurons will be generated in each ganglion, and that single neuron consistently survives and retains its original serotonergic phenotype.

In embryos of the leech and nematode, where particular founder cells can be traced by ancestry back to the fertilized egg, the cases of developmental equivalence that have been described to date involve groups of cells which are either lineally homologous or closely related. For example, the o and p blast cells of the leech embryo are the daughters of a sister pair of lineally equivalent O/P teloblasts, and many of the developmentally equivalent cell pairs in the nematode arise at identical points on bilaterally homologous branches of the overall lineage tree (Kimble, 1981; Sulston et al., 1983).

Homologous sublineages such as those which found different segments or halves of a bilaterally symmetrical structure very likely arose in some ancestral species by the duplication of an embryonic founder cell. Founder cell interactions which lead to a diversification of elements within such homologous sublineages could then be selected for secondarily. However, in the absence of a full understanding of the biochemical basis of developmental commitments, one must entertain the possibility that it may be inherently unstable for pairs of cells which are in intimate communication to share certain states of commitment. If this were the case, an embryonic cell duplication could entail the interaction and subsequent divergence of two initially equivalent sister cell lines. In this scenario the occurence of the duplication event itself would automatically engender the appearance of a novel cell type.

ALTERNATE LINES OF DESCENT FOR DIFFERENTIATED PATTERN ELEMENTS

The observation that an o/p blast cell which has undergone one or two partial commitments to the O pathway can still be induced by neighbor cell ablation to generate P pattern elements demonstrates that specific differentiated pattern elements can arise from at least two alternate sequences of embryonic cell division, one involving a purely p-type blast cell division pattern and the other including at least a few distinctively o-type divisions. The existence of alternate lines of descent would be readily explicable if uncommitted o blast cell daughters, such as cell o.ap in Fig. 5, are homologous to particular p blast cell daughters and therefore able to recapitulate a branch of the normal p blast cell lineage in response to neighbor cell ablation. This model would predict that those subsets of P pattern elements produced by a partially committed o blast cell clone should be the same as those produced by specific daughter cells in the p blast cell lineage, a prediction which can be tested experimentally by lineage tracer analysis of the descendant fates of p blast cell daughters.

The results summarized in Fig. 6 show this hypothesis to be incorrect. During normal development the primary p blast cell divides to yield two daughter cells, p.a and p.p, which give rise to complementary subsets of the six unambiguous P pattern elements. These subsets are not the same as those produced by a partially committed o blast cell clone following p bandlet ablation. For example, an o blast cell clone which has undergone two partial commitments to the O pathway will give rise to neuron pz7 and cell floret 3, but not any of the other unambiguous P pattern elements, following p bandlet ablation. These same two pattern elements are segregated at the very first division of the p blast cell lineage, revealing that the partially committed o blast cell clone is not recapitulating a single branch of the p blast cell lineage.

These findings indicate that the sequence of blast cell divisions which comprise normal development is not a prerequisite for the formation of specific groups of differentiated cell types. This conclusion is not surprising when considered from a phylogenetic viewpoint, since the alternative proposition--that each type of differentiated cell can only arise via a particular sequence of founder cell divisions--would render developmental cell lineages largely refractory to evolutionary diversification. Cell lineage descriptions of different species of nematode (Sternberg and Horvitz, 1981, 1982), as well as a comparative analysis of annelid embryology (Anderson, 1973), reveals that phylogeny has entailed considerable changes in species-specific patterns of cell lineage. One must therefore conclude that no matter how invariant a cell lineage may appear when studied in a single species, it possesses a functional plasticity sufficient to accomodate phylogenetic adaptation.

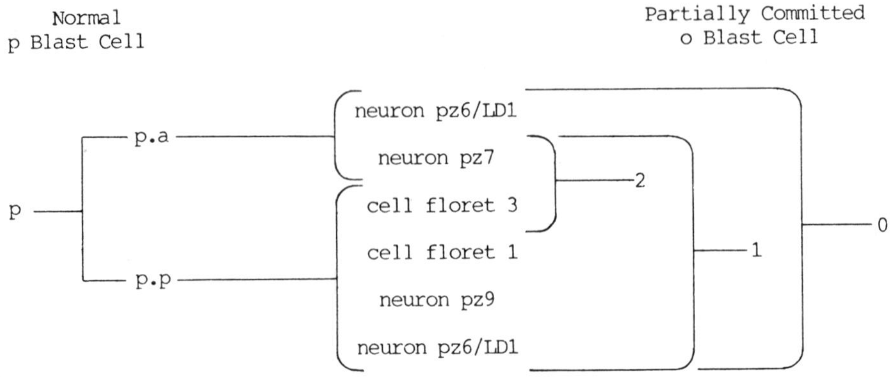

Fig. 6. Clonal relationship of six unambiguous P pattern elements. On the left is shown the normal segregation of pattern elements at the first division of the primary p blast cell. On the right is shown the subsets of P pattern elements which are produced in response to p bandlet ablation by an o blast cell clone which has undergone 0, 1 or 2 partial commitments to the O pathway. The two patterns of segregation are incongruent, indicating that the partially committed o blast cell clone does not produce P pattern elements by simply recapitulating a single branch of the normal p blast cell lineage. This finding suggests that the genealogical relationship between P pattern elements differs depending upon whether these elements are generated by their normal or alternate lines of descent.

The fact that a partially committed o blast cell clone will give rise to sets of P pattern elements which overlap but are not congruent with those produced by individual p blast cell daughters suggests an alteration in the genealogical relationship of the various P pattern elements. Thus, cell floret 3 and neuron pz7 would seem to be more closely related to one another than to other P pattern elements in one line of descent, and but distantly related in the other.

This conclusion may help to serve as a conceptual link between the development of the leech embryo and that of vertebrates and insects. Fate-mapping experiments performed in these other organisms have revealed that their developmental cell lineages vary considerably between individuals of the same species, with the result that neighboring elements of the mature body pattern do not bear a fixed genealogical relationship (Garcia-Bellido and Merriam, 1969; Crick and Lawrence, 1975; Jacobson, 1981). Such observations have long been cited as evidence that these other embryos must possess spatial patterning mechanisms which can bring about the production of a stereotyped array of differentiated cell types independent of the genealogical relationships between cells. If one accepts that an o/p blast cell is able to produce recognisable P pattern elements even though the genealogical relation between those elements has been changed, then one interpretation would be that the leech embryo possesses similar patterning mechanisms which only become evident when the normally regular sequence of embryonic cell divisions is experimentally perturbed.

ACKNOWLEDGEMENTS

The research described in this article was supported by National Research Service Award 5 F32 NS06814-02, research grants NS12818 and HD17088 from the National Institutes of Health, BNS79-12400 from the National Science Foundation, I-738 from the March of Dimes Birth Defects Foundation, and the Rowland Foundation.

REFERENCES

Anderson, D.T., 1973, "Embryology and phylogeny in Annelids and Arthropods," Pergamon Press, Oxford.

Blair, S.S., 1982, Interactions between mesoderm and ectoderm in segment formation in the embryo of a glossiphoniid leech, Dev. Biol., 89:389-396.

Blair, S.S., 1983, Blastomere ablation and the developmental origin of identified monoamine-containing neurons in the leech, Dev. Biol., 95:65-72.

Blair, S.S. and Weisblat, D.A., 1984, Cell interactions in the developing epidermis of the leech Helobdella triserialis, Dev. Biol., 101:318-325.

Conklin, E.G., 1905, The organization and cell lineage of the ascidian egg, J. Acad. Nat. Sci. Philadelphia, 13:1-119.

Crick, F.H.C., and Lawrence, P.A., 1975, Compartments and polyclones in insect development, Science, 189:340-347.

Fernandez, J., 1980, Embryonic development of the glossiphoniid leech Theromyzon rude: characterization of developmental stages, Dev. Biol., 76:245-262.

Garcia-Bellido, A., and Merriam, J.R., 1969, Cell lineage of the imaginal discs in Drosophila gynandromorphs, J. Exp. Zool., 170: 61-67.

Gimlich, R.L., and Braun, J., 1985, Improved fluorescent compounds for tracing cell lineage, Dev. Biol., in press.

Goodman, C.S., Bastiani, M.J., Doe, C.Q., du Lac, S., Helfand, S.L., Kuwada, J.Y., and Thomas, J.B., 1984, Cell recognition during neuronal development, Science, 225:1271-1279.

Illmensee, K., and Mahowald, A.P., 1974, Transplantation of posterior polar plasm in Drosophila. Induction of germ cells at the anterior pole of the egg, Proc. Nat. Acad. Sci. USA, 73: 549-553.

Jacobson, M., 1981, Rohon-Beard neuron origin from blastomeres of the 16-cell frog embryo, J. Neurosci., 1:923-927.

Kimble, J., 1981, Alterations in cell lineage following laser ablation of cells in the somatic gonad of Caenorhabditis elegans, Dev. Biol., 87:286-300.

Kramer, A.P., and Weisblat, D.A., 1985, Developmental neural kinship groups in the leech, J. Neurosci., in press.

Nicholls, J.G. and Baylor, D.A., 1968, Specific modalities and receptive fields in CNS of the leech, J. Neurophysiol., 31:740-756.

Nicholls, J.G. and Purves, D., 1970, Monosynaptic chemical and electrical connexions between sensory and motor cells in the central nervous system of the leech, J. Physiol., 209:647-667.

Schleip, W., 1936, Ontogenie der Hirudineen, in: "Klassen und Ordnungen des Tierreichs," Part 2, H.G. Bronn, ed., Akad. Verlagsgesellschaft, Leipzig, pp. 1-121.

Shankland, M., 1986, Positional interaction of embryonic founder cells during the development of the leech nervous system, Current Topics in Dev. Biol., in press.

Shankland, M. and Weisblat, D.A., 1984, Stepwise commitment of blast cell fates during the positional specification of the O and P cell lines in the leech embryo, Dev. Biol., 106:326-342.

Stent, G.S., Weisblat, D.A., Blair, S.S., and Zackson, S.L., 1982, Cell lineage in the development of the leech nervous system, in: "Neuronal Development," N.C. Spitzer, ed., Plenum Press, New York, pp. 1-44.

Sternberg, P.W. and Horvitz, H.R., 1981, Gonadal cell lineages of the nematode Panagrellus redivivus and implications for evolution by the modification of cell lineage, Dev. Biol., 88: 147-166.

Sternberg, P.W. and Horvitz, H.R., 1982, Postembryonic nongonadal cell lineages of the nematode Panagrellus redivivus: description and comparison with those of Caenorhabditis elegans, Dev. Biol., 93:181-205.

Stuart, D.K., Glover, J.C., Blair, S.S., and Weisblat, D.A., 1983, Development of leech serotonin neurons examined with serotonin antibody, cell lineage tracer and cell killing, Soc. Neurosci. Abstr., 9:605.

Sulston, J.E., and Horvitz, H.R., 1977, Postembryonic cell lineages of the nematode Caenorhabditis elegans, Dev. Biol., 56:110-156.

Sulston, J.E., Schierenberg, E., White, J.G., and Thomson, J.N., 1983, The embryonic cell lineage of the nematode Caenorhabditis elegans, Dev. Biol., 100:64-119.

Sulston, J.E., and White, J.G., 1980, Regulation and cell autonomy during postembryonic development of Caenorhabditis elegans, Dev. Biol., 78:577-597.

Taghert, P.H., Doe, C.Q. and Goodman, C.S., 1984, Cell determination and regulation during development of neuroblasts and neurones in grasshopper embryos, Nature, 307:163-165.

Weisblat, D.A. and Blair, S.S., 1984, Developmental indeterminacy in embryos of the leech Helobdella triserialis, Dev. Biol., 101: 326-335.

Weisblat, D.A., Harper, G., Stent, G.S. and Sawyer, R.T., 1980a, Embryonic cell lineage in the nervous system of the glossiphoniid leech Helobdella triserialis, Dev. Biol., 76:58-78.

Weisblat, D.A., Kim, S.Y. and Stent, G.S., 1984, Embryonic origin of cells in the leech Helobdella triserialis, Dev. Biol., 104:65-85.

Weisblat, D.A., Sawyer, R.T. and Stent, G.S., 1978, Cell lineage analysis by intracellular injection of a tracer enzyme, Science, 202:1295-1298.

Weisblat, D.A., Zackson, S.L., Blair, S.S. and Young, J.D., 1980b, Cell lineage analysis by intracellular injection of fluorescent tracers, Science, 209:1538-1541.

Whitman, C.O., 1878, The embryology of Clepsine. Quart. J. Micro. Sci. 18:215-315.

Whittaker, J.R., 1980, Acetylcholinesterase development in extra cells by changing the distribution of myoplasm in ascidian embryos, J. Embryol. Exp. Morph., 55:343-354.

Zackson, S.L., 1984, Cell lineage, cell-cell interaction, and segment formation in the ectoderm of a glossiphoniid leech embryo, Dev. Biol., 104:143-160.

Zur Strassen, O., 1896, Embryonalentwickelung der Ascaris megalocephala, Arch. Entwicklungsmech. Org., 3:27-105 & 133-190.

STRUCTURE AND REGULATED TRANSCRIPTION OF DIRS-1, A NOVEL *DICTYOSTELIUM DISCOIDEUM* TRANSPOSABLE ELEMENT

Joe Cappello, Stephen M. Cohen, Karl Handelsman, and Harvey F. Lodish

Whitehead Institute for Biomedical Research
Nine Cambridge Center, Cambridge, MA 02142
and Department of Biology, Massachusetts Institute of Technology, Cambridge, MA 02139

I. Introduction:

Transposons and transposon- and retrovirus-like DNA segments form a major part of the repetitive DNA in many eukaryotes. They are a cause of many "spontaneous" mutations in *Drosophila* and other organisms, and many catalyze the rearrangements of large segments of DNA that have occurred during evolution (Spradling and Rubin, 1981; Kleckner, 1981; Flavell et al., 1981; Roeder and Fink 1983; Fedoroff, 1983). In the course of cloning segments of *Dictyostelium discoideum* genomic DNA that are expressed preferentially during early differentiation, we isolated the DIRS-1 transposon (Chung et al., 1983; Zuker and Lodish, 1981). DIRS-1 is unusual in that its transcription is induced either by heat shock or by the high cell density that occurs during the early stages of differentiation. It is also novel in that its structure is unlike any transposon in prokaryotes or eukaryotes that has been identified to date. DIRS-1 has been independently isolated by Rosen *et al.* (1983) who call it Tdd-1.

II. Structure of DIRS-1.

The DIRS-1 genomic family has several structural features in common with well-characterized prokaryotic and eukaryotic transposable elements. There are multiple copies of DIRS-1-related sequences in the genome. The major component of these (about 40 copies) is a homologous 4.7 kb element which consists of 4.1 kb of internal unique sequence flanked by 330 bp inverted terminal repeats (ITRs). The elements are inserted at dispersed genomic locations, and the genomic locations of the elements vary in different Dictyostelium genetic stocks. Elements that are inserted at different chromosomal locations have similar terminal repeats and internal 4.1 kb segments (Chung et al., 1983).

DIRS-1 is an unusual eukaryotic transposon because it has long terminal repeats - 330 nucleotides - that are inverted (Fig. 1) (Zuker et al., 1984). The great majority of eukaryotic transposons and transposable-like elements (e.g. retroviruses) has direct terminal repeats (Spradling and Rubin, 1981; Majors et al., 1980). Only four eukaryotic transposable elements are known to bear inverted terminal repeats: the P and FB elements of Drosophila (Spradling and Rubin, 1981; O'Hare and Rubin, 1983), the σ element of yeast (Del Rey et al., 1982), and the Tc1 element of Caenorhabditis elegans (Rosenzweig and Hirsh, 1983). The inverted repeats flanking P and σ elements are very short (31 or 15 bp and 8 bp, respectively), whereas those at the ends of FB contain periodic tandem repeats of a 10-, a 20-, and a 31-bp sequence (Potter, 1982). The inverted terminal repeats of Tc1 are 54 bp in length (Rosenzweig and Hirsh, 1983). DIRS-1 is also unusual in that the lengths of the left and right repeats differ by 28 bp; to our knowledge, no other transposon has left and right repeats of different lengths.

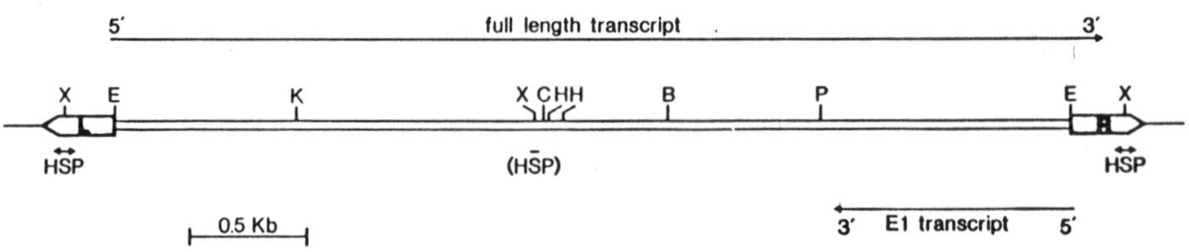

Fig. 1. Structure of DIRS-1 and its transcripts. This restriction map is a consensus of 6 cloned DIRS-1 elements. DIRS-1 is 4.7 kb in length and contains 330 bp inverted terminal repeats (shaded boxes). The unique internal 4.1 kb sequence is indicated by the open box. The terminal repeats contain heat shock promoters (HSP). The heat shock promoter in the left repeat is responsible for directing rightward transcription of the full length 4.5 kb RNA transcript. The HSP in the right terminal repeat directs leftward transcription of the E1 RNA. The extent and direction of these transcripts are indicated by arrows above and below the element. An internal heat shock promoter has been shown to be

functional in yeast (unpublished data). Restriction enzymes: B, BglII; C, ClaI; E, EcoRI; H, HindIII; K, KpnI; P, PvuII; and X, XbaI.

Both the left and right repeats flanking a single DIRS-1 element frequently contain nucleotide sequence alterations that are found in no other cloned copies of the repeats that we have sequenced; this suggests that the left and right repeats of a DIRS-1 element may be copied or corrected from a single DNA sequence during transposition (Zuker et al., 1984). The direct repeats of TY-1 and *copia* elements may be copied or corrected by an analogous process (Eibel et al., 1980).

III. Insertion of DIRS-1.

The majority of DIRS-1 related sequences in the genome are not part of an intact 4.7 kbp DIRS-1 element (Chung et al., 1983). Most of these vary in size and pattern of restriction enzyme digestion from the canonical DIRS-1. It is possible that some of these fragments were derived by deletions within DIRS-1. However, recent data suggest that many of these partial DIRS-1 sequences are generated by a strikingly novel preference for insertion sites exhibited by the transposition of DIRS-1: DIRS-1 inserts preferentially into itself but the target site within DIRS-1 is more-or-less random (Cappello et al., 1984a).

To establish this, the structure of several genomic clones containing intact 4.7 kb DIRS-1 elements was examined. Except for Cp19-5 (not shown in Fig. 2) all five cloned genomic DIRS-1 elements are flanked by DIRS-1 internal sequences (Fig. 2). Cloned cDNA pCCA5 is a partial transcript of an apparent DIRS-1 into DIRS-1 insertion (Cappello et al., 1984a). The 50 to 100 nucleotides immediately flanking the terminal repeats of each cloned DIRS-1 element (except one) are 80 to 100% homologous with internal sequences of DIRS-1. By blot hybridization, the homology of the flanking regions to DIRS-1 was shown, in several cases, to be longer than these immediate flanking nucleotides (Zuker, 1983).

In one clone, SB41, the DIRS-1-related flanking sequences are those that would result from a precise DIRS-1-into-DIRS-1 insertion: the EcoRI fragments flanking the intact DIRS-1 element in SB41 equal the total 4.1-kb internal EcoRI fragment of DIRS-1 (Cappello et al., 1984a).

At least for the DIRS-1-into-DIRS-1 insertion in SB41, there is no duplication of the target nucleotide

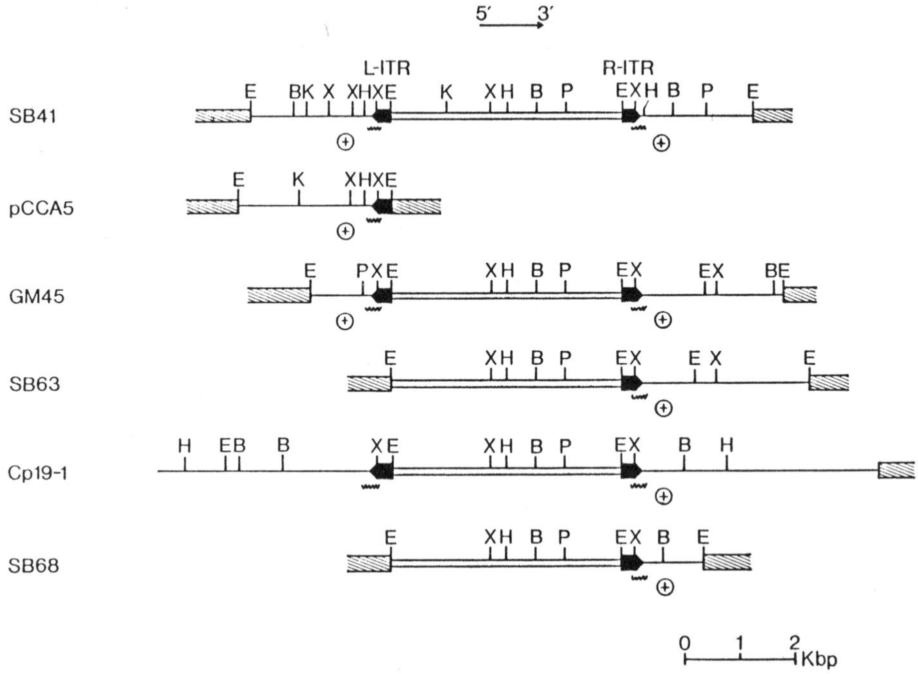

Fig. 2. Structure of DIRS-1 clones. The diagram displays the locations of the DIRS-1 flanking sequences homologous to DIRS-1. The boxed regions of each clone are the intact DIRS-1 elements. L-ITR and R-ITR refer to the positions of the ITRs at both ends of DIRS-1. The wavy lines below the restriction maps denote the positions of the flanking nucleotides we have sequenced that are homologous to internal DIRS-1 sequences. Additional segments of flanking regions which have been shown to be homologous to DIRS-1 by hybridization are indicated by a + below the maps.

sequence at either side of the inserted DIRS-1 element. Flanking sequences on both sides of the DIRS-1 element can be rejoined and aligned perfectly with the internal sequence of DIRS-1. The alignment accounts for every nucleotide present in the DIRS-1 sequence before insertion except one. This possible deletion of a T residue could be explained merely by an ambiguity in the definition of the final residue of the terminal-repeat sequences, which are also T's. The transposable elements Tc1 from _C. elegans_ and Tn554 from _S. aureus_ also transpose without an apparent duplication of host sequences at the insertion site (Rosenzweig et al., 1983; Murphy and Lofdahl, 1984).

The locations of the flanking-region homologies, which define the proposed sites of insertion in each of our clones, are distributed throughout DIRS-1 (Fig. 3). This result indicates that within the DIRS-1 sequence there is no absolute preference for insertion sites. Also, because the extent of DIRS-1 sequence in each of the flanking regions is variable, ranging from a full, uninterrupted complement of DIRS-1 sequences in SB41 to probably no more than a few hundred bases on either side of the intact DIRS-1 element in another genomic clone, CP19-1, it is possible that a preexisting DIRS-1-related sequence need not be a complete DIRS-1 in order function as a target for insertion by a second DIRS-1.

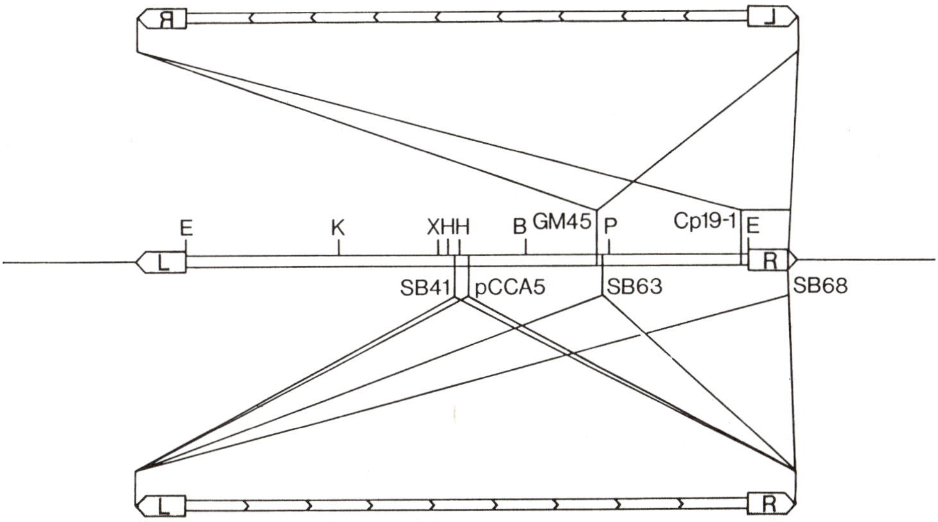

Fig. 3. Model for the generation of the sequences present in cloned DIRS-1 containing genomic segments. Intact DIRS-1 elements are flanked by internal DIRS-1 sequences. We interpret this as the result of an insertion of an intact DIRS-1 element into a second DIRS-1 or DIRS-1-related sequence. This is illustrated for five cloned genomic DIRS-1 sequences and for the cDNA clone pCCA5. The position along the DIRS-1 element denoted by the homology present in the flanking sequences is the putative point of insertion of the incoming element. In this model, the preexisting target element (the center element of the

(continued)

diagram) becomes the DIRS-1-related flanking sequences on both sides of the incoming element (those elements depicted above and below). As a generalization, the preexisting target sequence is depicted as an intact element. However, in at least two cases (GM45 and Cp19-1), the DIRS-1-related sequences flanking the intact element are not those expected from insertion at at single point in an intact DIRS-1 element. Only clone SB41 is known to contain both parts of the intact 4.1 kb EcoRI fragment of DIRS-1 flanking the central element. The insertion events associated with clones GM45 and Cp19-1 are depicted as inversions, that is, the incoming element is inverted with respect to the DIRS-1-related sequence at the target site. Those events depicted for SB41, pCCA5, SB63 and SB68 are depicted as colinear insertions.

The orientation of the incoming element can be either colinear or inverted with respect to the preexisting DIRS-1-related sequence. The DIRS-1-related flanking sequences in four cases are in the same orientation as the central complete DIRS-1 element. In the remaining two cases they are inverted with respect to the central complete element.

Several hypotheses can explain the preference of DIRS-1 for self-insertion. The simplest involves the recognition of internal DIRS-1 sequences as insertional "hot spots". Since the exact location of insertion sites within a DIRS-1 sequence can vary, a level of recognition more complex than the actual nucleotide sequence must be involved. The secondary structure of an integrated element, possibly in conjunction with specifically bound proteins, might serve as a recognition signal. It is possible, alternatively, that a specific nucleotide sequence in DIRS-1 is initially recognized by the incoming element; this could define a region into which the incoming element might insert that could span several kilobases of target DIRS-1 sequence. A possible analogy is provided by the KB restriction system of *Escherichia coli*. The K and B restriction nucleases recognize a specific target sequence within the DNA, but, after binding to the DNA, move a random distance along it before cleaving (Yvan, 1981).

IV. DIRS-1 transcription

A rather heterogenous set of polyadenylated transcripts of DIRS-1-related sequences accumulate in response to a variety of cellular stresses (Fig. 4) (Zuker et al., 1983). DIRS-1 transcripts are almost undetectable in low density, exponentially growing cells. Their abundance increases very rapidly after the cells are plated for development. The trigger for this is the high cell density required for differentiation (Zuker et al., 1983). Growing cells express DIRS-1 RNAs if subjected to a heat shock, or if grown to very high cell densities. Individual species of DIRS-1 RNAs that are induced very early in development (from 0.5 to 2.5 hr after plating) are also induced by heat shock (Zuker et al., 1983). Both during differentiation and heat shock, the increase in level of DIRS-1 RNAs is due, in part, to a four- to fivefold increase in the rate of RNA transcription (Zuker, 1983). The maximum accumulation of DIRS-1 RNA can be achieved by any single stress; it is not enhanced by the application of a second stress condition (i.e. development and heat shock together).

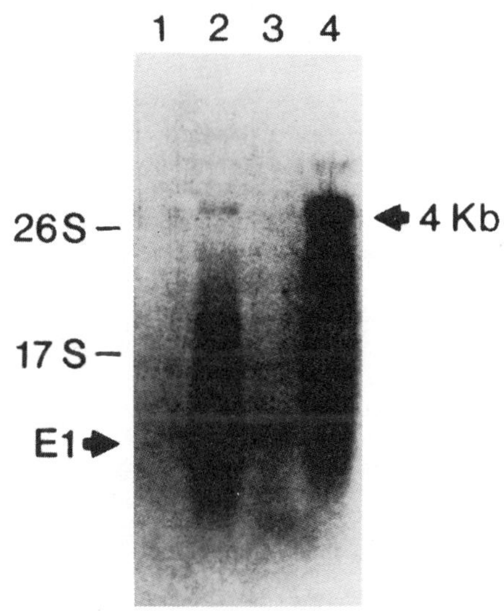

Fig. 4. Developmental regulation of DIRS-1 RNAs. RNA filter hybridization of Dictyostelium discoideum RNAs with labeled DIRS-1 DNA. Four micrograms of poly (A)$^{+}$ RNA isolated from vegetative cells (lane 1) heat shock-treated vegetative cells (lane 2) and 1 hour (lane 3) and 15 hour (lane 4) filter-developed cells were size fractionated on a 1% agarose-formaldehyde gel and transferred to Gene

(continued)

Screen. The filter was hybridized for 48 hour at 65C with 10- cpm of nick-translated DNA from clone pB41.6 per ml (Zuker and Lodish, 1981). The positions of the ribosomal RNAs are indicated as 26S and 17S. The positions of the ca. 4.5 kb and E1 RNAs are indicated.

Transcription of DIRS-1 and related sequences generates two predominant discrete polyadenylated RNA species: a "full length" 4.5 kb transcript and E1 (Zuker et al., 1983). As depicted in Fig. 1, E1 is transcribed from right to left on the "standard" DIRS-1 (Zuker et al., 1984; Cohen et al. 1983). In contrast, the majority of DIRS-1 RNAs including the 4.5 kb transcript are transcribed from the opposite strand, from left to right (Fig. 1). As is discussed below, the 4.5 kb RNA is transcribed from an intact DIRS-1 element.

In several respects E1 is unusual. It is transcribed exclusively during early development and, unlike the majority of DIRS-1 RNAs, reaches its maximal accumulation by 1 hour and is almost undetectable by 2 hours of development (Zuker et al., 1983). E1 is transcribed from the opposite strand than are the other DIRS-1 RNAs, initiating near the right terminal repeat and extending leftward into DIRS-1 (Cohen et al., 1984). In addition, transcription of E1 is induced to a much greater extent by heat shock than are other DIRS-1 RNAs (Fig. 4). The reasons for the differences in the regulation of E1 and the other DIRS-1 RNAs are not known. In particular, we do not know whether E1 is transcribed from complete copies of DIRS-1 or from some of the large number of partial copies present in the genome. As noted above, other heterogeneously sized DIRS-1 related RNAs may also be transcripts of partial copies of DIRS-1. However, we cannot exclude the possibility that some of these may be processed from the "full length" 4.5 kb transcript. Most cytoplasmic DIRS-1 RNAs are polyadenylated and are associated with polysomes, suggesting that they may encode protein(s) (Zuker, 1983). However, the longest reading frame in E1 is only 294 bases, suggesting that E1 is not translated (see below).

As is the case for many developmentally induced *Dictyostelium* mRNAs (Mangiarotti et al., 1982), the expression of DIRS-1 RNAs is regulated at the levels of RNA transcription and also at the stability of the transcripts. DIRS-1 transcripts are in very low abundance in vegetative cells grown at low density, but appear immediately upon plating at high density for development. They continue to accumulate throughout differentiation. DIRS-1 transcription rates, determined by in vitro run off transcription in isolated nuclei, indicate that DIRS-1 transcription increases 4 to 5 fold both in early developing cells and in heat shocked cells (Zuker, 1983). Some DIRS-1 transcripts are synthesized in vitro by nuclei isolated from vegetative cells; however proportionally much less DIRS-1 RNA is found in the polyadenylated nuclear or cytoplasmic RNA isolated from growing cells. Thus DIRS-1 RNAs transcribed in growing cells may be rapidly degraded. Alternatively, the apparent transcription of DIRS-1 RNA in nuclei from growing cells may reflect an artifactual induction of DIRS-1 transcription caused by stress to the cells during the preparation of nuclei.

The initial burst of DIRS-1 transcription at about 3 hours of development is followed by a drop to the approximate rate observed in vegetative cells. However, cytoplasmic DIRS-1 RNAS continue to accumulate to increasing levels until about 15 hours of development. This accumulation is due, in large part, to an increase in the processing or stability of DIRS-1 transcripts.

Processing and nuclear transit times of DIRS-1 transcripts in developing cells was assayed by continuous labeling of cells with [P]PO$_4$ followed by measuring the accumulation of DIRS-1-related poly (A^+) and poly (A^-) RNAs in the nucleus and the cytoplasm. Results indicate that polyadenylation of DIRS-1 RNA occurs within 45 min after synthesis, as is the case with other *Dictyostelium* RNAs (Mangiarotti et al., 1983). However, DIRS-1 RNAs begin to accumulate in the cytoplasm only after a lag of 135 min. This nuclear transit time is much longer than that of constitutively expressed mRNAs or most other developmentally induced *Dictyostelium* RNAs, which range from 25 to 60 minutes, and from 50 to 100 minutes, respectively. DIRS-1 RNAs are also unusual in that only 25% of the nuclear transcripts appear in the cytoplasm. The remaining 75% are apparently degraded within the nucleus, or immediately after export to the cytoplasm (Mangiarotti et al., 1983). The processing efficiency of most other *Dictyostelium* mRNAs ranges from 60 to 70%, although some RNAs unrelated to DIRS-1 are processed at about 20% efficiency.

Transcription and processing of DIRS-1 RNAs has several features in common with those of the *Drosophila* copia-like transposable elements (Stanfield and Lengyel, 1980; Falkenthal et al., 1981). Of copia transcripts 90% are also rapidly degraded in the nucleus. The 10% of poly (A^+) copia RNA that is exported to the cytoplasm is stable. In both cases the reasons for the low efficiency of processing and transport of nuclear RNAs is not known.

V. Sites of transcriptional initiation

Using S-1 nuclease mapping and primer extension we have mapped several sites within DIRS-1 where transcription initiation occurs. These are located at or near the junctions of both inverted terminal repeats with the internal sequence (Cohen et al., 1984).

Transcription of the majority of the DIRS-1-related RNAs is directed inward (rightward) from the left repeat. Transcription initiates predominantly at a site 9 bases upstream from the center of the EcoRI site that defines the end of the terminal repeat (Cohen et al., 1984). This initiation site is located within a short oligo(dT) sequence. An oligo(dT) sequence preceding transcriptional initiation sites is a characteristic feature of *Dictyostelium* protein-coding genes and may be required for precise initiation (Kimmel and Firtel, 1983). A sequence identical to the *Dictyostelium* consensus TATA box begins at positions -27.

Initiation of transcription of E1 RNA is more heterogeneous than that of the 4.5-kb RNA: E1 transcripts initiate over a range of 300 bases (Cohen et al., 1984). The longest E1 RNAs initiate near the boundary of the right ITR and the internal DIRS-1 sequence. This position in the right ITR is analogous to that of the initiation of 4.5 kb transcript in the left ITR, and may be directed by a TATA box located at position -42 within the right terminal repeat. The sequence of the first 120 nucleotides downstream (leftward) from the EcoRI restriction site defining the right ITR is unusual in that it contains a 100-base stretch consisting almost entirely of adenine and thymine residues. Consequently, other prefect matches for the consensus TATA box are located within this sequence.

Thus the locations of the sites for initiation of transcription support the idea that DIRS-1 contains within its terminal repeats promoters or enhancers for RNA synthesis. All ITRs contain a 25 base pair sequence which possesses two tandem and near-perfect matches with the consensus *Drosophila* heat shock promoter sequence (Zuker et al., 1984; Cohen *et al.* 1984; see below). We presume that this putative heat shock promoter sequence in DIRS-1 is responsible for regulating DIRS-1 transcription during both heat shock and development in *Dictyostelium*. This suggestion is supported by the finding that the same transcription initiation sites are used during both heat shock and developmental induction of DIRS-1 transcription (Cohen et al., 1984).

VI. DIRS-1 transcription in yeast.

Although the major initiation sites of DIRS-1 transcription are located an appropriate distance downstream from the putative promoters contained in the terminal repeats, the actual role of these sequences in the transcription of DIRS-1 is difficult to assess in *Dictyostelium*. For this reason, we transformed *S. cerevesiae* (yeast) cells with a plasmid containing an isolated terminal repeat (Cappello et al., 1984). This ITR induced transcription of flanking DNA sequences; as in *Dictyostelium*, transcription increased several-fold after a heat shock. An essential component of the DIRS-1 heat shock promoter, identified by deletion mapping, is the 18 base pair palindromic sequence (Fig. 5) located 120 base pairs upstream of the initiation sites of transcription. Each half of this palindrome contains an apparent functional heat shock promoter, as judged by the similarity of 9 of 10 and 10 of 10 bases with the consensus *Drosophila* heat shock promoter (Fig. 5; Zuker et al., 1983; Cohen et al., 1984; Pelham 1982; Pelham and Bienz 1982). Deletion of the entire palindrome (mutant A X18 and B X29) abolished heat shock inducible promoter activity. Deletions which remove either half of the palindrome, leaving intact only a single copy of the consensus heat shock promoter, achieved about one third of the normal heat shock induction (Fig. 5, mutants B RX34 and B XH41). These results establish that the 18 base palindrome containing two near-perfect matches with the *Drosophila* consensus heat shock promoter is an upstream activator of transcription that responds to heat shock. These data suggest that both halves of the palindrome function additively to produce full heat shock induction. Whether this cooperativity is due to the structural symmetry of the palindrome or to the divalence of the promoter sequence is not clear.

Although both terminal repeats contain functional heat shock promoters, transcription of the intact DIRS-1 element in yeast occurs almost exclusively from the left promoter (unpublished data). As in *Dictyostelium*, a transcript of 4.5 kb is produced which spans most of the element. In contrast to *Dictyostelium*, very little E1-like RNA (transcription inward from the right terminal repeat promoter) is evident. Clearly the transcription of DIRS-1 is complex and involves more than merely the function of the heat shock promoters that are contained in both

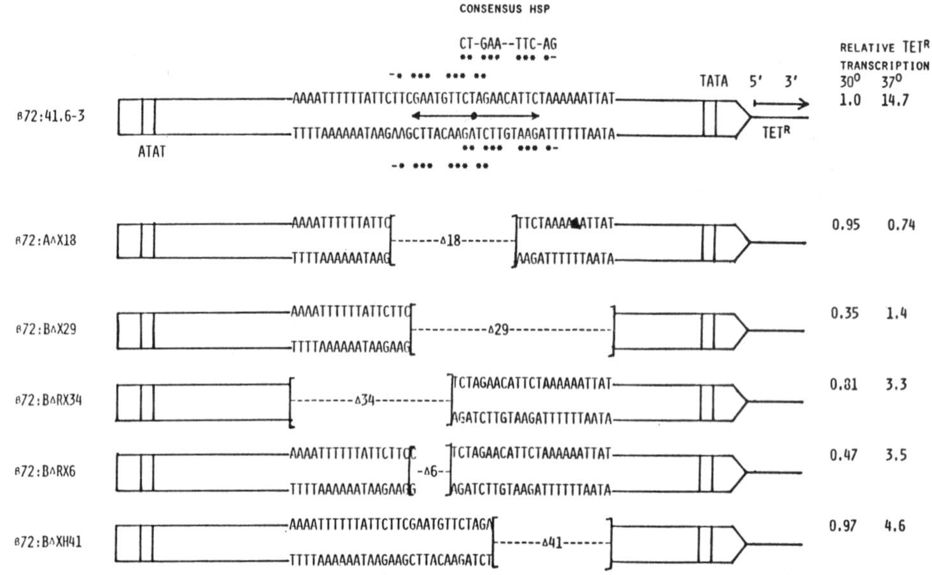

Fig. 5. Organization of the DIRS-1 terminal repeat. A schematic representation of the organization of the heat shock promoter sequences and TATA box sequences contained in the terminal repeats is presented. The consensus sequence of the region of the terminal repeat containing the *Dictyostelium* heat shock promoters is shown. Homology to the *Drosophila* consensus heat shock promoter (HSP) "CT-GAA--TTC-AG" is indicated by asterisks. Dashes indicate mismatched nucleotides. Two tandem overlapping copies of the HSP spanning the 18 bp palindrome (indicated by divergent arrows) are present on each strand of the terminal repeat. The positions of TATA boxes at each end of the repeat are indicated. Previously, we showed that in yeast the right ITR of clone SB41 was both capable of and sufficient for directing the heat shock inducible transcription of plasmid flanking sequences - in this case the tetracycline resistance gene of pBR322 (Cappello et al., 1984). In order to identify the sequences reponsible for this regulation, we generated a series of deletions of the ITR. By S-1 nuclease protection and by Northern blot analysis, we assayed

their ability to regulate transcription of the Tet^R gene in yeast (unpublished data). The wild type DIRS-1 ITR (72:41.6-3) and five of the deletion constructs are shown. The relative accumulation of transcripts with respect to the wild type ITR 30 C level was measured both at 30 C and after a 1 hour heat shock at 37 C. HIS3 RNA which remains constant throughout the one hour heat shock (unpublished data) was used as an internal control for the amount of total RNA per sample.

left and right inverted terminal repeats. Although a full understanding of DIRS-1 transcription will eventually arise only from its study in Dictyostelium, the functional data acquired from the mapping of these promoters in yeast cells has provided the first definition of a Dictyostelium promoter sequence.

VII. Nucleotide sequence of DIRS-1

As mentioned above, DIRS-1 RNAs are polyadenylated and associated with polysomes (Zuker, 1983). These observations are consistent with the notion that at least some DIRS-1 transcripts are mRNAs that encode polypeptides. These protein products might function in the transposition of DIRS-1. In order to characterize these products, we have determined the complete nucleotide sequence of DIRS-1.

The unique internal sequence of DIRS-1 (between the EcoRI sites) is 4158 base pairs long and contains three long open reading frames (ORF's; Fig. 6)(J. Cappello, K. Handelsman, and H.F. Lodish, in preparation). ORF-1 extends through the first 1150 bases of DIRS-1 and could encode a polypeptide of 38,000 daltons. ORF2 overlaps the final 170 nucleotides of ORF1 and extends for 2800 bases. Using the first ATG codon of ORF2 for initiation, a translation product of over 100,000 daltons is predicted.

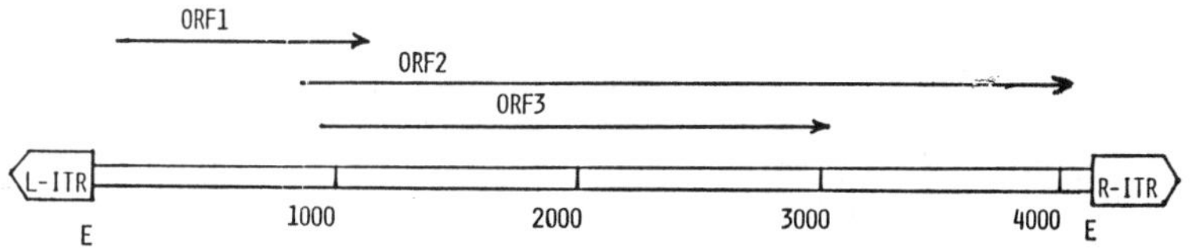

Fig. 6. DIRS-1 open reading frames. The nucleotide sequence of the internal 4.1 kb EcoRI fragment of DIRS-1 is diagrammed. The extent of (continued)

all open reading frames (ORF) longer than 100 amino acids is shown. The direction of transcription of the 4.5 kb DIRS-1 RNA is from left to right.

The most surprising and remarkable feature of the sequence of DIRS-1 is ORF3. This reading frame also overlaps the final 55 nucleotides of ORF1, extends for 2000 bases, and could encode a polypeptide of about 69,000 daltons. Consequently, ORF3 completely and codirectionally overlaps ORF2. For more than 2000 bases, DIRS-1 potentially encodes two different polypeptides simultaneously using one nucleotide sequence. To our knowledge, the utilization of the coding potential of the genetic material in this way is unprecedented.

The polarity of all three open reading frames of DIRS-1 is such that they could be encoded by the 4.5 kb RNA. No open reading frames longer than 100 amino acids are encoded on the opposite strand; thus it is doubtful that the DIRS-1 transcript E1 encodes a protein. Because ORF1 initiates at the first AUG codon from the 5' end of the 4.5 kb transcript, initiation of its translation is straightforward. However, the utilization of ORF2 and 3 would require either the use of an internal AUG codon, or the elimination of the 5' region of the 4.5 kb RNA, including part of ORF1. This could be accomplished by RNA splicing or by transcriptional initiation within DIRS-1. Evidence for or against both of these possibilities is under continued investigation.

By using antibodies prepared against the hypothetical polypeptides encoded by ORF1, 2 and 3, we are trying to detect these proteins in *Dictyostelium*, and also in DIRS-1-transfected yeast. A functional assay for DNA transposition developed in yeast containing DIRS-1 sequences will facilitate the study of the DIRS-1 translation products and their functions.

The occurrence of the two long overlapping reading frames is totally unexpected. All three long, overlapping open reading frames are found in a second genomic copy of DIRS-1 that we have sequenced (manuscript in preparation). The reason for the conservation of the coding capabilities for two very different polypeptides in such a small space is of great interest. Conservation of these unusual DIRS-1 ORF sequences in *Dictyostelium* might indicate that DIRS-1 transposition or some other function of its protein products is important for growth and/or differentiation of this organism.

References

Cappello, J., Cohen, S. M., and Lodish, H. F., 1984a, Dictyostelium transposable element DIRS-1 preferentially inserts into DIRS-1 sequences, Mol. Cell. Biol., 4:2207-2213.

Cappello, J., Zuker, C., and Lodish, H. F. 1984b, Repetitive Dictyostelium heat-shock promotor functions in Saccharomyces cerevisiae, Mol. Cell. Biol., 4:591-598.

Chung, S., Zuker, C., and Lodish, H. F., 1983, A repetitive and apparently transposable DNA sequence in Dictyostelium discoideum associated with developmentally regulated RNAs, Nucl. Acids Res. 11:4835-4852.

Cohen, S. M., Cappello, J., and Lodish, H. F., 1984, Transcription of Dictyostelium discoideum transposable element DIRS-1, Mol. Cell. Biol., 4:2332-2340.

Del Rey, F. J., Donahue, T. F., and Fink, G. R., 1982, sigma, a repetitive element found adjacent to tRNA genes of yeast, Proc. Nal. Acad. Sci. USA., 79:4138-4142

Eibel, H., Gafner, J., Stotz, A., and Philippsen, P., 1980, Characterization of the yeast mobile element Ty1, Cold Spring Harbor Symp. Quant. Biol., 45:609- .

Falkenthal, S., Graham, M. L., Korn, E. L., and Lengyel, J. A., 1981, Transcription, processing and turnover from the Drosophila mobile genetic element copia, Devel. Biol., 92:294- .

Fedoroff, N. V. 1983, Controlling elements in cell lineages, in: "Mobile Genetic Elements," J. A. Shapiro, ed., Academic Press, New York, pp. 1-57.

Flanell, R. B., O'Dell, M., and Hutchinson, J., 1981, Nucleotide sequence organization in plant chromosomes and evidence for sequence translocation during evolution, Cold Spring Harbor Symp. Quant. Biol., 45:501- .

Kimmel, A. R., and Firtel, R. A., 1983, Sequence organization in Dictyostelium: unique structure at the 5' ends of protein coding genes, Nucl. Acids Res., 11:541-552.

Kleckner, N., 1981, Transposable elements in prokaryotes, Annu. Rev. Genet., 15:341- .

Majors, J. E., Swanstrom, R., DeLorre, W. J., Payne, G. S., Hughes, S. H., Ortiz, S., Quintrell, N., Bishop, J. M., and Varmus, H. E., 1980, DNA intermediates in the replication of retroviruses are structurally (and perhaps functionally) related to transposable elements, Cold Spring Harbor Symp. Quant. Biol., 45:731- .

Mangiarotti, G., Zuker, C., Chisholm, R., and Lodish, H. F., 1983, Different mRNAs have different nuclear transit times in *Dictyostelium discoideum* aggregates, *Mol. Cell. Biol.*, 3:1151-1517.

Mangiarotti, G., Lefebvre, P., and Lodish, H. F., 1982, Differences in the stability of developmentally regulated mRNAs in aggregated and disaggregated *Dictyostelium discoideum* cells, *Devel. Biol.*, 89:82-91.

Murphy, E., and Lofdahl, S., 1984, Transposition of Tn554 does not generate a target duplication, *Nature (London)*, 307:292-294.

O'Hare, K., and Rubin, G. M., 1983, Structures of P transposable elements and their sites of insertion and excision in the *Drosophila melanogaster* genome, *Cell*, 34:25-35.

Pelham, H. R. B., 1982, A regulatory upstream promoter element in the *Drosophila* HSP 70 heat shock gene, *Cell*, 30:517-528.

Pelham, H R. B., and Bienz, M., 1982, A synthetic heat-shock promoter confers heat inducibility on the herpes simplex thymidine kinase gene, *EMBO J.*, 1:1473- .

Potter, S. S., 1982, DNA sequence of a foldback transposabale element in *Drosophila*, *Nature (London)*, 297:201-204.

Roeder, G. S., and Fink, G. R., 1983, Transposable elements in yeast, *in:* "Mobile Genetic Elements," J. A. Shapiro, ed., Academic Press, New York, pp. 250-326.

Rosen, E., Sivertsen, A., and Firtel, R. A., 1983, An unusual transposon encoding heat shock inducible and developmentally regulated transcripts in *Dictyostelium*, *Cell*, 35:243- .

Rosenzweig, B., Liao, L. W., and Hirsh, D., 1983, Target sequences for the *C. elegans* transposable element Tcl, *Nucl. Acids Res.*, 11:7137-7140.

Spradling, A. C., and Rubin, G. M., 1981, *Drosophila* genome organization: conserved and dynamic aspects, *Annu. Rev. Genet.*, 15:219- .

Stanfield, S. W., and Lengyel, J. A., 1980, Small circular deoxyribonucleic acid of *Drosophila melanogaster*: homologous transcripts in the nucleus and cytoplasm, *Biochemistry*, 19:3873- .

Yvan, R., 1981, Structure and mechanism of multifunctional restriction endonucleases, *Annu. Rev. Biochem.*, 50:285- .

Zuker, C., 1983, Repetitive developmentally regulated genes in *Dictyostelium*, Ph.D. Thesis, Massachusetts Institute of Technology, Cambridge, MA.

Zuker, C., Cappello, J., Chisholm, R. L., and Lodish, H. F., 1983, A repetitive *Dictyostelium* gene family that is induced during differentiation and by heat shock, *Cell*, 34:997-1005.

Zuker, C., Cappello, J., Lodish, H. F., George, P., and Chung, S., 1984, *Dictyostelium* transposable element DIRS-1 has 350-base-pair inverted terminal repeats that contain a heat shock promoter, *Proc. Natl. Acad. Sci. USA*, 81:2660-2664.

Zuker, C., and Lodish, H. F., 1981, Repetitive sequences cotranscribed with developmentally regulated *Dictyostelium discoideum* mRNAs, *Proc. Natl. Acad. Sci. USA*, 78:5386-5390.

DEVELOPMENTAL CONSTRAINTS AND EVOLUTIONARY SALTATIONS: A DISCUSSION AND CRITIQUE

Jeffrey S. Levinton
Department of Ecology and Evolution, State University of New York
Stony Brook, New York 11794

"Fashion me, therefore, one form of a many-colored and many- headed beast. There is a ring of heads both of tame and wild beasts, and it can change and produce them out of itself at will.
'That is clever molder's work,' he said." *The Republic* of Plato

CONSTRAINT AND SALTATION

Developmental biology has long been a focus for evolutionary theory (von Baer, 1828; Haeckel, 1866; Garstang, 1922; deBeer, 1958; Goldschmidt, 1938; Waddington, 1940; Riedl, 1978; Gould, 1977; Bonner, 1982; Raff and Kaufman, 1983). Time and again, the concepts of *constraint* and *saltation* have been (re)formulated in developmental terms, but, surely, Goldschmidt's (1938) "Physiological Genetics" is the classic in this field. A major group of *constraints* are non-random channelizations of evolutionary direction due to limitations imposed by epigenetic interaction in the developing organism. In the context of development, *Saltations* are rapid evolutionary fixations of phenotypic discontinuities; many might be governed by developmental constraints which do not permit continuity of form in polymorphic populations. It is the purpose of this essay to discuss these two concepts critically.

PHYLOGENETICISTS AND DEVELOPMENTALISTS

The significance of ontogeny in evolutionary theory has been emphasized from what I shall call the *phylogeneticist* and *developmentalist* viewpoints. The phylogeneticists view ontogeny as the source of information of an organism's evolutionary history. Ernst Haeckel, that most devoted German disciple of Charles Darwin, combined a materialistic view of evolution with a belief that man's ancestry was recorded in the sequence of embryonic development. His association between the order of ontogeny of derived taxa and the order of acquisition of ancestral character states was unduly rigid and invited later strong criticism

(Garstang, 1922; deBeer, 1958; Gould, 1977). But why should development be indicative of ancestry at all? Why do primitive structures that are degenerate or absent in the adult often make an appearance albeit brief in the embryo only to later appear as vestigial structures or even to disappear? How can we explain the occasional reappearance of ancestral structures, as developmental abnormalities, that supposedly disappeared millions of years before in evolution? Why, for example, do some whales have hindlegs (Andrews, 1921), do some horses develop ancestral toes instead of splints (Marsh, 1892)? Some level of order in development survives an evolutionary process characterized by divergence.

A feeling that the process of development is crucial to an understanding of evolution comes also from the nature of development itself. The developmentalists claim that "the diversity of structures that have been formed through the process of evolution is constrained by the rules which govern pattern formation during development" (Stock and Bryant, 1981; p. 432). As such, evolutionary change is involved intimately with developmental sequences. The individual, therefore, is treated in terms of its entire ontogeny and development is therefore both the constraint and target of selection. The developmentalist claim should be the key to the phylogeneticist claim. If we can understand whether development really does provide constraints, and if we can understand how developmental programs are shaped by evolutionary processes, then we might come to understand the limits to which phylogeny can be expressed in ontogeny.

DEVELOPMENT: CONSTRAINTS AND DISCONTINUITY OF FORM

The notion of constraint and discontinuity of form in populations has been long appreciated on both genetic and morphological grounds. Wright's (1934, 1935a) classic analysis of digit number in guinea pigs (*Cavia cobaya*) recognized the constrained nature of development, despite complex polygenic effects on traits. Quantitative geneticists working on mutations affecting vertebrate skeletons and teeth have long been aware of the presence of limited and discrete variation (e.g., Hadorn, 1961; Green, 1962; Grewal, 1962; Gruneberg, 1965; Garn et al., 1963). In their classic review, Sinnott and Dunn (1935) showed that plant development is strongly regulated by genes that affect rates, and by genetically controlled quantum changes in morphology (see also Haldane 1932). Extreme fruit-shape differences are already established in the very earliest primordia of squash (*Cucurbita*). In one case, two genes are known to contribute to a disc shaped form. Many of the strong differences among genotypes can be attributed to phenotypic changes early in the developmental process (e.g., Atchley et al., 1984). In plants, mutations affecting early stages of development often have the most significant final effects on overall form. Form differences are usually not the result of " mere continuations or extensions of the later stages of a growth period" (Sinnott and Dunn, 1935, p. 140).

Why might there be constraints? The developmentalists provide at least some keys to the solution. Consider the quantum nature of development. It is not continuous but discontinuous. Despite the potential continuous nature of additive gene effects and the continuous possible range of concentrations of morphogens, hormones, etc., morphological structures often come as completed structures or not at all (Maynard Smith, 1960). Of equal interest is the importance of localization in development. Embryos develop only as the result of a complex series of timing events that bring different cells into contact, or place cells or even molecules of restricted developmental potency in a proper environment for induction. The spatial position of cell groups seems crucial in the generation of morphological patterns due to postulated localized intercellular movement of dissolved substances (Turing, 1952; Wolpert, 1969; Garcia-Bellido, 1975; Summerbell, 1981), transcellular electric fields (Jaffe and Stern, 1979; Nuccitelli, 1983), or due to mechanochemical interactions (e.g. Odell et al., 1981; Oster et al., 1983). Must this not influence the direction of evolution? These two phenomena - *integrity of structure* and *topological restriction of development* suggest that an embryo can only be transformed in a limited number of directions during the process of development and evolution. That is the fundamental message that Richard Goldschmidt's book "Physiological Genetics" (1938), derived from Spemann (1938), underscored so well.

Some examples of developmental mutants show the discontinuous and often spectacular nature of possible structural change. Consider the *cyclops* mutant (Bowen et al., 1966) in males of the brine shrimp, *Artemia salina.* After the fourth instar, the lateral eyes move forward and fuse together as a single large compound eye by the ninth instar. During this fusion the ganglia and nerves of the two optic stalks fuse; the resultant eye is rather similar to the normal medial eye of the cladoceran *Leptodora*. A quirk of development has caused a structure to change from that characteristic of one taxonomic order to another! Previous work on the eyes of *Artemia* and *Triops* suggests that the organogenesis of the protocerebral sense organs involves close connections between compound and nauplius eyes (Dahl, 1959). The upper lateral nauplius eye nerve is the remnant of a connection between the nauplius eye and the ganglion opticum of the compound eye. The development of the observed structures, both normal and mutational, involves extensive integration and quantum spatial reorganization of already complex structures.

The development of the vertebrate limb shows similar quantal steps. Alternative symmetries and structures, demonstrated by manipulations of development and repair, can be spectacularly different. An excision of the posterior part of the chick (*Gallus gallus*) wing bud results in the elaboration of only a single skeletal element: a humerus, or a humerus fused with a reduced radius. If, however, the anterior part of the bud is excised the posterior half develops nearly normally, elaborating part of the humerus, the ulna, and digits 2, 3, and 4 (Hinchliffe and Gumpel- Pinot 1981). Clearly, developmental mechanisms are well organized, depend upon tissue interactions, and often involve discrete steps (Maderson, 1975; Alberch, 1980).

A developmental notion of macromutation springs from the nature of development described above. After all, if a simple transplant places toes on wings or replaces scales with feathers, why couldn't evolution occur in major steps? Thus some have either directly interpreted developmental change, via such phenomena as tissue interactions, as the vehicle for major evolutionary jumps (Schindewolf,

1936; 1950; Goldschmidt, 1940; Lovtrup, 1974; Maderson et al., 1982), or at least see them as the possible stuff of major saltations (Frazzetta, 1970; Gould, 1980; 1982; Alberch et al., 1979).

DEVELOPMENT, GENES, AND SELECTION: COMPONENTS OF THE EVOLUTIONARY RATCHET

The developmentalist point of view presented above leads to a model, which I call the *evolutionary ratchet*, generating some of the constraints in evolution. The model characterizes an organism as being a product of a long evolutionary history, where the evolution of timing, rates, and localization leads to a complex developmental process that can be disrupted less and less easily as time progresses. Any spatiotemporal interactions in the developing phenotype which, when accumulated, cannot be disrupted due to integration are accumulated by the *epigenetic ratchet*, and lead to *epigenetic constraints* (see Rachootin and Thomson, 1981). Shifts in tissue interactions are well known in evolution (Hall, 1984) but such transitions may be difficult to achieve due to interdependencies between different germ layers. Riedl (1978) thought of such complex organisms as carrying a high *burden*. "By burden I mean the responsibility, carried by a feature or decision. ....with systemization [i.e., the evolution of development] the functional burden carried by decisions increases and with this a new lack of freedom called canalization also increases" (Riedl, 1978, p. 80). He argues that "..patterns of decisions that have a certain degree of burden have no real prospect of being fully dismantled.." (p. 93).

Aside from the developmental ratchet, two more factors contribute to the overall evolutionary ratchet. The *genetic ratchet* refers to new genes which, when incorporated, cannot easily be lost by random processes such as genetic drift. We are not, however, referring to a simple contribution to the organism's survival in a changing environment. Rather, the evolutionary change of a gene affects the integration of the phenotype during development. Pleiotropy is one likely explanation. Incorporated genes having widespread effects on the overall phenotype may not be easily lost in evolution (Lande, 1978; Kollar and Fisher, 1980). Ohno (1973) has characterized incorporations into the genetic ratchet as "frozen accidents". The x-chromosome in mammals and dosage compensation may be an example. Evolution after the development of dosage compensation (inactivation of one of the two x chromosomes in the female) would select against translocations of x-linked genes to other chromosomes. It is of interest that the x chromosome seems to always be about five percent of the total mammalian genome.

Such a model is evocative of the reverse of the usual metaphor of topographic highs, connoting high fitness. Instead, one thinks of evolution as leading the organism to cut channels, forming the developmental landscape. Waddington's *epigenetic landscape* was conceived as a reflection of the limited number of stable developmental pathways possible from the combined actions of different genes involved in development. "The reactions of these genes with one another (and with the environment) interlock so as to define a developmental track which will always be followed by the antennae of flies of a certain genotype. ....the normal developmental track is one towards which a developing system tends to return after disturbance" (Waddington, 1940, p. 93). The pictorial representation of this is the steepness of the walls of the valleys, which reflect the tendency to return to the developmental track, if a deviation from the central stream occurs. Such a theory acknowledges that the evolution of internal organization is as important as the effects of the external environment. As Riedl (1978) puts it (p. 244): "..selection does not work from the outside only." Internal organization may become more efficient as the result of natural selection.

The evolutionary process can therefore be understood as a dialectical interaction between the external conditions of the environment and the internal organization of the organism (Schmalhausen 1949). With the passage of time and evolution, the joint actions of both internal and external controls may lead to still more internal organization, a portion of which must influence or constrain future evolutionary directions.

The *selection ratchet* refers to features that, once acquired, are not easily lost due to functional integration within the developing phenotype. For example, it would appear that the vertebrate humerus is essential in the function of the radius and ulna, since one observes that the latter is lost or modified before, or more frequently, than the former. Such functional complexes interact with the external milieu.

In the construction of functional morphological models, all three types of constraints - *epigenetic, genetic, and selectional* - will have to be incorporated into the study of functional morphology and adaptation. Stebbins (1969, 1974) has pointed out the connection between the evolution of functioning complexes and the rise of gene sets that are strongly interactive because of their collective effect on fitness. "The adaptive value, and hence the acceptance or rejection by natural selection of most new mutations, depends not upon direct interactions between these mutations and the external environment, but upon their interaction with other genes, and their contribution to the adaptiveness of the genotype as a whole" (Stebbins 1974, p. 302). Thus, "the direction that it [evolution] will take in later stages of evolution is influenced only in part by direct interactions between populations and their environment" (p. 303).

The actions of the evolutionary ratchets complement neoDarwinian theory; we only acknowledge that evolution tends to organize the phenotype and genotype so that it begins to resist some directions of evolutionary change for reasons of internal organization. Such presumptions are implicit in the works of the founders of the Modern Synthesis (e.g., Stebbins, 1969; Mayr, 1963; Dobzhansky, 1970), although genetic, rather than epigenetic, interections have been recognized. I would term this the hardening of the developmental ground plan. We have no real evidence that such hardening does occur, but it is a very plausible expectation from developmental considerations. I find it especially unfortunate that Maderson et al. (1982) take such points about development and evolution to be major challenges to the Modern Synthesis. They are not. As Riedl (1978, p. 235) notes: "In terms of feedback it is a *selectional theory*. It assumes the correctness of the Neodarwinian synthesis as a precondition, but supplements this .." The concern for constraints influences our conception of the nature of variation.

The constraints generated by the ratchets define the genetic and phenotypic organization of development and allow us to construct appropriate specific models of evolution. This is not a radical concept, as Waddington noted in 1940 (p. 145): "..the concept..provides a terminology in which it is easy to admit the recognition that phenomena do not present themselves as being all of the same kind." As evolution progresses, the "kinds" increase in number and the constraints become more important. With time, form changes by reorganizing the basic patterns established by the evolution of development. Thus, in any move into a major new habitat, morphological changes will show to varying extents, the stamp of evolution in all previous environments. Thus, the archetype concept of Richard Owen, so denigrated by many Darwinians, thus survives, without its Platonic stamp. It has a rather obvious materialistic basis. It seems unlikely to me that the notion of constraint will be easy to understand completely in all biological taxa. The most progress will be made where the genotype and the phenotype are understood as completely as is possible. This leaves us with very few useful biological systems, including those traditional genetic and developmental workhorses such as the fruitfly *Drosophila melanogaster* and the nematode *Caenorhabditis elegans*. Unfortunately, neither of these species can be said to be understood well from an ecological point of view.

BUILDING AND INTERNALIZATION OF THE NEWLY EVOLVED PROGRAM

The three ratchets only describe the difficulty of losing a gene or phenotype at the present time, and do not indicate how the traits or genes might have been accumulated by the ratchet. A possible example of the combined action of the genetic and selective ratchets is the sequence of incorporations of evolutionary changes underlying a functioning phenotype. Consider an enzyme. To ensure function, the enzyme must have a sequence of amino acids that determines the proper conformation for reactions with substrates and cofactors. The change of an amino acid at a crucial site might eliminate function. Such a change would be selected against, and the loss or gain of function would be directly exposed to natural selection, as the individual would suffer without the appropriate residue. Consider, however, the case of tryptophan synthetase A, a protein of 267 amino acids in the bacterium *Escherichia coli* (Yanofsky et al., 1981; and references). At position 210, a substitution of serine for glycine (the wild type) causes no change of activity. A mutant with glutamic at this site eliminates function. If, however, this non-functioning mutant encumbers a further change, of tyrosine to cystine at site 174, full function is restored. The same change, at site 174, of the wild type protein eliminates function. Thus the same mutant (tyr>cys at site 174) will destroy or save the phenotype's full function, depending upon the order of incorporation. The sequential incorporation of mutants, moreover, may result in a frozen ensemble of changes which cannot be removed except by rare double mutants. The phenotypic and genotypic basis of a functioning phenotype is therefore as much a matter of the order of genetic incorporation of change as it is a matter of function *per se*.

The evolution of an ordered developmental program subject to the ratchets can be described as *building*. As this process continues, order becomes important and internal relationships become as important as interactions with the external environment. If the program is organized as a unit, it may then become *internalized*, that is, a genetic-epigenetic mechanism seals it off from the external environment and the program is then a unit, interacting and evolving dependencies with other such units. Of course, it is unlikely that total independence will evolve, but, the internal order may be more important than the forces of the external environment. Once such units evolve, two important changes may occur in the evolution of the organism. First, internal constraints will direct evolution. Second, the units may be shuffled as organized entities with other such units. Evolution would no longer consists of infinite molding by the external milieu, as might have happened during the evolutionary building of a given unit.

Potential complex relationships between genotype, functional considerations, and developmental order can be illustrated by the formation of the developmentally arrested dauer larva stage of the nematode *Caenorhabditis elegans* (Riddle et al., 1981). Normally the larvae pass through four molt stages, but overcrowding or starvation causes entry into the dauer larvae in the second. The duration of the dauer stage does not affect subsequent longevity. Studies of mutants suggest that the genes responsible for induction of the dauer stage are ordered in a pathway, and that the order corresponds to neural processing of environmental stimuli necessary to stimulate the developmental switch. Two of the mutant genes are correlated with chemotaxis defects, and both mutants exhibit ultrastructural alterations in specific neurones which have previously been implicated in the chemosensory response to salts. This case shows the intimate relationship between genes, developmental order, and selection. We do not know how the sequence was built up in evolution, but the disruption of a gene's function probably affects the internal organization of the order of gene action, interaction between neurosensory substances, the establishment of neuronal pathways, and the ability to monitor the external environment for the proper cue to enter the dormant state.

It is of interest that some dauer mutants can sidestep any interaction with the external environment. In *dauer constituative* mutants the dauer larvae stage is switched on irrespective of crowding and food conditions. This suggests that once a developmental pathway is integrated, probably by natural selection in this case, it can then be incorporated into the internal organization of the genotype and phenotype. One can imagine a harsh seasonal environment where the dauer larva is no longer an option; the developmental pathway becomes fixed by selection and incorporated into development.

CRITIQUE OF THE RATCHET THEORY

The notion of hardening of developmental programs has been seen by many as a fundamental challenge to the notion that natural selection is unrestricted in its bounds. The evolution of development is thought by many to bias evolutionary direction (Davenport, 1979; Gould, 1982; Alberch, 1981; 1982; Maderson et al., 1982). These arguments, furthermore, point to the creative power of combining units that are relatively independent to produce novel structures (e.g., Lovtrup, 1974). While these arguments are useful and have been underemphasized in the past, it is worrisome that they will be applied uncritically to construct a new model of evolution that tends to think of change in the developmental program as a necessary or at least common progenitor to major morphological change in evolution. Minor changes in temporal arrangment of developmental timing would under this argument be the cause of large macroevolutionary jumps (Gould 1982). Differences between major bauplans, moreover, would be conceived as due mainly to differences in embryogenesis that may have arisen by accident or for reasons of internal organization, and not selection to function in the external environment. This reasoning is entirely premature and may lead us too far in the opposite direction. Gould (1982) creates a mythical community of strict Darwinians who cannot accept the notion of developmental channelization in evolution. This tactic only separates us into artificial intellectual camps and obscures issues which must be addressed without rhetoric.

A caricature of such a point of view would state that the reason a pussycat differs from a platypus is that their basic embryological ground plan evolved independent of adaptation and any disruption might lead to severe disruptions of integration during development. Thus an evolutionary line descending from a platypus is condemned always to resemble its ancestor and arose due to nonadaptive rearrangements of developmental blocks. In my view, this point of view is a motivation for a general belief in nonadaptive origins of bauplans and long-term stasis (e.g. Gould, 1983). The relative degree of fixation of developmental programs causes the rates of evolutionary change to be dampened, and deviations from the archetypal form become unlikely.

Such major restrictions of evolutionary change may well be true at some levels of development. But it must be subject to empirical investigation. An example of major differences in early development among related species, with strong similarity in final morphological outcome, would provide evidence supporting the notion that different developmental pathways can be eventually molded into the same overall adult phenotype, which may have been selected to function optimally in the environment. Similarly, major changes in developmental pattern, despite similar final adult morphologies, that can be related to relatively minor aspects of natural selection, would indicate that the notion of burden may be a relatively unimportant concept, at least in the systems considered.

Frogs are notable for their overall adult morphological homogeneity, despite great antiquity and among-species genetic distances, relative to other groups such as mammals (e.g. Cherry et al., 1978). Is this adult homogeneity determined by a singular developmental ground plan that has been hardened to the extent that no fundamental change is possible? Apparently not. Early development in the hylid frog genus *Gastrotheca* is quite different from that of other frogs, yet the adult form looks like a typical frog (del Pino and Elinson, 1983). Almost all of the body forms from an embryonic disk, a group of cells that forms from fusing cells of the blastoporal lips. In two other unrelated genera, similarly located yolk-poor cells are spread around the blastopore, instead of being concentrated in a disk. The difference in *Gastrotheca* development seems due to a temporal separation of the times of closing of the blastopore and the anterior extension of the archenteron. The embryo of *Gastrotheca* looks bird or reptile like as it initially develops into several sheets of tissue, which secondarily form tubes after folding. This disk, however, is formed by gastrulation and is not homologous with those of other vertebrates.

Why early development should be so different in this genus is not clear, but it might be due partially to extraordinary slow development (two weeks from fertilization to end of gastrulation, as opposed to one day in *Xenopus laevis*). This might cause the temporal separation of the closing of the blastopore and the anterior movement of the archenteron and fundamentally rearrange tissue interactions, relative to other frog embryos. Interestingly, this slow development is related to an overall syndrome of egg brooding, maternal incubation of embryos, multinucleate embryogenesis, and very large eggs. Some of these may have ecological significance; the environment may have selected for the rather unusual parental traits. The novel development of *Gastrotheca* organizes the developing embryo as an elongate disk lying above a large amount of yolk. The alternative development of *Xenopis laevis* and other anurans results in an elongate embryo that is important for early swimming (del Pino and Elison, 1983). But no macromutation changing adult development has occurred; the adult is still basically a frog. As del Pino and Elison (1983, p. 589) note: "the embryos of *Gastrotheca* illustrate that it is possible to modify greatly the pattern of early development without altering the basic adult morphology." In this case, adult stasis cannot be said to be due to a rigidity of developmental program.

Development in the freshwater clam *Unio* provides further evidence that natural selection can break up seemingly fundamental embryological spatial interactions yet a typical adult is produced. Its larval stage is highly modified relative to its closest marine relatives. Larvae typically are brooded until the glochidium stage, which bears a pair of hooked valves capable of attaching to the gills of fishes. It completes development on the gills and eventually drops off to finish development in soft sediments. *Unio* would be expected to have spiral cleavage, a characteristic mode of early cell division common throughout annelids, mollusks, polyclad flatworms and other phyla. At the third cell division typical groups with spiral cleavage show a differentiation between 4 macromeres and 4 micromeres. After the next few cell divisions the spiral arrangement of cells is apparent. The embryo at this stage is mosaic, that is, cells have specific fates which cannot be easily reversed. In most mollusks, the first generation of micromeres divides and forms eight cells, which eventually forms the apical region of the embryo and the prototroch, a larval feeding organ. But these structures are lacking in *Unio* larvae, due to a retardation of division in the first tier of micromeres. The second tier of micromeres, arising (as is normal) from cleavages from the macromeres, gives rise to the bulk of the larva. The importance of other micromeres has been greatly exaggerated relative to the typical pattern of spiral trochophore larval development in most related groups (Lillie, 1898). The 2d micromere gives rise to the shell gland; this differs from other mollusks who derive their shell gland from cell 2D (Kniprath, 1981). This seems related to the development of structures of ecological relevance to the unique functioning *Unio* larvae, such as the powerful adductor muscle used by the larva to snap the valves closed on prospective fish gills. As in the amphibian case, major adjustments can be made in early development to satisfy ecological necessities, with little consequence for the subsequent adult phenotype. There is little reason to believe, therefore, that ontogeny is necessarily a tightly bound sequence of developmental changes, with major changes in the adult inevitably springing from major alterations of early ontogeny.

ATAVISTIC FEATURES: THE WHISPERINGS WITHIN

Atavisms are coordinated, often incomplete, structures that appear as developmental anomalies and resemble ancestral character states of the taxon to which the individual belongs. We presume that such developmental anomalies often induce already evolved features that have been suppressed, rather than creating newly coordinated structures that are prospective for this branch of a clade. The efficacy of this assumption rests on consistency with other evidence, since one cannot completely exclude the prospective nature of developmental anomalies, given the degree of interaction in development.

The degree of organization and apparent antiquity of atavistic features is often surprising. Consider the atavistic partial limbs of sirenians and cetaceans. The fossil record of the Cetacea is spotty, but at least early representatives (e.g. *Protocetus*, Middle Eocene) had a well-developed pelvis and sacrum, indicating the likely presence of almost complete hind limbs. Early Eocene forms have a dentition resembling both carnivorous mesonychid terrestrial mammals (the presumed ancestors of whales) and middle Eocene forms (Gingerich et al., 1983). Modern whales lack hind limbs; evidence of an apparent terrestrial ancestry has been suppressed. Andrews (1921) described a humpback whale, *Megaptera nodosa*, with hind limbs over a meter long. The femur was nearly complete, even though it is normally an internal and diminutive cartilaginous element. Specimens of the sperm whale *Physeter catodon* have been discovered with femur and partial phalanges, even though the femur is normally a rudiment (Lande, 1978). If the atavisms of whales and sirenians are any indication, the capacity to reexpress limb elements long lost in evolution suggests that some measure of developmental organization has been retained. A chain reaction, something like that in control of amphibian metamorphosis, must have been altered during the course of evolutionary limb reduction and loss, rather than the genes or epigenetic subprograms specifying the structures themselves.

Two most spectacular atavisms relate birds to their reptile-like ancestors. Kollar and Fisher (1980) performed a typical tissue interaction experiment by combining chick epithelium with mouse (*Mus musculus*) molar mesenchyme. The result was a great surprise. A variety of dental structures appeared, including perfectly formed crowns with differentiated ameloblasts, depositing enamel matrix. For some of the structures to have been produced, the bird epithelium could not have lost - after at least sixty million years - the appropriate mechanism to induce the production of dentin in mouse mesenchyme. This result is grander in scale, but similar to Kurtén's (1963) discovery of the reappearance of an associated group of dental structures, sometimes including the second molar, in the lynx (*Lynx*

lynx), otherwise unknown in the Felidae since the Miocene. The absence of complete teeth could be explained either by loss of genes to completely specify the structures, or by some difference in the inducing mesenchyme. But we must conclude that the genes to specify the organized structures have not been completely lost.

Perhaps the most celebrated inference of an atavism was reported in the chick (Hampé, 1959). Avian evolution involved a reduction of the fibula, relative to presumed reptile-like ancestors. During the development of the avian leg, the fibula is shortened relative to the tibia, while the metatarsals fuse with the fibula. These adult features are typical characteristics of the avian limb. Hampé altered the course of development with a series of experiments, such as grafts of fibular primordial tissue, or by insertion of a barrier to equalize blastema size between the tibia and the fibula. Two remarkable things resulted from these manipulations. First of all, the chick formed a structure like a complete fibula. But when the fibula was completed, it apparently induced contact distally to the joint, reminiscent of *Archaeopteryx*. In turn, *Archaeopteryx*, seems intermediate to the reptilian condition, where a series of ankle bones exist instead of two fused elements at the ends of the tibia and fibula, respectively. Again, the evidence supports the notion that the genes to specify structures have not been lost, only a switch mechanism which controls their expression.

The common reexpression of such traits suggests a large degree of organization in either the genome or the epigenetic programs that have been retained, subsequent to the loss of the determined phenotypic structure. If it is the same genes that are retained, this would be superficially surprising to anyone expecting that a relaxation of selection should produce a corresponding relaxation and eventual elimination of the genes for the structure (Kollar and Fisher, 1980). The result indicates that there really are no genes specifying only one structure. The genes, rather, must have pleiotropic effects on other characters of selective advantage for the organism (Lande, 1978). It suggests that some of the organization among genes may be used in yet other structures.

The order apparent in reexpression may not be due to the very same genes, throughout the history of the evolution and loss of the trait. As Waddington (1940, 1942) notes, the elaboration of the trait, including the tissue interactions involved, might eventually become so important that the genes involved might shift with no disruption of the epigenetic outcome. Genes that influence and help to effect the same epigenetic program might come and go. He refers to this process as canalization of development; the developmental program assumes an importance in itself. In cases where a structure is suppressed in evolution, but reexpressable, it is still possible that the reexpressed epigenetic programs might still be used in the determination of other structures. The preservation of potential for reexpression of the phenotype therefore may not necessarily indicate that the genotype has been exactly preserved. There must therefore be an analogous situation to pleiotropy which has only genetic implications. To distinguish between preserved genes and developmental programs which have widespread influence, we should establish the terms genetic pleiotropy and epigenetic pleiotropy. Evidence from *Drosophila*, however, suggests considerable autochthony of genes involved in development. For example, a controlling sequence, the homeobox, has been found to be widespread in animals (North, 1984).

DEVELOPMENTAL AND EVOLUTIONARY PATHWAYS - JUMPS ACROSS THE BREACH?

The preceding sections indicate that the developmental phenotype is a strongly interactive unit, and that evolutionary direction is very much channeled by the previous evolution and internalization of developmental programs. As Williams' (1950) study of cervical articulations in turtles demonstrates, this generalization is compatible with the notion of adaptation. The nature of variation, however, is constrained by development and, as a consequence, the response to selection must deal with available developmental programs. The question arises: can major new phenotypes spring into being by simple rearrangments of developmental interactions?

Atavisms have recently been a center of attention among evolutionary biologists (e.g. Lande, 1978; Riedl, 1978; Gould, 1982; Alberch et al., 1979; Raff and Kaufman, 1983). This interest reflects the feeling that something important must spring from the fact that major traits, apparently long lost in evolution, can be reexpressed by simple manipulations such as rearranging tissue interactions (Kollar and Fisher, 1980). As Gould (1982, p. 343) notes: "But the genome embodies an extensive set of latent capacities, some the echoes of distant ancestors..Small changes in rates often activate these potentials, and the result is not only a surprise, but often a major one". The proximate phenomena of tissue interactions are therefore believed to represent the ultimate limitations that development imposes on evolutionary directions.

The major morphological changes induced in the form of atavisms seems consistent with the argument that changes in early development can rearrange morphology and may produce major discontinuous macroevolutionary jumps. After all, if an increase in calcium can induce a chick to make ankle bones, the long lost structures of reptile-like ancestors, then why can't minor changes cause major saltational jumps in evolution? This impression is only reinforced by major mutants, such as the magnificant *cyclops* mutant of *Artemia salina* discussed above.

Expecting major jumps stems from an important assumption, which is best phrased by Maderson et al. (1982, p. 307): "..developmental constraints are the basis of many discontinuities and clumpings in organic morphospace. The sparse and clumped distribution of morphology does not represent a set of optima constructed by natural selection from a set of unbounded possibilities." If this is true, then the major problem of morphological discontinuity among taxa is solved readily. Maderson et al. go on to say that such a point of view is "not congenial" with the traditional notions of the Modern Synthesis, due to the inferred rapidity of origin of novel structures and the non-adaptive nature of evolutionary change under developmental constraints.

The major problem with this argument is the fact that the structures themselves, while determined as units, must be integrated functioning units to permit the organisms to survive in the external environment. The evolution of the unit is a different process from the determination of number, symmetry or position of the units. The evolution of the unitized switch of the dauer larva, as discussed above, may have involved a gradual accumulation of the actions of many genes. A developmental switch gene would arise only after the gradual evolution of the developmental unit is complete.

The same can be said for any structure. Gradual evolution will form the structure, but only then can other mechanisms switch on or off the developmental program invoking the structure's presence. The quantal nature of development is thus somewhat deceiving. A long process of gradual evolution may be behind the current presence of a quantally determined developmental program. Once the program is evolved, then its burden, in the sense of Riedl (1978), will determine whether it can be shuffled out of the organism's total development or rearranged to cause the combinatorial evolution of already extant developmental units. The same type of misinterpretation was behind the belief that butterfly mimicry consisted of single step mutations (Punnett, 1915; Goldschmidt, 1945). Because the entire mimetic morph seemed to segregate as one unit, Punnett believed incorrectly that one gene, accumulated in one step, determined the mimicry. But, gradual evolution had resulted in eventual linkage of many genes whose aggregate effect, in combination with unlinked modifiers, produced the mimetic morph. Current quantal status does not indicate quantal evolutionary steps.

The significance of atavisms stems from this same consideration. It is often easy to induce novel structures by simple experiments. But this act only induces what evolution has already created. It does not, and could not, propel the organism forward into a new morphological realm. This is not to say, that a possible reshuffling might lead to occasional novel combinations and a rather distinctive organism - a minor hopeful monster, so to speak. But the reinduced structures have already evolved, and this evolution may have been through the cumulative action of many genetic modifications. The major work, in other words, has already been done. That is why the new structure appears so suddenly.

The importance of atavisms is overstated because of the fallacy of looking at a highly evolved system, backwards towards ancestral states that have been suppressed in the evolution of new taxa, and then mistakenly inferring a creative force. The complete suppression of these states may have taken millions of years and many intermediates. But a bird is not just a reptile with suppressed features. The evolution of the derived states characterizing the class Aves involve the origin of many new structures, though certainly preexisting materials and developmental programs must have been modified. If lizard epidermis is placed in contact with chick dermis, it does not produce feathers, only scales! Although it is of great interest that the genes to specify ancestral structures are often not lost mere suppression does not create organized new structures. Looking backwards is not the same as moving forwards.

The prospect for the single-step origin of wholly new developmental descendants is also falsely enhanced by examining the results of embryological manipulations. Because the formation of so many structures is initiated by tissue contacts, it is possible to create novel structural rearrangements with the appropriate experiments. But the presence of claws on wings comes only from an isolated manipulation; the wider effects of a mutation with such an effect on the entire phenotype cannot be understood properly with such experiments. Broader genetic- phenotypic investigations must be done to understand whether such changes can be isolated, or whether they have other deleterious effects. The evidence from developmental mutants, cited below, fails to support the optimism accorded major single step changes in evolution.

Despite these arguments, one cannot exclude the possibility that major developmental mutants might be the source of large scale saltational jumps in evolution. Even if no new units were evolved in the process, combinations of already existent units might generate startling new creatures. The potential certainly exists in the homoeotic mutants that cause switches of determination in overall developmental fate. We could then imagine evolution as a reshuffling of basic units, to produce the beast of Plato's *Republic* comprised of many heads and colors that can materialize as any of a number of forms. Raff and Kaufman (1983) note that some simple developmental events, such as duplication of cells giving eventual rise to entire structures, might be the source of evolutionary novelty. This is well taken. But these same authors overemphasize the possible importance of homoeotic mutants in evolutionary change.

As a general rule, major developmental mutants give a picture of hopeless monsters, rather than hopeful change. Epigenetic and genetic pleiotropy both impart great burden to any major developmental perturbation. Thus it is unlikely that mutants affecting any fundamental prepattern in development are likely to produce a functional organism. Genes that activate switches in prepatterns are not sufficiently isolated in effect on other parts of the phenotype to expect major saltations. The cyclops mutant of *Artemia* is lethal. The homoeotic mutants of *Drosophila melanogaster* suffer similar fates. After all, the bithorax complex must function perfectly since it controls, at the least, the fate of segments posterior to the second thoracic (Lewis, 1978). The *ultrabithorax* mutant converts the first abdominal and third thoracic segments to second thoracic-type segments. But the gene is always lethal in the homozygote. Disruptions, i.e., mutants, have drastic effects on other parts of the phenotype. The *apterous* mutation, for example, causes reduction of wings and halteres. But it also is the basis for non-vitellogenic oocyte development, failure of development of the larval fat body and precocious death, though these effects are alleviated somewhat at lower temperatures (Wilson 1981 a,b). The *engrailed* mutant strongly changes the prepattern for the normal formation of the *Drosophila* sex comb. A secondary sex comb is formed. But the engrailed gene also affects the formation of wing veins and development of the scutellum. Pleiotropy reduces the likelihood of major switches that escape tremendous drops in viability due to correlated changes.

The problems with the evolutionary potential of atavisms is illustrated by a relatively minor atavism in the guinea pig (Wright, 1934; 1935,a,b). A polydactyly mutant, (*Pollex*), increases the number of toes in the hind limb of the heterozygote from four to five, the presumed ancestral pentadactyl vertebrate limb condition. The homozygote for this allele dies, usually before birth, and usually has even more toes. Other abnormalities are also apparent in a large number of embryonic organs and incoherent development is especially prominent in the limb buds, brain, and visceral arches. The introduction and fixation of such an "atavistic gene" into a population would be highly unlikely. Highly complex syndromes of this sort are common in the mutations of major effect studied in other vertebrates. As in *Drosophila* they arise from the obviously complex pleiotropic effects of the mutant genes (Hadorn, 1961, p. 140-148, 182-202).

The organization in development is a two-edged sword. On the one hand the presence of organization indicates that evolution must move along the valleys of Waddington's epigenetic landscape. But organization also is so intricate that major jumps are precluded simply because of the high probability of producing a major disruption of proper development. Even Waddington's theory of canalization of development (1942) requires gradual reorganization of the developmental process. Hadorn's (1961) monograph of lethal developmental mutants puts it well (p. 304):

"It must also always be kept in mind that normal development requires the co-ordination in space and time of a large number of individual processes. A mutation which causes the slowing down or the speeding up of a single process may easily induce a standstill of development. In particular, if the development of a primordium depends on the organising function of another rudiment and the inducing stimulus is delayed, the competence of a reacting system which follows its own rhythm may have disappeared."

"In addition, the process of development must make enormous demands on the harmonious co-operation of the numerous individual processes which are originated in the genetic substances of the chromosomes."

Thus, the accumulated evidence suggests that major developmental mutants are of little potential for evolution. The side effects are drastic. Moreover, the impression created by atavisms and other homoeotic mutants are misleading. The sudden appearance of change only indicates current developmental organization, not a necessary path to macroevolutionary jumps.

It might be argued that hopeful developmental monsters are significant in evolution, but nevertheless too rare to be seen in the laboratory. This may be true and is a frustrating problem since special environmental circumstances could indeed have made such monsters frequent and subject to rapid selection during unusual times in the past. It is of interest, however, that any geneticist interested in major developmental mutants would be delighted to find viable hopeful monsters in the laboratory, given the various tricks usually necessary to keep developmental mutants in laboratory cultures. But, alas, major developmental mutants are invariably sickly and show pervasive deformities. From both theoretical and empirical points of view, hopeful monsters have led only to hopeless mooting.

Although major jumps are probably excluded from the arguments presented above, evolution by quantal, if rather small, jumps, is far more probable. The phylogenetic and developmental analyses of Alberch (1980, 1981, 1983) working with salamanders, suggested that the previous evolution of developmental programs does result in constraints on the pattern and arrangment of already evolved units. Thus if reduction of the head or the limb is to occur in evolution, certain bones will be lost before others (Alberch and Gale, 1983), and most probably as units. But the between population and among species changes found, for example in *Bolitoglossa*, are rather easy extrapolations of within species variants (Alberch, 1980, 1983). The stuff of phenotypic evolution is small developmental units of relatively low burden on fitness.

ALLOMETRY AND THE PROBLEM OF DEVELOPMENTAL CONSTRAINT

In his important primary works on allometry and growth, Huxley (1931, 1932) discussed a possible link between developmental constraint and form, and concluded that it may have important implications for the interpretation of form in terms of function and evolutionary trends. If we consider a longitudinal study of shape, the value of the allometric exponent and intercept should be explained partially by mechanisms of gene action and development. If this is so, then trends among taxa may be due to the same genetic and developmental mechanisms. Are size changes in phylogeny simply an extrapolation of ontogeny (Gould, 1977)? To the degree that this is true, any evolutionary trend under the influence of functional differences among phenotypes of different shape must be constrained by developmental-genetic effects. Allometry has therefore been a primary battleground where proponents of non-adaptive constraints have argued with supporters of entirely adaptive explanations for shape differences in phylogeny.

An excellent example comes from the work of Huxley (1932), and, more recently, Gould (1974) on the relationship between antler size and skull size (or body height) in different species of deer (Cervidae). All of the measurements yield an approximate straight line on a log- log plot. This is true to the degree that the spectacular extinct Irish Elk (actually, a deer), *Megalocerus giganteus*, appears on the plot as a "typical" species. The question is thus raised: is the Irish elk a captive of its unusually large size via a genetic-developmental constraint, or is the overall relationship between antler size and skull size among the species simply regulated by natural selection on overall form, based upon some as yet poorly understood biomechanical or other selective constraint. Huxley (1932) argued forcefully for the former, though some of his later remarks may reflect an abandonment of this position (Huxley 1960).

A classic study of evolutionary relative growth in titanotheres (Osborn, 1929; Hersh, 1934) shows a typical interpretation of interspecific allometric relationships based upon developmental constraints. Using Osborn's data, Hersh found the following, when comparing horn length to skull length: (1) Species within a genus, consisting of an ancestor-descendant series, have the same values for b (allometric coefficient) and k (allometric exponent); (2) genera are distinguished from one another by differences in b and k; (3) b is a decreasing exponential function of k as phylogeny proceeds. Hersh (1938, p.548) notes that "If we were to suppose that the value of k for the relation between horn-length and skull-length, instead of being so unusually high, were more nearly a common value in the neighborhood of 1, then a titanothere with a skull length of 800 mm would possess a horn about 1.5 mm long. Even if k should have the quite high value of 2.75, such a specimen would have a horn not longer than about 5 mm, which, considering the broad oval base of the titanothere horn, would most likely fail to be detected as more than a change of contour in the fronto-nasal region of the skull."

He used this to reach conclusions of far-reaching importance: (1) The mammalian organization may have a potential mechanism for horn growth. In many groups, however, the horn does not develop because of the allometric relationship, which reflects a developmental process. Therefore (2) "the titanotheres of early Eocene times did not have horns because they were not large enough. As mutations for larger size occurred and were selected..the ontogenetic mechanisms for the production of horns was able to produce first incipient horns, and with the passage of generations horns of progressively increasing size were produced." (Hersh 1934, p. 548). Finally (3): "..the horns in their incipiency were not directly adaptive. But once the horns had reached a sufficient size to be used as organs of offense and defense, we might reasonably conclude that the animals with larger horns, as a consequence of the presence of horns, were selectively favored at the expense of smaller-horned animals." (p.550). At this point selection for larger horns might drag along a correlated selection for larger body size. Hersh concluded that the process would end when the newborn titanothere has a horn so large that it would damage its mother during birth!

I reproduce this argument in some detail as it illustrates beautifully the concept of developmental constraint in evolution, so popular again, a half a century later (Alberch, 1982; Gould 1982). Hersh and Huxley conceived of evolutionary trends as regulated by developmental programs, to the degree that they might be thrust by a correlated evolutionary change into a new realm that is incidentally adaptively significant.

Can geometrical comparisons be so easily translated into developmental and evolutionary change? If development does constrain proportions within developmentally united blocks, then one should be able to dissect an allometric relationship within a clade and see an overall similarity in trends over great and small taxonomic distances. After all, the constraint hypothesis presumes that development locks in a certain overall pattern of growth. Unfortunately, variation is expectable enough that it would be difficult to quantify the magnitude of a deviation from a given allometric exponent that would be sufficient to falsify the notion of developmental constraint.

Let us return to the antlers of deer. Is the allometric relationship between body size and antler size a product of a genetic-developmental constraint, or is it shaped by natural selection? Arguments by Gould (1977) and Lewontin (1978) support the former. Small deer such as the muntjac have small antler size relative to overall body size. But larger species such as the reindeer, or the well-endowed Irish

Elk, had proportionally far larger antlers. Clutton-Brock et al. (1980) demonstrated that large deer species tend to be more polygynous (more females mated per male) than smaller species. Larger species, with proportionally larger antlers, tend to form larger breeding groups, than do smaller species. This suggests a role in intermale combat. But there is still a significant allometric relationship between shoulder height and antler length within groups identified for similar overall breeding size. The slope of the within-group relationship is less than the one established when species from all groups are considered together. Among other possible explanations, this may be due to a general overall advantage of larger deer and their antlers in mating success over smaller deer. Of course, the developmental constraint hypothesis may also be invoked, but it is surely not an hypothesis that seems any more likely than a number of others.

More instructive examples come from cases where evolutionary trends among closely related species seem to cross the overall allometric trend. Such cases, when explained in ecological terms, would appear to falsify the developmental constraint hypothesis. In the primates, testis weight follows an overall allometric relationship with body weight. Significant deviations occur, however, from the overall trend. These can be related to breeding system (Harcourt et al., 1981). Monogamous and harem genera have a smaller testis than those where several males compete for the same female. In the latter case, the contribution of greater volumes of sperm is selectively advantageous, as it dilutes the chances of contributions from matings by other males. Species with larger testes also have a proportionally larger volume of seminiferous tubules as opposed to merely increased supportive tissue. The relationship cuts across taxonomic borders. These results suggest that there is an adaptive reason for an overall correlation between body size and testis size, but that species- specific differences have selected for significant deviations from the overall trend.

A similar case can be made for deviations from the overall brain- size body-size allometric relationship found in small mammals and primates. Folivores tend to have smaller brains relative to body weight than do frugivores (Harvey and Bennett 1983). This may be related to the more elaborate ensemble of behaviors required to locate dispersed and clumped sources of fruit, as opposed to more homogeneously distributed leaves. The trends seem to freely cross the overall allometric relationship, suggesting that developmental constraints cannot be that strong. Indeed, the breakdown of such overall allometric curves into distinct and virtually non-overlapping subgroups suggests that the impressive linearity of larger-scale groupings is deceiving. The primates have notably larger brains for their body weights, which further suggests that the allometric relationship is not frozen by an overall developmental program (Lande 1979). The relatively small brain relative to body weight of the gorilla (*Gorilla gorilla*) results probably from its folivorous habits, as opposed to the better endowed and frugivorous chimpanzee, *Pan troglodytes* (e.g., Harvey and Bennett, 1983). In other cases, however, such a prediction does not hold terribly well. On the basis of such reasoning, carnivores would be expected to have larger brains for their body sizes than ungulates, but this is not the case (Radinsky, 1978). The overall slope for all species may have other selective explanations, but may have some developmental restrictions as well.

The problem of the overall relationship of brain weight to body weight in the mammals has been a subject of controversy, even over the value of the exponent. Based upon recent evidence, the value of the exponent when comparing all species, is 0.75 (Martin, 1981). This suggests the possibility of a relationship with metabolic rate which also scales on body weight with the same exponent. It is possible that the brain acts as a coordinator of bodily activity correlated with overall metabolic rate.

Allometric exponents relating brain to body size among species at different taxonomic levels often differ. It is the usual case that the value of the exponent, comparing a given measurement with overall body size, decreases, as the species compared come from lower taxonomic levels (Count, 1947; Gould, 1975; Clutton-Brock and Harvey, 1979). This has been investigated systematically by Bennett and Harvey (1983). This relationship is curious and cannot be easily explained with either a selectionist or developmental constraint hypothesis. If some singular selective or developmental factor tightly regulated the relationship, then the exponent should not differ with taxonomic level. Lande (1979) and Harvey and Bennett (1983) relate this to a difference in rate of response of brain and body size in the same evolutionary series. If body size responds more rapidly than brain size in an overall trend of size increase, then we would expect a lag time until brain size adjusts to the overall scaling value.

As Lande notes, the study of trait relationships should be best understood in the context of quantitative genetic analyses that establish the degree of genetic correlation among traits. Cheverud et al. (1983) studied the quantitative genetics of age- specific trait values in the mouse and found that ontogenetic gene effects most commonly caused an individual to be larger or smaller throughout ontogeny. It was much less common to find genetic effects which were opposite at different ontogenetic stages, or to find effects unique to a given stage.

Lande (1979) used the literature on genetic correlations between brain and body weight to conclude that short term selection experiments would involve mostly change in overall body size with changes in brain size largely a genetically correlated response. The allometrical relationships of closely related species and populations within a species suggest that the genetic regression of brain on body size permits a successful prediction of the magnitude of the brain response when size is increased either by selection or genetic drift. If the laboratory data can be used to extrapolate to longer term evolution, Lande concludes that during the long-term allometric diversification within most mammalian orders there has been more net directional selection of brain size than body size.

More recent studies suggest that the genetic correlation between brain and body size may vary, depending upon the ontogenetic stage upon which selection acts. The genetic correlation between brain and body size is far higher in early postnatal growth in mice than in later growth (Atchley et al., 1984). The genetic correlation between brain size and body size decreases in mice as brain size growth in mice slows and eventually stops. Riska and Atchley (1984) suggest that this ontogenetic change in correlation may explain the increased value of the allometric exponent with increasing level of taxonomic comparison mentioned above. Body size evolution among higher taxa may include rapid shifts involving selection on earlier stages of growth which share more genetic growth determinants with brain size. Evolution among closely related species may involve adjustments of body size, which can be accomplished by selection on genetic components influencing later growth. Although this is an attractive hypothesis, the predicted allometric slope from selection on the earlier stages of growth can range from 0.43 - 0.76, using an estimate of one standard error. Unfortunately, this range embraces the slopes seen among all taxa (0.75) and those observed among more closely related taxa (0.4 and up, Harvey and Martin, unpublished). The analysis therefore may suggest that pleiotropy is indeed important as an evolutionary constraint. It does not prove, however, that the overall brain body size relationship is frozen by a long-standing genetic correlation that cannot be overcome. Riska and Atchley interpret the gorilla's large body size relative to brain size (Martin 1981) as a continuation of later postnatal growth relative to the chimp, whose relative brain size is considerably greater (see also Shea 1983). If the chimp were the ancestor of the gorilla, this reasoning may be justified, but the phylogenetic relationships of these two forms is in hot dispute, and one is certainly not the ancestor of the other. The fact that in such closely related forms body size can change so much relative to brain size suggests a degree of evolutionary lability that would permit fine adjustments of brain to body size to satisfy functional considerations. Moreover, it is not always possible to predict genetic correlations based upon developmental expectations. In the fowl, no genetic correlation exists between length of the tarsometacarpus and carpometacarpus, despite the serial homology (Cock, 1969). Intercorrelations among skeletal elements are often difficult to understand and impossible to predict (Cock, 1969; Cheverud et al., 1983).

The extension of laboratory measurements of genetic correlations among traits to interpretations of field distributions and even interspecific studies is tenuous at best. In *Drosophila*, extrapolations from laboratory measurements are possible to a degree (Robertson, 1962). Lines of *D. melanogaster* selected for change in wing, leg, and thorax length showed genetic correlations between wing and leg, but not between either and thorax length. In *D. subobscura* and *D. robusta* geographic races, wing length and leg length tend to covary, whereas head width and thorax length do not. The similarities in experiments and geographic variation may be due to genetic mechanisms whose effects are translated through developmetal apparatus of the imaginal disks.

In conclusion, the available evidence is firm enough to cast considerable doubt on the developmental constraint hypothesis as a universal explanation for allometric relationships in form. While no one would doubt that developmental constraints must have a strong influence on short term selective processes it seems likely that most of these constraints are breachable over the long term. The intriguing evidence presented by Riska and Atchley may, however, lend credence to a hypothesis of constraint through pleiotropy, but much more information is required before a quantitative assessment can be made.

THE CUMULATIVE AND GRADUAL ASSEMBLY OF COMPLEX BAUPLANS

The major phyla, and sometimes different taxonomic classes or orders, are often thought of as a series of bauplans, with no intermediates between them. Gould (1977, 1983) conceives of bauplans as sufficiently constrained by development that further evolutionary change is unlikely. Gaps between major forms are often explained in terms of current observations of the discrete jumps in phenotypic difference that can be induced as developmental anomalies (Gould 1982). These considerations raise the question of whether bauplans arose gradually or in discrete major steps. I will focus on the mammal bauplan but use other examples in this discussion.

It may be sufficient to document change through a geological section to justify the phyletic nature of single-character evolution, but an attribute such as "mammalness" cannot be restricted to one character, even if the definition of the group can be made in this way (Simpson, 1960; Crompton and Jenkins, 1973). The success and ecological position of mammals owes as much to homeothermy, reproduction, and a complex central nervous system, as it does to the acquisition of a new jaw joint and more complex teeth. This is after all the crux of any macroevolutionary question. (1) What is the pattern of acquisition of the many features? (2) What controls the pattern and rate of acquisition? (3) Once acquired, what holds the plexus together, if anything? These macroevolutionary questions can in principle be independent of the issue of speciation. The pattern of acquisition of various mammalian traits can be entirely coincident with speciation events or may consist of entirely within-species phyletic evolution. But as long as the mammals, with all of their unique characters fully formed, did not arise in a single speciation event, the questions still have relevance.

Three end member patterns can be imagined for the acquisition of the characters making up a complex group. (1) *Saltational Hypothesis*: First, all characters could have been acquired at once. I don't believe that anyone takes Schindewolf's (1936) claim that the first bird hatched from a reptile's egg very seriously, without accepting some clever definitions of bird and reptile. Gould (1980) has revived the notion of hopeful monster, but certainly not at the level of wholly new organized body plans. (2) *Independent blocks Hypothesis*. Secondly, we might imagine that the features that characterize a group such as the mammals can be ascribed to distinct blocks, and that these blocks evolved independently. As time progressed, interactions between the blocks occurred but only because of coincidental response to the same new environment, although later functional interdependencies among the blocks may have arisen. This has also been termed *mosaic evolution* (see DeBeer, 1958; Stebbins, 1984), but here the notion has developmental connotations, as discussed above. (3) Correlated Progression Hypothesis. Finally, we could argue that different morphological complexes can be distinguished, but they have a sufficient interrelationship such that each system appears to "..evolve in a loose correlation with all the other systems" (Kemp 1982, p. 313).

I support Kemp's (1982) hypothesis that correlated progression is a major component of mammalian evolution. Over time, the system has congealed from both a functional and developmental point of view, to the point that evolution to maintain the whole is probably as important as evolution designed to adapt independently any particular block to a change in the environment. I emphatically do not imply that this will be the case for all groups. Indeed, since theory does not predict unique solutions in most cases, our approach to such questions must be largely inductive.

The evolution leading to mammals is unique in that a remarkable gap in morphological organization seen in living forms between mammals and reptiles, is bridged in the fossil record with an array of intermediates (Crompton and Jenkins, 1968, 1973; Kemp, 1982; Parrington, 1971). This is particularly interesting, given the considerable evidence that additional discoveries are likely to be made with increased sampling (e.g. Jenkins et al., 1983). As it now stands, the fossil record of the mammal-like reptiles (synapsids) documents approximately 130 million years of change from their earliest appearance in the early Upper Carboniferous to the first appearance of the "true" mammals in the Upper Triassic. Because the continents were not separated during this period, faunas were geographically relatively homogeneous and incomplete sections in one location can therefore be filled in, at least at a coarse scale, with collections from other localities. The earliest mammals, found in South Wales and northeastern Arizona, already show a diversity of jaw structure (Jenkins et al., 1983).

The synapsids made their first appearance in the Late Carboniferous. They dominated the reptilian faunas until the end of the Triassic, but then declined gradually when the dinosaurs replaced them and dominated the second age of reptiles in the Jurassic-Cretaceous. During this period their apparent descendants, the mammals, played a minor role in terrestrial faunas and were typically diminutive in size. The gradual shift from dominance by obvious reptilians such as the pelycosaurs in the Permian to groups much more mammalian in character in the Triassic is believed to be associated with a general adaptation to terrestrial life (Kemp, 1982). This would imply that the groups extant in the Triassic were superior to their predecessors at surviving the exigencies of terrestrial life. We must remember, however, that this superiority did not guarantee a place of dominance in the terrestrial world. Indeed, the dinosaurs seem to have forestalled the rise of the mammals for a considerable period of time. This suggests that trends seen within a group do not imply superiority to anything other than their own antecedents and the relatives of the latter. It is unfortunate to think of the mammals as superior to the dinosaurs, or, for that matter, to think of the dinosaurs as inherently superior to the mammal-like reptiles.

The distinction between advancement within a clade and interclade superiority is crucial. The former can be argued on functional morphological grounds, while the latter can only be argued from historical evidence of interclade competition, which is often weak or nonexistent (Benton, 1983a,b). Even within the synapsid clade, "advancement" can only be considered within the context of the characters that are associated with terrestriality *per se*. Many unique features acquired by members of the clade may have momentarily been more important in survival than the features associated with the overall trend towards "mammalness."

Unfortunately, many characters that we associate with "mammalness" - heart and double circulation, sweat glands, hair, lactation and parental care, an impermeable skin, production of a hypertonic urine, and a larger number of mitochondria per cell - cannot be documented meaningfully at this time from the fossil record. We are thus left with those characteristics that can be inferred directly and indirectly from the cranial and postcranial skeleton. This is unfortunate as it may focus unduly our attention at only part of the picture. Skeletal characters can reveal much about physiology and activity, but not everything. We have for example, a very incomplete picture of temperature control. Even among the extant homeothermic mammals, large differences exist in the absolute value of temperature, even if the temperature is maintained relatively constant. In fossils, there is a strong case for homeothermy in advanced mammal-like reptiles and dinosaurs, based upon the lack of seasonal growth rings in bones, other features of bone structure, and predator- prey ratios characteristic of endotherms (e.g. Bakker, 1974, 1977). Homeothermy could have been accomplished at first by increased body size and by the evolution of mechanisms to retard loss of the heat gained in exersize (McNab, 1978). An increase of metabolic rate could have occurred separately. In the cynodonts, a group of advanced mammal-like reptiles, the presence of a secondary palate and the complex mastication apparatus both point to a high metabolic rate, requiring concomitant features enhancing the rate of food acquisition.

Most research (e.g. Crompton, 1963; Crompton and Jenkins, 1968, 1973) has focused upon the changes in the jaw and dentition. This was partly due their relatively good preservation and to the importance of dentition and mastication to the efficient processing of food required to service the mammalian life style which includes a high metabolic rate. The record of transitions - from the Pelycosauria, to the Therapsida, to the Cynodonta, to the Mammalia - is impressive, as it includes a complete graded series ranging from the primitive reptilian skull with (1) alternate tooth replacement and relatively simple homodont teeth, (2) articular-quadrate jaw joint, and (3) relatively simple jaw musculature, to the mammalian condition of (1) single replacement of complex heterodont teeth, (2) dentary-temporal jaw joint, and (3) complex jaw musculature which reduces the mechanical load on the articular-quadrate joint (Crompton 1963). A host of important features evolve in concert with these three major systems, most notably the various muscular support structures such as the temporal fenestra (Frazzetta, 1969).

The most interesting transition can be seen from the perspectives of jaw articulation and mastication. The evolution of efficient food processing requires complex occlusion and the ability to move the lower jaws in several directions (Crompton, 1963). In the advanced cynodont, *Thrinaxodon*, the postcanine teeth did not occlude in the sense of extensive intermolar contact. This was precluded by the reptilian tooth pattern of alternate tooth replacement (Crompton and Jenkins, 1968). In more advanced cynodonts, occlusal facets essentially characteristic of the later mammals were established by gradual wear. Later, the faceting was genetically determined.

Such changes in occlusion were accompanied by a change in the jaw musculature that permitted greater latitude in movement and increased mastication strength. Such strength would have exerted significant forces on the reptilian jaw joint, which functioned as a third class lever. The evolution of the jaw musculature increasingly reduced the load on the reptilian jaw joint, and the opposing action of the temporalis and masseter muscles permitted the lateral control necessary for complex chewing. In the cynodonts, a trend for reduction of the post dentary bones eventually led to the possibility for the mammalian style jaw joint, which, in several transitional forms, existed in line with the older reptilian joint. This is the most unlikely transition of all: the coincidence of two fulcra (Jaanusson 1981). In modern reptiles and mammals, it would seem like an impossible transition for the location of a jaw articulation to switch from one fulcrum to another without a saltation.

In conjunction with this already remarkable series of changes, the evolution of the mammalian jaw joint is intimately involved with the evolution of the bones of the middle ear. The articular and the quadrate jaw bones in reptiles are the homologues of two of the mammalian middle ear bones, the malleus and the incus (Hopson 1966). As the new mammalian dentary-temporal jaw articulation arose, the post dentary bones decreased in size and were freed for cooption as part of the middle ear apparatus. Allin (1975) argues that, along with the trends in synapsid evolution towards masticatory efficiency, some features just anterior to the reptilian jaw joint indicate an auditory function. If this is true, then the evolution of hearing and chewing is an excellent example of Kemp's (1982) concept of correlated progression. From the point of view of functional morphology, this series of changes demonstrates that functioning structures cannot be predicted "from the ground up." In order to assemble a model of functional change, one must start with the reptilian joint, and the constraint of a requisite for efficient hearing.

Our account of the changes occurring in the evolution of mammals demonstrates the complex interrelationships among the various parts of the skull, most notably muscular supports, dentition, and jaw articulations. The evolution of the mammalian middle ear (at least as we see it now) could not progress without significant change in the various features associated with the evolution of efficient chewing. It seems unlikely that any part of the skull could have evolved very much without an effect on the functioning of another. The best example is the need to reduce the mechanical load on the reptilian jaw joint, as masticatory forces become more intense and complex (Crompton and Jenkins, 1968). Changes in the postcranial skeleton, the evolution of dorso-ventral flexibility for example, also helped contribute to the general need for an increase in food gathering efficiency. The latter was encumbered by still another evolutionary change incorporating higher metabolic rates which generate the need for a higher rate of food intake. As Kemp (1982) noted, it is difficult to take a modern mammal and dissect the crucial mammalian feature; they are all so interdependent. It is therefore not surprising and rather gratifying to see the fossil record show a correlated progression of change.

This general account of transitional change can be made more precise by defining the phylogenetic relationships of the various synapsid groups involved in the evolution of mammals with a cladistic analysis. We can then ask the question: were the various mammalian characters acquired in independent groups or can we speak of a correlated progression of features. Kemp (1982) has associated the features acquired by the various groups with a cladogram depicting genealogical relationships. Current data can only articulate the big picture; we only have a general cladistic skeleton, and cannot say with the evidence whether transitions consist of minute quantitative changes or major qualitative changes. Intermediate morphological stages are usually absent below the family level, but the continual discovery of new forms suggests that missing data, rather than saltations, are the likely explanations. As an example, evidence until recently would have suggested a possible major quantum change, perhaps based upon a developmental mutant, to explain the origin of the angular region of the mammalian jaw, from the reptilian pseudangular process. Recent evidence, however, suggests that the angular region arose as a neomorphic structure based upon an apparent incipient presence in an early mammal (Jenkins et al. 1983).

The pattern of acquisition of mammalian characters illustrated in the cladogram (Fig. 1). shows the gradual nature of mammalian evolution. Significant incremental changes towards mammalness in the size of the jaw adductor musculature, complexity of the dentition, and the postcranial skeleton, all occur in the majority of internodal steps in the cladogram. In the case of the interaction of jaw musculature, dentition, nature of the jaw joint, and the middle ear, these changes could not help but be interactive. Intermediate stages in the process are therefore functionally harmonious, the result of adaptation, but not optimal from the point of view of function. To use Darwin's wording, they were the best in their "country." Some other features, however, may have been acquired according to the model of mosaic evolution. The evolution of the kidney, for example, may have been in response to similar selective pressures posed by the terrestrial environment on feeding, but the acquisition need not be intimately correlated with changes in the skull.

As the record of the evolution of mammals is a story filled with intermediates, two important and unresolved questions come to mind. (1) Why did the process take so long - ca. 130 million years? (2) Is there any functional problem posed by intermediate forms? The case of the mammals demonstrates that both questions, even if they have been major concerns in the history of evolutionary biology, may have little meaning when considering organizational changes of this magnitude. The question of tempo is nearly impossible to solve. We often forget that the complex of taxa involved in mammalian evolution had important and unique features other than those upon which we focus in understanding the transitional stages leading to the mammals. In some cases these unique features may have interacted significantly with intermediate character complexes and may have either led members of the taxon in question down a dead end path, or at least decreased the chances for advancement towards the mammalian condition.

For example, the dinocephalians constituted the majority of the early Late Permian therapsid faunas. They consisted of very large animals with a number of features that functionally may have precluded an easy advancement to the mammalian condition. Unfortunately, we have no idea as to how such conflicts

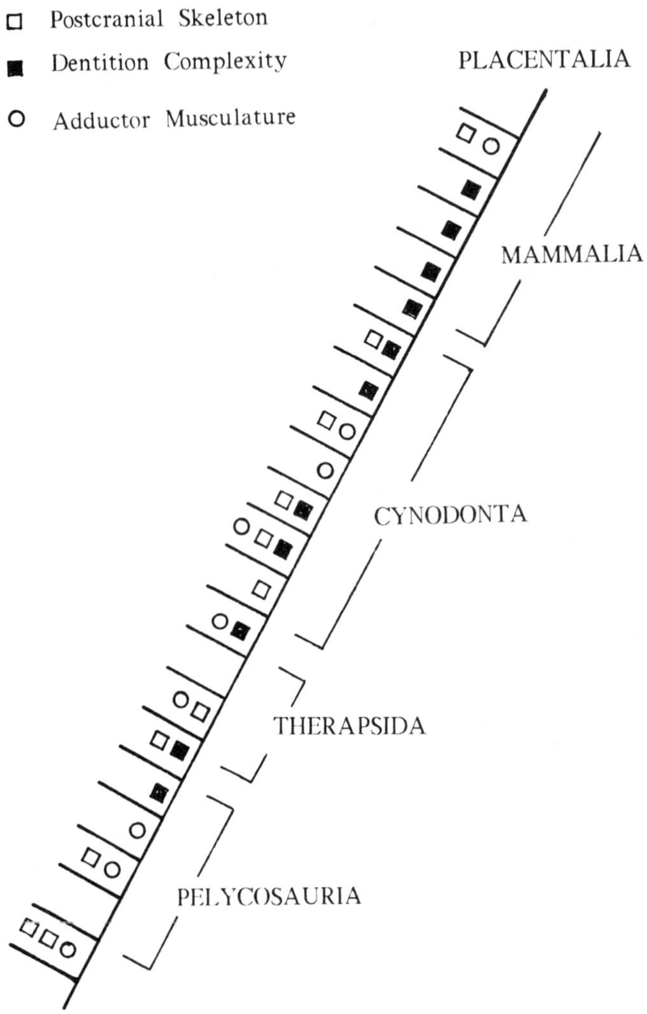

FIGURE 1. A CLADOGRAM FOR THE GENERA IN THE LINEAGE LEADING TO THE THERIAN MAMMALS: Major new acquisitions of skeletal features leading to the mammalian condition are indicated (after Kemp, 1982).

might influence whether a group can be a progenitor of further advanced forms. In the context of terrestriality, there seems no doubt that the overall path towards the mammals involved adaptation, but other conflicting adaptations may have been present in many of the groups. Unless some clear model of selective value of these diversions and retardants can be established, it becomes fruitless to argue why the rate was so slow (or fast, for that matter). In many ways, the diversions are more interesting than the main mammalian line itself. Too often, we ignore the fact that much of the diversity of life simply is precluded from being ancestors by being burdened withone or many features that restrict evolutionary direction and potential. It is therefore equally fruitless to view the evolutionary trend towards mammalness as "progress"; mammal-like reptiles didn't know what they were going to be, and true mammals do not remember what their ancestors were! Once mammalness was completed, however, it does make sense to ask why they did so well with the array of features they acquired over the last 130 million years.

We might ask if this sort of gradual accumulation of characters, found in the broad picture of mammalian evolution, can be seen at lower levels, such as orders or families. As noted above, interfamily transitional forms are often absent, probably as the result of inadequate finds rather than saltation. But sufficient evidence exists to examine the question of whether evolution within a mammalian order is of the same type as on the level of Mammalia.

The mammalian order Carnivora first appeared in the middle Paleocene and was represented by members of the family Miacidae until the end of the Eocene, some 20 million years later. After the extinction of some coexisting carnivorous mammals, the miacids radiated into what produced most of the modern families of the Carnivora (canids, felids, viverrids, mustelids, ursids) as well as two extinct families. Using factor analysis, Radinsky (1982) demonstrated that the differences that now distinguish modern viverrids, canids, felids and mustelids were less pronounced in the Oligocene, when these families make their first appearance in the fossil record. Cranial differences in representatives of the living families can be related to differences in prey-killing, but not necessarily to differences in prey type or size. But only some of these distinctive features can be found in Oligocene ancestors of the modern families. Radinsky suggested that body size differences established early in the history of the group may have been an important factor in the radiation. As body size is correlated to prey size in carnivores (Rosenzweig, 1966), adaptations to prey handling may have been triggered by or coincided with body size differentiation. These adaptations may have been entrenched with the accumulation of further traits, when the miacids survived the late Eocene extinction of many early carnivores, perhaps even as a matter of chance. From a quite different specific context, size variation in spores may have been the initial variation that was later evolved into morphologically distinct gametes, and ultimately, seeds in plants (Tiffney, 1982). Heterospory is represented initially in the fossil record simply by spore size variation within a single sporangium in the Devonian *Chaleuria*. Thus later complex morphological diversification may have been initiated by simple differences in size.

At a still lower taxonomic level, Rose and Bown (1984) have examined dental evolution within the Eocene primate family Omomyidae, sampled from the excellently preserved faunas of the Willwood formation in the Bighorn Basin, Wyoming. This formation was the source of evidence for gradual evolution collected by Gingerich (1976). Two cases of transitional evolution at the generic level have been elucidated. In one transition, from *Tetonius homonculus* to *Pseudotetonius ambiguus*, modifications that characterize the latter genus do not appear synchronously and patterns of transformation differ strongly among characters. The mosaic nature of change and the strong variability of some characters makes the point of demarcation between genera largely arbitrary. This is apparently the general case in mammals and in other groups (Levinton 1983).

The evolution of the angiosperms provides another example of the piecemeal evolution common in the evolution of major new groups. As Tiffney (1982) noted, many of the features thought to be characteristic of the angiosperms were already present in their remote ancestors. Thus, if there is a feature such as "angiospermness," it did not arise suddenly, but rather by a combination of correlated progression and mosaic evolution. Tiffney outlined a scheme of potential correlated progression remarkably similar in approach to that of Kemp's analysis for mammals. Many of the individual features of angiosperms appeared in various fossil groups, but only one line seemed to accumulate them all. Stebbins (1984) briefly summarized the literature and reached the same conclusion for the origin of birds from a reptilian ancestor, amphibians, and several other groups. Indeed, Schindewolf (1936) is exactly correct. The first bird did hatch from a reptile's egg. But this is only true when we use some particular feature to define the birds. The class Aves was nearly already fully formed the moment it was "born." In some cases, the birth of a new group can be marked with a key innovation that must have permitted an extensive radiation, as in the development of selenodont teeth in the Artiodactyls. But an examination of the Carnivora reveals no keystone innovations that can readily define "carnivoreness" in one step (Radinsky, 1982). The same can be said in the evolution of the mammals.

This discussion demonstrates, in some well-documented cases, that what we now consider a fixed bauplan in given groups of organisms arose gradually by accumulation of features that may or may not have arisen by correlated progression. In some groups, the doors to success were opened by key innovations, but this seems not to be the case for the mammals. This perspective, obtainable only from the fossil record, undermines the nearly essentialist viewpoint voiced, for example, by Gould which treats different bauplans as integrated entities that resist change and tends to ignore their gradual origins. Gould states (1983, p. 80): "..for surely the cluster of cats exists primarily as a result of homology, and historical constraint. All felines are alike because they arose from a common ancestor shared with no other clade. .... All feline species have inherited the unique cat *Bauplan*, and cannot deviate far from it as they adapt." Genealogically, of course, all felids have descended from a common ancestral species. But this omits the essential nature of the evolutionary process. The "cat bauplan" was assembled gradually, first as "mammalness," and later as "catness." After this set of adaptations was congealed by both functional interactions and incorporation of traits into developmentally linked complexes, it is no surprize that current conditions do not now permit much of a deviation from the basic organization that we use to define cats. The entire process of adaptation and crystallization of the cat Bauplan involved a presence in the same ecological milieu, that is, hunting, seizing, and eating usually mobile prey. In writing of present constraints, Gould leads the reader away from the likely adaptive origin of cat traits. It is thus meaningless for Gould (1983, p. 80) to dismiss correlations of cat morphology with the present environment and to argue that "..genealogy, not current adaptation, is the primary source of clumped distribution in morphological space." How does one distinguish between genealogy and adaptation? The dichotomy is false.

The fossil record reveals a very clear pattern. From the lowest level of resolution of morphological change in phyletic lineages at the species level, to the highest level, we see the gradual accumulation of novelties which contribute to form the distinctive and morphologically separate clades which exist today, or have existed at any time in the past. The unique genetic properties of various traits strongly influences the pattern of change seen in a geological column. On higher levels, we see traits assembled over time and trait groups gradually congealing into total organisms with the evolution of various developmental programs. Some of these programs make any change very costly to the organism, but it is a matter for empirical investigation to determine this cost. But it is clear that the evolution of bauplans are often gradual and that the present distribution of groups tightly clustered in morphospace is in large part a reflection of this gradual congealing. Developmental constraints may regulate the congealing, but adaptation may often be a major element in the evolution of developmental constraints. Since the organism evolves mechanisms to stay in its favored environment, the gradual congealing is a reciprocal process of (1) culling of variants arising as mutants and (2) the increasing commitment of the organism to a given lifestyle which keeps it locked into the same selective regime for long periods of time. It is this very reciprocal process which weakens the claims for the isolated importance of historical or developmental constraints on evolution. Organisms adapt partially by sealing their own fates through a continually increasing commitment to a given lifestyle. As in the mammals, there is patent evidence that this increasing commitment is associated with improved performance. Adaptation is not divorced from history; it is an integrated part of history.

CONCLUSIONS

(1) Developmental constraints are part of a larger picture of temporally increasing commitment of the organism's morphology to a bounded set of evolutionary directions. I characterize this increasing commitment as the evolutionary ratchet, which has genetic, epigenetic, and functional components.

(2) Adaptation may be intimately involved in the evolution of constraints; it is therefore incorrect to speak of constraints as necessarily nonadaptive or a unique nonadaptive component in evolution.

(3) A notion of saltational evolution arises naturally from examination of developmental variation. But this can often be fallacious; a long period of gradual evolution and congealing of a developmental program, may seem at present to be saltational. Once developmental programs are congealed and separated as a block, alterations by mutation are usually phenotypically discontinuous, but are also strongly pleiotropic in effect and greatly diminish fitness. In saying this, one should not overlook that the developmental mechanism can and does create discontinuity.

(4) Even if developmental commitment does occur, one should not presume so without direct empirical evidence. Cases are known where strong changes in early developmental programs have little effect on the final adult form.

(5) The major bauplans seem not to have arisen suddenly, nor is there much evidence that major form differences are nonadaptive. The fossil record of mammals and other groups show that the traits characteristic of the gestalt of the group have accumulated piecemeal. At every level, evolutionary transitions seem to be fragmentary, and not of major bauplan shifts.

(6) One should not exclude developmental constraints, nor should one ignore the possibility of discrete differences being generated by developmental-genetic sources of novelties. But this recognition is long-standing among evolutionary biologists. It is now time for an empirical program to document the interactions of genetics, development, and natural selection on phenotypic evolution.

ACKNOWLEDGEMENTS

I am grateful to Bill Atchley, Mike Bell, Stefan Bengtson, John Bonner, Paul Harvey, Paul Maderson, Joan Miyazaki, Ledyard Stebbins, and David Wake, who read versions or parts of this draft and provided many useful comments and criticisms. I appreciate the kind hospitality of the Station Zoologique, Villefranche-Sur-Mer, where this paper was mainly prepared. This paper is contribution 534 to the Graduate Studies in Ecology and Evolution, State University of New York at Stony Brook.

REFERENCES

Alberch, P., 1980, Ontogenesis and morphological diversification, *Amer. Zool.*, 20:653-667.

Alberch, P., 1981, Convergence and parallelism in foot morphology in the Neotropical salamander genus *Bolitoglossa.* I. *Evolution*, 35:84-100.

Alberch, P., 1982, Developmental constraints in evolutionary processes, J.T. Bonner, ed., *in:* "Development and Evolution, Springer-Verlag, Berlin, pp. 313-332.

Alberch, P., 1983, Morphological variation in the Neotropical salamander genus, *Bolitoglossa*, *Evolution*, 37:90

Alberch, P., and Gale, E., 1983, Size dependence during the development of the amphibian foot. Colchicine-induced digital loss and reduction, *J. Embryol. Exp. Morph.* 76:177-197.

Alberch, P., Gould, S.J., Oster, G.F., Wake, D.B., 1979, Size and shape in ontogeny and phylogeny, *Paleobiology*, 5:296-

Allin, E.F., 1975, Evolution of the mammalian middle ear, *J. Morph.*, 147:403-438.

Andrews, R.C., 1921, A remarkable case of external hind limbs in a humpback whale, *Am. Mus. Novitates* 9:1-16.

Atchley, W.R., Riska, B., Kohn, L.A.P., Plummer, A.A., and Rutledge, J.J., 1984, A quantitative genetic analysis of brain and body size associations, their origin and ontogeny: data from mice, *Evolution*, 38:1165-1179.

Atchley, W.R., and Rutledge, J.J., 1980, Genetic components of size and shape. I. Dynamic components of phenotypic variability and covariability during ontogeny in the laboratory rat, *Evolution*, 34:1161-1173.

Baer K.E. v, 1828, Entwickslungsgeschichte der Thiere: Beobachtung und Reflexion. Konisgberg: Borntrager.

Bakker, R.T., 1974, Experimental and fossil evidence for the evolution of tetrapod bioenergetics, *in*: "Perspectives in Biophysical Ecology," D. Gates and R. Schmerl, eds., Springer- Verlag, New York, pp. 365-399.

Bakker, R.T., 1977, Tetrapod mass extinctions - a model of the regulation of speciation rates and immigration by cycles of topographic diversity, *in*: "Patterns of Evolution as Illustrated by the Fossil Record," A. Hallam, ed., Elsevier Scientific, Amsterdam, pp. 439-468..

Benton, M.J., 1983a, Large-scale replacements in the history of life, *Nature*, 302:16-17.

Benton, M.J., 1983b, Dinosaur success in the Triassic: a noncompetitive ecological model, *Q. Rev. Biol.*, 58:29-55.

Bonner, J.T., 1982, "Evolution and Development," Springer-Verlag, Berlin.

Bowen, S.T., Hanson, J., Dowling, P., and Poon, M.-C., 1966, The genetics of *Artemia salina* VI. Summary of mutations, *Biol. Bull.*, 131:230-250.

Cherry, L.M., Case, S.M., and Wilson, A.C., 1978, Frog perspective on the morphological difference between humans and chimpanzees, *Science*, 200: 209-211.

Cheverud, J.M., Rutledge, J.J., and Atchley, W.R., 1983, Quantitative genetics of development: genetic correlations among age-specific trait values and the evolution of ontogeny, *Evolution*, 37:895-905.

Clutton-Brock, T.H., Albon, S.D., and Harvey, P.H., 1980, Antlers, body size and breeding group size in the Cervidae, *Nature*, 565- 567.

Clutton-Brock, T.H., and Harvey,P.H., 1979, Comparison and adaptation, *Proc. Roy. Soc. Lond.*, B205:547-565.

Cock, A.G., 1969, Genetical studies on growth and form in the fowl, *Genet. Res.*, 14:237-247.

Count, E.W., 1947, Brain and body weight in man: their antecedents in growth and evolution, *Ann. N.Y. Acad. Sci.*, 46:993-1122.

Crompton, A.W., 1963, On the lower jaw of *Diarthrognathus* and the origin of the mammalian lower jaw, *Proc. Zool. Soc. London*, 140:697-753.

Crompton, A.W., and Jenkins, F.A., 1968, Molar occlusion in Late Triassic mammals, *Biol. Rev.*, 43:427-458.

Crompton, A.W., and Jenkins, F.A., 1973, Mammals from reptiles: a review of mammalian origins, *Ann. Rev. Earth Planet. Sci.*, 1:131- 153.

Crompton, A.W., and Jenkins, F.A., 1979, Origin of mammals, *in*: "Mesozoic Mammals," Lillegraven, J.A., Kielan-Jaworska, A., and Clemens, W.A., eds., Univ. of California Press, Berkeley, pp.

Dahl, E., 1959, The ontogeny and comparative anatomy of some protocerebral sense organs in notostracan Phyllopods, *Quart. J. Micr. Sci.*, 100:445-462.

Davenport, R., 1979, "An Outline of Animal Development," Addison- Wesley, Reading.

DeBeer, G., 1958, "Embryos and Ancestors," 3rd edition, Clarendon Press, Oxford.

DelPino, E. M., and Elinson, R.P., 1983, A novel developmental pattern for frogs: gastrulation produces an embryonic disk, *Nature*, 306:589-591.

Dobzhansky, T., 1951, "Genetics and the Origin of Species," 3rd. ed. Columbia Univ. Press, New York.

Dobzhansky, T., 1970, "Genetics of the Evolutionary Process," Columbia Univ. Press, New York.

Frazzetta, T.H., 1969, Adaptive problems and possibilities in the temporal fenestration of tetrapod skulls, *J. Morph.*, 125:145-158.

Frazzetta, T.H., 1970, From hopeful monsters to Bolyerine snakes, *Am. Nat.*, 104:55-70.

Frazzetta, T.H., 1975, "Complex Adaptations in Evolving Populations," Sinauer, Sunderland, MA.

Garcia-Bellido, A., 1975, Genetic controll of wing disc development in "Drosophila", *in*: "Cell Patterning," *CIBA Foundation Symp.* 29, pp. 161-182.

Garn, S.M., and Lewis, A.B., 1963, Third molar polymorphism and its significance to dental genetics, *J. Dent. Res.*, 42 suppl.:1334-1363.

Garstang, W., 1922, The theory of recapitulation. A critical restatement of the biogenetic law, *J. Linn. Soc. London, Zoology* 35:81-101.

Gingerich, P.D., 1976, Paleontology and phylogeny: patterns of evolution at the species level in Early Tertiary mammals, *Amer. J. Sci.*, 276:1-28.

Gingerich, P.D., Wells, N.A., Russell, D.E., and Ibrahim Shah, S.M., 1983, Origin of whales in epicontinental remnant seas: new evidence from the early Eocene of Pakistan, *Science*, 220:403-406.

Goldschmidt, R., 1938, "Physiological Genetics," McGraw-Hill, New York.

Goldschmidt, R., 1940, "The Material Basis of Evolution," Yale Univ. Press, New Haven.

Goldschmidt, R.B., 1945, Mimetic polymorphism, a controversial chapter of Darwinism, *Quart. Rev. Biol.*, 20:147-164, 205-330.

Gould, S.J., 1974, The evolutionary significance of 'bizarre' structures: antler size and skull size in the 'Irish Elk', *Megalocerus giganteus, Evolution*, 28:191-220.

Gould, S.J., 1975, Allometry in primates with an emphasis on scaling and the evolution of the brain, *Contr. Primatol.*, 5:244- 292.

Gould, S.J., 1977, "Ontogeny and Phylogeny," Harvard Univ. Press, Cambridge, MA.

Gould, S.J., 1980, Is a new and general theory of evolution emerging? *Paleobiology*, 6:119-130.

Gould, S.J., 1982, Change in developmental timing as a mechanism in macroevolution, *in*: "Development and Evolution," Bonner, J.T., ed., Springer-Verlag, Berlin, pp. 333-346.

Gould, S.J., 1983, Irrelevance, submission, and partnership: the changing role of palaeontology in Darwin's three centennials and a modest proposal for macroevolution. in: "Evolution From Molecules to Men," Bendall, D.S.,ed., Cambridge Univ. Press, Cambridge, pp. 347-366.

Green, E.L., 1962, Quantitative genetics of skeletal variations in the mouse. II. Crosses between four inbred strains (C3H, DBA, C57BL, BALB/c), *Genetics*, 47:1085-1096.

Grewal, M.A., 1962, The development of an inherited tooth defect in the mouse, *J. Embryol. Exp. Morph.*, 10:202-211.

Gruneberg, H., 1965, Genes and genotypes affecting the teeth of the mouse, *J. Embryol. Exp. Morph.*, 14:137-159.

Hadorn, E., 1961, "Developmental Genetics and Lethal Factors," (transl. from German) Methuen & Co, London.

Haldane, J.B.S., 1932, The time of action of genes, and its bearing on some evolutionary problems, *Am. Nat.*, 66:5-24.

Haeckel, E., 1866, "Generelle Morphologie der Organizmen," 2 vols., Reimer, Berlin.

Hampè, A., 1959, Contribution a l'etude du developement et de la regulation des deficiencies et des excedents dans la patte de l'embryon de poulet, *Arch. Anat. Microscop. Morphol. Exp.*, 48:347- 478.

Harcourt, A.H., Harvey, P.H., Larson, S.G., and Short, R.V., 1981 Testis weight, body weight and breeding system in primates, *Nature*, 293:55-57.

Harvey, P.H. and Bennett, P.M., 1983, Brain size, energetics, ecology and life history patterns, *Nature*, 306:314-315.

Hersh, A.H., 1934, Evolutionary relative growth in the Titanotheres, *Am. Nat.*, 68:537-561.

Hinchcliffe, J.R. and Gumpel-Pinot, M., 1981, Control of maintenance and anteroposterior skeletal differentiation of the anterior mesenchyme of the chick wing bud by its posterior margin (the ZPA), *J. Embryol. Exp. Morph.*, 62:63-82.

Hopson, J.A., 1966, The origin of the mammalian middle ear, *Am. Zool.*, 437-450.

Huxley, J.H., 1931, The relative size of antlers in deer, *Zool. Soc. Lond.*19-864.

Huxley, J.H., 1932, "Problems of Relative Growth," MacVeagh, London.

Huxley, J.H., 1960, The emergence of Darwinism, *in*: "Evolution After Darwin," v. 1., S. Tax, ed., Univ. of Chicago Press, Chicago, pp. 1-21.

Jaanusson, V., 1981, Functional thresholds in evolutionary progress, *Lethaia*, 14:251-260.

Jaffe, L.F., Stern, C.D., 1979, Strong electrical currents leave the primitive streak of chick embryos, *Science*, 206:569-571.

Jenkins, F.A. Jr., Crompton, A.W., and Downs, W.R., 1983, Mesozoic mammals from Arizona: new evidence on mammalian evolution, *Science*, 222:1233-1235.

Kemp, T.S., 1982, "Mammal-like Reptiles and the Origin of Mammals," Academic Press, London.

Kniprath, E., 1981, Ontogeny of the molluscan shell field: a review, *Zoologica Scripta*, 10:61-79.

Kollar, E.J. and Fisher, C., 1980, Tooth induction in chick epithelium: expression of quiescent genes for enamel synthesis, *Science*, 207:993-995.

Kuren , B., 1963, Return of a lost structure in the evolution of felid dentition, *Soc. Sci. Fenn.*, 26:4-12.

Lande, R., 1978, Evolutionary mechanisms of limb loss in tetrapods, *Evolution*, 32:73-92.

Lande, R., 1979, Quantitative genetic analysis of multivariate evolution, applied to brain: body size allometry, *Evolution*, 33:402-416.

Levinton, J.S., 1983, Stasis in progress: the empirical basis of macroevolution, *Annu. Rev. Ecol. Syst.*, 14:103-137.

Lewis, E.B., 1978, A gene complex controlling gene segmentation in "Drosophila", *Nature*, 276:565-570.

Lewontin, R.C., 1978, Adaptation, *Scientif. Amer.*, 239(3):156-169.

Lillie, F.R., 1898, Adaptation in cleavage, Biological Lectures of the Marine Biological Laboratory, Woods Hole. Ginn and Co., Boston pp. 43-67.

Lovtrup, S., 1974, "Epigenetics," John Wiley and Sons, New York.

Maderson, P.F.A., 1975, Embryonic tissue interactions as the basis for morphological change in evolution, *Amer. Zool.*, 15:315-327.

Maderson, P.F.A., Alberch, P., Goodwin, B.C., Gould, S.J., Hoffman, A., Murray, J.D., Raup, D.M., de Ricqles, A., Seilacher, A., Wagner, G.P. and Wake, D.B., 1982. The role of development in macroevolutionary change, *in*: "Evolution and Development," Bonner, J.T., ed., Springer-Verlag, Berlin, pp. 279-312.

Marsh, O.C., 1892, Recent polydactyle horses, *Am. J. Sci.* , 43:339- 355.

Martin, R.D., 1981, Relative brain size and basal metabolic rate in terrestrial vertebrates, *Nature*, 293:57-60.

Maynard Smith, J., 1960, Continuous, quantized and modal variation, *Proc. Roy. Soc. London*, 152B:397-409.

Mayr, E., 1963, "Animal Species and Evolution," Belknap Press, Harvard Univ. Press, Cambridge, MA.

McNab, B.K., 1978, The evolution of endothermy in the phylogeny of mammals, *Am. Nat.*, 112:1-21.

Murray, J.D., and Oster, G.F., 1984, Generation of biological pattern and form, *IMA J. of Mathematics Applied in Medicine and Biology*, 1:51-75.

North, G., 1984, How to make a fruitfly, *Nature*, 311:214-216.

Nuccitelli, R. 1983. Transcellular ion currents: signlas and effectors of cell polarity, *in*: "Modern Cell Biology," v. 2, J.R. McIntosh, ed., Alan R. Liss, New York, pp. 451-481.

Odell, G.M., Oster, G., Alberch, P., and Burnside, B., 1981, The mechanical basis of morphogenesis. I. Epithelial folding and invagination, *Devl. Biol.*, 85:446-462,

Ohno, S., 1973, Ancient linkage groups and frozen accidents, *Nature*, 244:259-262.

Osborn, H.F., 1929, The Titanotheres of ancient Wyoming, Dakota and Nebraska, U.S. Dept. Int., Geol. Survey., Monogr. 55, 2 vols., Washington, D.C.

Oster, G.F., Murray, J.D., and Harris, A.K., 1983, Mechanic aspects of mesenchymal morphogenesis, *J. Embryol. Exp. Morph.* , 78:83-125.

Parrington, F.R., 1971, On the upper Triassic mammals, *Phil. Trans. Roy. Soc. Lond.*, 261B:231-272.

Punnett, R.C., 1915, "Mimicry in Butterflies," Cambridge Univ. Press, Cambridge.

Rachootin, S., and Thomson, K.S., 1981, Epigenetics, paleontology, and evolution, *in*: "Evolution Today," Proc. Intl. Congr. Syst. Evol. Biol., Scudder, G.G.E., and Raveal, J.L., eds., Hunt Inst. for Botanical Documentation, Carnegie Mellon Univ., Pittsburgh, pp. 181-194.

Radinsky, L., 1978, Evolution of brain size in carnivores and ungulates, *Am. Nat.*, 112:815-831.

Radinsky, L.B., 1982, Evolution of skull shape in carnivores. 3. The origin and early radiation of the modern carnivore families, *Paleobiology*, 8:177-195.

Raff R.A. and Kaufman, T.C., 1983, "Embryos, Genes, and Evolution," MacMillan Publ. Co., New York.

Riddle, D.L., Swanson, M.M., and Albert, P.S., 1981, Interacting genes in a nematode dauer larva formation, *Nature* , 290:668-671.

Riedl, R., 1978, "Order in Living Organisms," (transl. from German by R.P.S. Jefferies), John Wiley and Sons, Chichester.

Riska, B. and Atchley, W.R., 1984, Genetics of growth predict patterns of brain-size evolution, unpublished.

Robertson, F.W., 1962, Changing the relative size of body parts of *Drosophila* by selection, *Genet. Res. Cambr.*, 3:169-180.

Rose, K.D., and Bown, T.M., 1984, Gradual phyletic evolution at the generic level in early Eocene omomyid primates, *Nature*, 309:250-252.

Rosenzweig, M.L., 1966, Community structure in sympatric Carnivora, *J. Mammal.*, 47:602-612.

Schindewolf, O.H., 1936, "Palaeontologie, Entwickslungslehre un Genetik," Borntrager, Berlin, 108 pp.

Schindewolf, O.H., 1950, "Grundfragen der Palaontologie," Schweizerbart, Stuttgart.

Schmalhausen, I.I, 1949, "Factors of Evolution, The Theory of Stabilizing Selection," Blakiston, Philadelphia.

Shea, B.T., 1983, Paedomorphosis and neoteny in the Pygmy chimpanzee, *Science*, 222:521-522.

Simpson, G.G., 1960, Diagnosis of the classes Reptilia and Mammalia, *Evolution*, 14:388-392.

Sinnott, E.W. and Dunn, L.C., 1935, The effect of genes on the development of size and form, *Biol. Rev. Cambr.*, 10:123-151.

Spemann, H., 1938, "Embryonic Development and Induction," Hafner, New York (reprinted 1967).

Stebbins, G.L., 1969, "The Basis of Progressive Evolution," Univ. of North Carolina Press, Chapel Hill.

Stebbins, G.L., 1974, Adaptive shifts and evolutionary novelty: a compositionist approach, *in*: "Studies in the Philosophy of Biology," Ayala, F.J. and Dobzhansky, T., eds., Univ. California Press, Berkeley, pp. 285-306.

Stebbins, G.L., 1984, Mosaic evolution: an integrating principle for the Modern Synthesis, unpublished manuscript.

Stock, G.B. and Bryant, S.V., 1981, Studies of digit regeneration and their implications for theories of development and evolution of vertebrate limbs, *J. Exp. Zool.*, 216:423-433.

Summerbell, D., 1981, The control of growth and the development of pattern across the anteroposterior axis of the chick limb bud, *J. Embryol. Exp. Morph.*, 63:161-180.

Tiffney, B.H., 1981, Diversity and major events in the evolution of land plants, *in*: "Paleobotany, Paleoecology, and Evolution," Niklas, K.V., ed., Praeger Publ., New York, pp. 193- 230.

Turing, A.M., 1952, The chemical basis of morphogenesis, *Phil. Trans. Roy. Soc. Lond.*, 237B:37-72.

Waddington, C.H., 1940, "Organizers and Genes," Cambridge Univ. Press, Cambridge.

Waddington, C.H., 1942, Canalization of development and the inheritance of acquired characters, *Nature*, 150:563-565.

Williams, E.E., 1950, Variation and selection in the cervical central articulations of living turtles, *Bull. Amer. Mus. Nat. Hist.*, 94:509-561.

Wilson, T.G., 1981a Expression of phenotypes in a temperature- sensitive allele of the apterous mutation in *Drosophila melanogaster*, *Devl. Biol.*, 85:425-433.

Wilson, T.G., 1981b, A mosaic analysis of the Apterous mutation in *Drosophila melanogaster*, *Devl. Biol.*, 85:434-435.

Wolpert, L., 1969, Positional information and the spatial pattern of cellular differentiation, *J. Theoret. Biol.*, 25:1-47.

Wright, S., 1934, An analysis of variability in number of digits in an inbred strain of guinea pigs, *Genetics*, 19:506-551.

Wright, S., 1935a, Polydactylous guinea pigs, *J.Hered.*, 25:359-362.

Wright, S., 1935b, A mutation of the Guinea pig, tending to restore the pentadactyl foot when heterozygous, producing a monstrosity when homozygous, *Genetics*, 20:84-107.

Yanofsky, C.T., Platt, T. Crawford, I.P., Nichols, B.P., Christie, G.E., Horowitz, H., van Cleemput, M., and Wu, A.M., 1981, The complete nucleotide sequence of the tryptophan operon of *Escherichia coli*, Nucleic Acids Res., 9:6647-6668.

INTRASPECIES GENOMIC VARIATION

Roy J. Britten*

Division of Biology, California Institute of Technology, and Carnegie Institution of Washington

Genomic variation is obviously the primary source of evolutionary change, however at present two major issues stand in the way of understanding its full significance: first, we know little yet about the kinds and amounts of variation; second, we need to identify the minor fraction of the variation which affects genes and their regulatory systems. Nevertheless genomic variation is fascinating, both because so little is known and because the data we do have shows unexpectedly large variation. In addition silent or non-coding variation is very much larger than amino acid replacement variation in coding regions of the DNA of the three species for which comparisons can now be made, indicating that selection retards the change of most amino acids in most proteins.

Direct measurements of overall single copy DNA sequence variation exist primarily for sea urchins. The Adh gene of *Drosophila melanogaster* has been sequenced in five population samples (supplying a measure of silent and replacement substitutions in one region). Restriction fragment length polymorphism studies (RFLP) in some human (*Homo sapiens*) genomic regions permit estimates of sequence variation. Mobile element variation is known mostly from *Drosophila*. There are hints of genome size variation in some plants and one mammal. Some local unstable regions have been turned up near a few genes. The number of ribosomal genes and the spacer sizes vary.

*Correspondence: Caltech, 101 Dahlia, Corona Del Mar, Ca. 92625

Here, the focus is on the nuclear genome and its DNA sequence structure. Thus mitochondrial or chloroplast variation is ignored and little said about variation of karyotype or chromosomal morphology. As a measure of DNA sequence variation Nei and Li's (1979) definition of sequence diversity (pi) appears most useful. It is the fraction of nucleotides that differ in a comparison of a DNA region between two genomes averaged over all of the comparisons in a set. Wherever a numerical value of variation is used (usually as percent) this is the precise meaning. When DNA sequences of individuals are hybridized the Tm reduction will be proportional to the sequence divergence. As it happens the proportionality factor is 100. In other words a 1% sequence diversity will give a Tm reduction of 1°C (Britten et al., 1974).

SINGLE COPY DNA SEQUENCE VARIATION IN SEA URCHINS

The first measurements of overall single copy DNA sequence variation showed an astonishing 4% difference in the average DNA sequences between individuals from the same population of *Strongylocentrotus purpuratus* (Britten et al., 1978). This result was based on measurements of the reduced melting temperature of hybrids of the DNA of different individuals. Labeled DNA from one individual was reassociated with driver DNA from the same or another individual and the resulting duplexes were melted in a chaotropic solvent (2.4M tetraethyl ammonium chloride) which reduces the effect of base composition on the melting temperature. The single strand specific nuclease sl was used to assay the remaining duplexes. With this method the melting curves for precise duplexes are very narrow (1.5°C) and high accuracy can be achieved. The size of the nuclease resistant duplexes was measured and all temperatures were corrected for the effect of length. Thus it was possible to directly compare the measured melting temperatures with those of perfect duplexes. Calibration using inbred laboratory mice (*Mus musculus*) demonstrated the accuracy of this comparison.

Several other sea urchin species have been studied (Grula et al., 1982) and the results are shown in Table 1. In addition some measurements with starfish DNA (*Pisaster ochraceus*, *Evasterias troscheli* and *Dermasterias imbricata*) indicate single copy DNA polymorphism polymorphism of 3-5% (Smith et al., 1982).

In every measurement that we have made, very wide melting curves are observed for interindividual hybrids suggesting that a few sequences show very large variation, with Tm reductions of up to 20%. Further a significant fraction, much more than that expected to be in coding regions, showed little variation. To prove that there was a wide range of variation the labeled DNA was separated into more and less variable fractions which were hybridized with total DNA driver. The thermal stability reductions of these fractions listed in Table 1 show that highly variable regions and less variable regions exist. We do not know if the large range of variation present in the sea urchin genome is due to selection. If it were the genetic load would be large. The rate of change of the single copy DNA (mutation rate) could vary from region to region depending on unknown aspects of chromatin structure and repair processes (Britten, 1985; and Bohr et al., 1985).

Table 1. Intraspecific Single Copy DNA Sequence Variation

S. franciscanus	3.2 (a)
S. intermedius	3.0 (a)
S. drobachiensis	2.0 (a)
S. purpuratus	4.5 (b)
(variable fraction)	8.9 (a)
(less variable ifraction	2.8 (a)
Mus musculus	<0.5 (c) (single measurement)

(a) Values are average reduction in thermal stability below perfect duplexes and are equal to the percent variation. Accuracy is estimated to be ± 0.5 (Grula et al., 1982). The latter figure is larger due to the inclusion of a greater quantity of the more variable sequences.
(b) Compromise between the values of Britten et al. (1978) (4°C); and Grula et al. (1982) (5°C).
(c) Britten et al. (1978).

Table 2. S. *purpuratus* RFLP

probe(a)	individ-uals	enzyme	number of occurrences (b)	H (c)
Cyl 3'	3	hind3	3,1,1,1	.70
Cy2a 3'	3	"	2,1,1,1,1	.78
Cy2b 3'	3	"	2,1,1,1,1	.78
10HB	6	"	5,2,1,1,1,1,1	.76
16pB	6	"	3,2,2,2,1,1,1	.83
10HB	6	EcoR1	4,2,1,1,1,1,1,1	.82
16PB	6	"	4,2,2,1,1,1,1	.81
pG6	7	"	7,5,1,1	.61
pG6	8	BamH1	9,3,2,1,1	.51
pG6	8	Hind3	9,4,3	.46

				.71

(a) Each probe is strictly single copy, that is each gives only one or two bands on a genome blot of individual DNA. The first three are from Lee et al. (1984) and the second four are from Posakony et al. (1983), while the last three are unpublished results (J.W. Roberts, and R.J. Britten).

(b) These are minimum estimates of the variation since in each case two bands were considered distinct only if there were convincingly large differences in mobility, and in several cases small differences that are probably real were ignored.

(c) The heterozygosity was calculated in the usual fashion as $1- \text{sum}(x_i)^2$, where x_i is the fraction of all of the bands which have each observed size.

RESTRICTION FRAGMENT LENGTH POLYMORPHISM IN SEA URCHINS

The large variation of average single copy DNA in sea urchins is corroborated by RFLP measurements for a few regions. Table 2 summarizes measurements made with several probes. The estimates of heterozygosity have considerable statistical uncertainty, due to the small number of individuals examined. It is worth noting that one of the probes used (pG6) apparently is in a region with less than average variation, while the five others all apparently fall in regions of average or above average variation. There is also RFLP data for starfish actin gene regions which corroborate the large single copy DNA polymorphism mentioned above (Smith et al., 1982; and Kovesdi et al., 1984).

Using the variation listed in Table 1 for S. purpuratus the expectation that a morph will be held in common between two individual haploid genomes is $(1\text{-pi})^{3s}=(.955)^{18}=0.44$, where s is the number of bases in a restriction site. This formula assumes that for an average length fragment there is an equal probability $(1\text{-pi})^{s}$ of a change affecting either terminal restriction site or of a new site appearing within its length. This simple formula has been checked by computer modeling. The heterozygosity to be expected was calculated assuming one principal morph and a set of less frequent morphs as the data indicate. On this basis the heterozygosity to be expected is approximately $H= 1-(0.44)^2 =0.81$, which is in adequate agreement with the average shown in Table 2.

COMPARISON OF SINGLE COPY DNA VARIATION TO PROTEIN VARIATION.

If the measured sea urchin single copy DNA variation occurred in the coding sequences of genes, the resulting protein variation would be very much larger than has ever been observed. A difference of 4% in two coding sequences would lead to a difference in about every tenth amino acid in the protein coded for, and thus dozens of differences per protein molecule. Marine invertebrates typically exhibit a heterozygosity of 0.15, implying that one out of seven proteins exhibit a difference in electrophoretic mobility between a pair of individuals. Thus, the average single copy DNA variation is 100 fold greater than amino acid replacement variation in these coding regions. For a convincing comparison both the enzyme polymorphism and DNA variation were measured in the same species and population. Alexander Pudovkin of the Institute of Marine Sciences, Vladivostok, USSR had studied the protein variation of Strongylocentrotus intermedius and found a typical value of 0.18 using standard methods (pers. comm.). Dr. Brykov of the Far East Science Center of Sciences of the USSR was kind enough to send us samples of the DNA of a number of individuals of the same population permitting us to make the measurements shown in Table 1, and to make a quantitative comparison.

The single copy DNA sequence variation of S. intermedius is 3%, and if this variation were present randomly in a coding sequence the amino acid sequence variation would be about 7% (allowing for 25% silent substitutions). However the observed amino acid sequence variation is about 0.06%, one hundred times smaller (calculated from the heterozygosity of 0.18 assuming an average protein contains 300 amino acids). To obtain an accurate estimate of the ratio of the variation between the coding regions and the average single copy DNA we would need to know both the average molecular weight and the amount of hidden variation, not revealed electrophoretically. No hidden variation was present in the Adh gene of five populations of D. melanogaster from a worldwide set of locations (Kreitman, 1983). It appears that neither the uncertainty in the average molecular weight nor in the hidden variation significantly affect the conclusion that the coding regions show much smaller variation than the average single copy DNA, although proteins need to be examined which are more variable.

This conclusion is very likely also true for the other sea urchin species we have studied, shown in Table 1. There is no reason to expect that their protein polymorphism is very much higher than that of typical marine invertebrates. We know that one member of the genus (S. intermedius) shows typical protein variation. Thus it appears that most coding regions are under intense selection pressure, presumably owing to the sensitivity of protein function to most possible amino acid substitutions. Evidence for selection has also been obtained from interspecies comparisons of DNA regions expressed in messenger RNA, primarily consisting of coding regions (Roberts et al., 1985). The divergence among sea urchin species was measured by thermal stability using cDNA probes made from egg or gastrula polysomal RNA. The evolutionary change is about half of what is observed with total single copy DNA probes. Some 3' flanking (non-coding) sequences are present in the probes and silent substitutions are measured. Thus it is clear that there is strong selection against changes of most amino acids in a very large class of proteins that are relatively prevalent in sea urchin eggs and embryos.

Do other eukaryotes besides sea urchins show greater silent or non-coding variation than replacement variation? It seems likely but data is apparently only available for Drosophila and human DNA. Stepping outside the self-imposed restriction (to variation data only) there are a number of interspecies sequence comparisons (e.g. Britten, 1985) which show much smaller differences for replacement substitutions than for non-coding and silent substitutions.

Barker et al. (1984) used a large set of supposedly randomly chosen single copy probes to examine a large number of restriction fragments of human DNA (equivalent to 447 sites) with 16 restriction enzymes. They found one polymorphic site that did not contain CpG. Nine polymorphic sites included CpG (out of a total 124 sites tested that contained CpG) suggesting 6% polymorphism for these "hotspots". Barker et al. have measured the frequencies in a larger group of subjects for each of the ten variants and pi can be calculated from their data to be .07% (though this value is dominated by the variation at CpG). While the high incidence of variation in the small number CpG containing sites suggests caution in the interpretation of RFLP, other regions of human DNA do not show this effect. The human growth hormone gene cluster has been examined for RFLP (Chakravarti et al., 1984) and they calculate pi= 0.2 to 0.3% for this region. They observe only two out six variable sites that contain CpG. Of more than 12 variable restriction sites in the beta globin cluster only one contains a CpG (Jeffreys, 1979; and Antonarakis et al., 1982) and pi for this region can be estimated to be 0.2%. Studies of primary DNA sequences of the beta globin region indicate a variation of about 0.2% for spacers and introns (Kazazian et al., 1983).

From this small amount of data we can only roughly approximate the average human single copy DNA variation but even if it were as low as 0.1% it would still be more than ten times greater than the implied variation of enzyme coding regions from polymorphism studies. An estimate of electrophoretic variation for 71 human enzyme loci (Harris and Hopkinson, 1972) gave an average heterozygosity of 6.7%. Taking an average protein to have 300 amino acids this implies a variation per amino acid of .022% and a coding region replacement variation of .008% per nucleotide. Hidden variation could raise this figure. Nevertheless variation which leads to amino acid replacement is much less than average single copy DNA variation for the proteins studied.

For Drosophila there is direct evidence (Kreitman, 1983) for strong selection on the great majority of the amino acids of Adh. The Adh gene was cloned and sequenced from five population samples of D. melanogaster (W. and E. coast of U.S.A., Africa, France and Japan). The slow and a fast allele were present in all samples, differing from each other by a thr-lys replacement substitution. Otherwise no replacement substitutions were present in any sample. Taking the observed 13 silent substitutions in the coding regions as a base, Kreitman calculated that in the absence of selection, 39 amino acid replacement substitutions would be expected whereas only one is observed.

Thus the variation of proteins in three species is lower than that expected from single copy DNA variation - by one or two orders of magnitude! Strong selective forces are required to suppress 99% of typical single copy variation in the coding regions. In order to reduce the variation by such a large factor selection must apply to almost all of the amino acids of practically all of the proteins that have been examined. For example say that 1/4 of the amino acids of a protein were relatively free from selection and their codons therefore shared the degree of variation of average single copy DNA. The calculated protein variation would be at least 1/4 of the value calculated from the single copy variation, which is still much larger than the observed protein variation and thus much longer than the observed variation.

MOBILE ELEMENTS AND REARRANGEMENT IN ANIMAL GENOMES

There are no measurements in any species which permit even an estimate of the total amount of rearrangement variation or mobile element variation in natural populations. In one genus, *Drosophila*, there is evidence that it could be very large (Strobel et al., 1979). The variation became important in "walking" along the genome (Bender et al., 1983) when it was observed that a repeat present in one strain was very likely to be absent in the same location in another strain. Recently Ananiev et al. (1984) compared the locations of 12 cloned transposable elements by in situ hybridization in Oregon RC and gt w^a (both stocks having been inbred for many generations). Most members of all 12 families of transposable elements were in different locations along the chromosomes of the two strains. Only about 10% of sites coincided in the two strains (true for each family). The frequencies of each of the families are similar in the two strains ($\pm$25%) except for Dm2068 which occurred only in gt w^a (43 sites).

If the genomes of these laboratory strains are still representative samples of populations, the data imply that the heterozygosity of wild *Drosophila* populations is very large. There are many puzzles, such as the lack of effect of the major DNA sequence variation on band or chromosomal morphology or apparently on synapsis. Further the frequencies of many families are similar in the two strains and thus appear to be somehow stabilized even when the locations of individual members are not. We must assume that transposition and multiplication are held in check by control mechanisms and it is a subtle question as to whether the host or the transposon or both benefit from or operate these controls.

The *D. simulans* genome has only about a third of the middle repetitive DNA of the *D. melanogaster* genome and individual families are less dispersed (Dowsett and Young, 1982). Comparisons of dispersed repetitive families among the *D. melanogaster* subgroup of species (Dowsett,1983) show enormous differences in frequency and many families of repeats are apparently species specific.

The positions of copia are not stable even for as short a time as the decades that Canton S stocks have been maintained separately (Junakovic et al., 1984), though copia has some aspects of a virus. There is some data on the occurrence of P-factor in wild populations and collections made at various times (Bregliano and Kidwell, 1983) suggesting that it may have recently spread through the species. The possibility that some transposable elements may be infectious and could have a wide host range needs to be kept in mind. Good evidence for such a concept is still non-existent and the *Drosophila* interspecies comparisons mentioned above suggest that a wide host range is not a general property of the set of repeats that were examined.

There is a homology between five nucleotides of the terminal inverted repeat of copia and those of the transposon "cin" of corn (Shepherd et al., 1984), and there are terminal repeat homologies between transposons of monocots and dicots (Doring and Starlinger, 1984), though sequence convergence due to function can not be ruled out. The population genetics of transposons and the influence of natural selection probably will remain difficult until we understand the specific replication control mechanisms and the rates of horizontal transfer, if they are significant. Transposons do land in *Drosophila* gene regions, are a major cause of mutations and sometimes lead to high frequency reversions. They influence gene regulatory processes, about which we know little in any case.

The role of these elements in *Drosophila* evolution appears to have been overwhelming. A possibly related fact is that large scale rearrangements (genomic revolutions) have occurred more than once in insect evolution. For example *Drosophila* (Manning et al., 1975; and Crain et al., 1976) has a small genome with a long period pattern of interspersion of repeats in the single copy DNA while the house fly (*Musca domestica*) has a much larger genome with a short period pattern (Crain et al., 1976b). Further a silk moth (*Antherae pernyi*, Efstratiadis et al., 1976) shows a short period pattern while the honey bee (*Apis mellifera*, Crain et al., 1976b) shows the *Drosophila* pattern. Cullis (this volume) has shown that flax (*Linum usitatissumum*) has a long period interspersion pattern and it has a relatively small genome. The garden pea (*Pisum sativum*) has a larger genome size and a short period interspersion pattern. Thus it sppears that very extensive genomic changes have also occurred during dicot evolution.

All that we know about these "revolutions" are the end results - strikingly different patterns of sequence organization - and the systematic relationships of the few species studied. Since, in each of the comparisons above, one species has a small genome and few interspersed short repeats while the second has a larger genome and many short interspersed repeats, it is possible that many thousands of short repeats were inserted (over a period of time) throughout the DNA in one lineage but not in the other.

However, it is equally possible that the changes occurred in the reverse direction with the systematic removal of repeats. The existence of mechanisms for both increase and decrease of numbers of interspersed repeats (possibly transposons) might lead to continuous drift, yielding over evolutionary time, a range of patterns including the extremes mentioned above. The evolutionary frequency change of interspersed families of repeats (Moore et al., 1978, 1981) is consistent with such a view. it is also probable that the instabilities of the type observed in flax under stress (Cullis, this volume) or in corn after out-breeding (Peterson, this volume) have long-term evolutionary effects that could be reflected in the quantitative aspects of patterns of genome organization. Observations of the genomic patterns of more species are required to learn the minimum time required to establish these large differences.

With this evidence for so much heterogeneity and variation it is worth inquiring if the concepts of "fluid genomes" and "genomic revolutions" are generally applicable to other eukaryotes, but we do not know the answer yet. Work on the human genome has only reached the stage of discovery of a few families of probable transposons. The study of human genomic variation by RFLP mentioned above (Barker et al., 1984) did not uncover any variation due to transposons or rearrangement, although many of the restriction fragments were large and small size differences would not always have been detected - such as the 300 nucleotides expected for Alu insertion. A transposon family has been observed to vary in location among individual sea urchins (Liebermann et al., 1983) and others probably exist (e.g. Johnson et al., 1984).

Some interspecies comparisons have been made that bear on the issue of variation in species other than Drosophila. Moore et al. (1978, 1980, and 1981) showed that families of interspersed repeats change rapidly in frequency in sea urchin genomic evolution, some by as much as a factor of 20 in a few tens of millions of years. Considering the number of families of repeats and their large frequencies this means that the incorporation into the species' genome of a newly inserted repeat occurs every few years on the average. The variation must be large unless these changes are concentrated in small populations at times of speciation.

There are no applicable modern studies of vertebrate repeat frequency changes although it can be argued (Britten, 1984) on the basis of older measurements that vertebrates have a comparable rate of insertion and/or deletion of repeated sequences. Recent measurements (J.W. Roberts and H. Hwu, unpublished data) indicate rapid evolutionary frequency change among hominids of both the Alu family of repeats and a high frequency element of the Kpn family of repeats. Considering the number of known, suspected and as yet undiscovered human transposons it is possible that the locations of individual transposons will be recognizable polymorphic markers of individual human genomes and populations.

MULTIPLE GENES

Several important structural genes occur in multiple copies and gene duplication or multiplication must be a significant evolutionary process. Histone genes, ribosomal RNA genes, tRNA genes and genes of the immune system occur in high numbers (large gene sets) while actin and hemoglobin are examples of genes with smaller numbers of copies (small gene sets). The underlying processes are probably quite distinct for the small and large sets since, for example, most of the individual actin and hemoglobin genes appear to have functionally diverged from each other, at least as far as the regulation of their expression, while for histone or ribosomal RNA there are at least subsets such that the function of many copies cannot be distinguished. Nevertheless in both large and small sets of genes there is evidence for coincidental evolution, that is, greater similarity is observed among members of a set within an individual genome (and species) than between these genes in different species. This result directly implies the production of copies by some mechanism (including conversion) as well as the probable loss of copies to retain the gene frequency. Any such mechanism, except conversion, requires variation in the number of copies, at least during some periods of the evolutionary history of each species.

This is a complex subject, and mathematical investigation has begun (e.g. Ohta, 1980). Observations of variation between individual genomes do exist. For example the evidence for conversion in the human A gamma and G gamma hemoglobin genes was first obtained with an individual that was heterozygous (Slightom et al., 1980). The number of ribosomal genes varies in *Xenopus* (Brown et al., 1971), *Drosophila* and in corn (*Zea mays*, Rivin et al., 1983), but significant wild population studies do not yet exist.

SPECIAL CASES

A large amount of what amounts to anecdotal information about variation has been published which may supply clues worth following up. For example, two highly variable regions flank the human alpha globin genes (Higgs et al., 1981) and a variable region made up of a tandem array of a simple sequence occurs 5' of the human beta globin gene (Moschonas et al., 1982). Satellites sometimes show surprisingly great interspecies homologies while others apparently are rapidly evolving. A recently studied example is the highly polymorphic deca-satellite of the African green monkey, *Cercopithecus aethiops* (Maresca and Singer, 1983; and Maresca et al., 1984).

There is apparently an unstable region a few hundred nucleotides 5' of the human insulin gene (Bell et al., 1981; and Rotwein et al., 1981). Several different inserts have been observed in different individual genomes ranging from 1.5 to 3.4 kb, and the presence of an insert appears to be associated with diabetes. A variable short segment is observed 345 bp 5' of the initiation codon of the human antithrombin III gene (Bock and Levitan, 1983). Such regions have turned up during initial studies of gene regions through the observation of RFLPs; it seems likely that many more of them must exist in the genome, and that the mechanisms involved may be distinct from those of transposons. The occurrences in the 5' flanking region suggest the possibility of gene regulatory polymorphism. Alexandraki and Ruderman (1985) have observed variation in the size of the alpha and beta tubulin mRNA apparently due to insertion/deletion polymorphism in the 3' untranslated region of the gene.

Genome size variation has apparently been observed in several plant species and one mammal (pocket gopher, Thomomys; Bachmann et al., 1972). Within a single species variation of up to 35% in haploid DNA content is observed, although cytological examination suggests that the change is primarily in heterochromatin and may be associated with a satellite. With this exception all mammals appear to have almost the same genome size, so it is surprising to find different species of this genus with a more than twofold ratio of DNA content. One might expect a similar phenomenon in the case of the African green monkey which has a satellite including 20 to 30% of its total DNA and is present in smaller quantities in closely related primates. The quantity of mouse satellite apparently differs in different strains and it probably changes rapidly in quantity since different individual members of the family have apparently been amplified as much as a million times in closely related species.

Evidence for plant genome size variation is primarily from cytochemical staining measurements. Consistency of observations to better than 5% is convincingly shown but it seems important to check the results with other biochemical methods. The reported variations are large. For example strains of sunflowers (Helianthus annuus) range in nuclear DNA content from 6.1 to 9.9 pg (Nagl and Capesius, 1976), while different species range over a factor of four in DNA content (all 2n=34). Microseris (Asteraceae) shows a 20% intraspecifc variation between strains and some studies of the DNA content of the progeny of crosses show intermediate values (Price et al., 1983; and Price et al., 1980). Some strains of Vicia also contain 20% more DNA than others (Raina and Rees, 1983) and different species range over a factor of 7 in DNA content. Allium also shows large intraspecific variation (Jones and Rees, 1968). If these results are correct there is more to DNA content and its variation than we know. Many of the large differences have little effect on chromosome morphology, principally a correlation of size with DNA content.

Along with this plant data it is worth noting that there is some evidence for significant single copy DNA sequence variation in tobacco (Nicotania tabacum, Okamuro and Goldberg, 1985). This is a single observation, and no other measurements are available for plants. In fact our ignorance for both plants and animals is overwhelming. There are no significant measurements of DNA content variation in animals (except for the rodent mentioned above). What hints there are suggest it would be worthwhile to look. Two species of Drosophila differ in genome size by almost a factor of two (Sober et al., 1968; and Bier et al., 1969). There is evidence that some of this genome size variation could be due to single copy DNA (Hunt et al., 1981). There must be a great deal of interesting variation in corn, the species in which mobile elements were first observed, but nothing I have seen gives any idea of the overall transposon variation, the single copy DNA sequence variation, or the genome size variation.

POSSIBLE POPULATION INSIGHTS

The sea urchin DNA sequence variation has implications regarding the population history. Since we have reasonably good estimates of the rate of evolutionary change (about 0.66%/MY, Britten, 1985) and variation (2 to 5%) of single copy DNA sequences, we may be able to test population models. For example the effective population can be calculated using the formula pi=4Nu/(1+16Nu/3) (Nei,1983; where u=mutation rate per nucleotide per generation=10^{-7}, and pi=.03). The result (N=500,000) is small as there must be more S. purpuratus in each of very many local beds of sea urchins. It is hard to believe that this is a good estimate of the effective breeding population size at their present stage of success. No greater differences are seen between individuals from populations 2000 kilometers apart and different individuals in one bed. The interpretation of these observations will require a deeper insight into the history of populations than is supplied by present models.

We may ask (but not decisively answer) an old question. Is the polymorphism observed for Drosophila Adh due to balancing selection? Only a tiny fraction of amino acid substitutions in the Adh gene can have been removed from the population by chance but there is no way to know if this conclusion applies specifically to the thr-lys substitution which is responsible for the slow and fast alleles. The presence of slow and fast alleles is apparently worldwide and occurs in some related species (e.g. Kreitman, 1983), showing that the thr-lys substitution was ancient. There is weak evidence against the thr-lys substitution being effectively neutral in wild populations. It has a significant effect on Adh function but selection has not been clearly shown even in vineyard flies (van Delden, 1982).

The work of uncovering the extent of genomic rearrangement has only begun but there is little doubt that an understanding of the processes of transposon multiplication and transmission and their effect on genes and their regulatory systems will deeply change our understanding of population genetics and the evolution of species.

Acknowledgements. I express my appreciation to Eric Davidson for continued intellectual support and for helpful criticism of the manuscript. This work was supported by NIH grant GM34031.

REFERENCES

Alexandraki, D., and Ruderman, J. V., 1985, Multiple polymorphic α- and β-tubulin mRNAs are present in sea urchin eggs, Proc. Natl. Acad. Sci. USA, 82:134-138.

Ananiev, E. V., Barsky, V. E., Ilyin, Yu. V., and Ryzic, M. V., 1984, The arrangement of transposable elements in the polytene chromosomes of Drosophila melanogaster, Chromosoma (Berl), 90:366-377.

Antonarakis, S. E., Boehm, C. D., Giardina, P. J. V., and Kazazian, H. H., Jr., 1982, Nonrandom association of polymorphic restriction sites in the β-globin gene cluster, Proc. Natl. Acad. Sci. USA, 79:137-141.

Bachmann, K., Goin, O. B., and Goin, C. J., 1972, Nuclear DNA amounts in vertebrates, Brookhaven Symp. Biol., 23:419-450.

Barker, D., Schafer, M., and White, R., 1984, Restriction sites containing CpG show a higher frequency of polymorphism in human DNA, Cell, 36:131-138.

Bell, G. I., Karam, J. H., and Rutter, W. J., 1981, Polymorphic DNA region adjacent to the 5' end of the human insulin gene, Proc. Natl. Acad. Sci. USA, 78:5759-5763.

Bender, W., Spierer, P., and Hogness, D. S., 1983, Chromosomal walking and jumping to isolate DNA from the Ace and rosy loci and the bithorax complex in Drosophila melanogaster, J. Mol. Biol., 168:17-33.

Bier, K., Kunz, W., and Ribbert, D., 1969, Insect oogenesis with and without lampbrush chromosomes, in: "Chromosomes Today," Vol. 2, D. Darlington and K. R. Lewis, eds., Plenum Press, New York, pp. 107-115.

Bock, S. C., and Levitan, D. J., 1983, Characterization of an unusual DNA length polymorphism 5' to the human antithrombin III gene, Nucl. Acids Res., 11:8569-8582.

Bohr, V. A., Smith, C. A., Okumoto, D. S., and Hanawalt, P. C., 1985, DNA repair in an active gene: Removal of pyrimidine dimers from the DHFR gene of CHO cells is much more efficient than in the genome overall, Cell, 40:359-369.

Bregliano, J., and Kidwell, M. G., 1983, Hybrid dysgenesis determinants, in: "Mobile Genetic Elements," J. A. Shapiro, ed., Academic Press, New York, pp. 363-410.

Britten, R. J., 1985, Rates of DNA sequence evolution differ between taxonomic groups, Science, submitted.

Britten, R. J., 1984, Mobile elements and DNA repeats, Carlsberg Res. Commun., 49:169-178.

Britten, R. J., Cetta, A., and Davidson, E. H., 1978, The single-copy DNA sequence polymorphism of the sea urchin Strongylocentrotus purpuratus, Cell, 15:1175-1186.

Britten, R. J., Graham, D. E., and Neufeld, B. R., 1974, Analysis of repeating DNA sequences by reassociation, in: "Methods in Enzymology," 29E, L. Grossman and K. Moldave, eds., Academic Press, New York, pp. 363-406.

Brown, D. D., Wensink, P. C., and Jordan, E., 1971, A comparison of the ribosomal DNAs of Xenopus laevis and Xenopus mulleri: the evolution of tandem genes, J. Mol. Biol., 63:57-73.

Chakravarti, A., Phillips, J. A., III, Mellits, K. H., Buetow, K. H., and Seeburg, P. H., 1984, Patterns of polymorphism and linkage disequilibrium suggest independent origins of the human growth hormone gene cluster, Proc. Natl. Acad. Sci. USA, 81:6085-6089.

Crain, W. R., Eden, F. C., Pearson, W. R., Davidson, E. H., and Britten, R. J., 1976, Absence of short period interspersion of repetitive and non-repetitive sequences in the DNA of Drosophila melanogaster, Chromosoma (Berl), 56:309-326.

Crain, W. R., Davidson, E. H., and Britten, R. J., 1976b, Contrasting patterns of DNA sequence arrangement in Apis mellifera (honeybee) and Musca domestica (housefly), Chromosoma, 59:1-12.

Döring, H.-P., and Starlinger, P., 1984, Barbara McClintock's controlling elements: now at the DNA level, Cell, 39:253-259.

Dowsett, A. P., 1983, Closely related species of Drosophila can contain different libraries of middle repetitive DNA sequences, Chromosoma (Berl), 88:104-108.

Dowsett, A. P., and Young, M. W., 1982, Differing levels of dispersed repetitive DNA among closely related species of Drosophila, Proc. Natl. Acad. Sci. USA, 79:4570-4574.

Efstratiadis, A., Crain, W. R., Britten, R. J., Davidson, E. H., and Kafatos, F. C., 1976, DNA sequence organization in the lepidopteran Antherae pernyi, Proc. Natl. Acad. Sci. USA, 73:2289-2293.

Grula, J. W., Hall, T. J., Hunt, J. A., Giugni, T. D., Graham, G. J., Davidson, E. H., and Britten, R. J., 1982, Sea urchin

DNA sequence variation and reduced interspecies differences of the less variable DNA sequences, Evolution, 36:665-676.

Harris, H., and Hopkinson, D. A., 1972, Average heterozygosity per locus in man: an estimate based on the incidence of enzyme polymorphism, Ann. Hum. Genet., 36, 9-20.

Higgs, D. R., Goodbourn, S. E. Y., Wainscoat, J. S., Clegg, J. B., and Weatherall, D. J., 1981, Highly variable regions of DNA flank the human α-globin genes, Nucl. Acids Res., 9, 4213-4224.

Hunt, J. A., Hall, T. J., and Britten, R. J., 1981, Evolutionary distances in Hawaiian Drosophila measured by DNA reassociation, J. Mol. Evol., 17:361-367.

Jeffreys, A. J., 1979, DNA sequence variants in the $^{G}\gamma$-, $^{A}\gamma$-, δ- and β-globin genes of man, Cell, 18:1-10.

Johnson, S. A., Davidson, E. H., and Britten, R. J., 1984, Insertion of a short repetitive sequence (D88I) in a sea urchin gene: A typical interspersed repeat?, J. Mol. Evol., 20:195-201.

Jones, R. N., and Rees, H., 1968, Nuclear DNA variation in Allium, Heredity, 23:591-605.

Junakovic, N., Caneva, R., and Ballario, P., 1984, Genomic distribution of copia-like elements in laboratory stocks of Drosophila melanogaster, Chromosoma (Berl), 90:378-382.

Kazazian, H. H., Jr., Chakravarti, A., Orkin, S. H., and Antonarakis, S. E., 1983, DNA polymorphisms in the human β globin gene cluster, in: "Evolution of Genes and Proteins," M. Nei and R. K. Koehn, eds., Sinauer Associates, Sunderland, Massachusetts, pp. 137-146.

Koveski, I., Preugschat, F., Stuerzl, M., and Smith, M. J., 1984, Actin genes from the sea star Pisaster ochraceus, Biochim. Biophys. Acta, 782:76-86.

Kreitman, M., 1983, Nucleotide polymorphism at the alcohol dehydrogenase locus of Drosophila melanogaster, Nature, 304:412-417.

Lee, J. J., Shott, R. J., Rose, S. J., III., Thomas, T. L., Britten, R. J., and Davidson, E. H., 1984, Sea urchin actin gene subtypes: Gene number, linkage and evolution, J. Mol. Biol., 172:149-176.

Liebermann, D., Hoffman-Liebermann, B., Weinthal, J., Childs, G., Maxson, R., Mauron, A., Cohen, S., and Kedes, L., 1983, An unusual transposon with long terminal inverted repeats in the sea urchin Strongylocentrotus purpuratus, Nature, 306:342-347.

Manning, J. E., Schmid, C. W., and Davidson, N., 1975, Interspersion of repetitive and nonrepetitive DNA sequences in the Drosophila melanogaster genome, Cell, 4:141-155.

Maresca, A., and Singer, M. F., 1983, Deca-satellite: A highly polymorphic satellite that joins α-satellite in the African green monkey genome, J. Mol. Biol., 164:493-511.

Maresca, A., Singer, M. F., and Lee, T. N. H., 1984, Continuous reorganization leads to extensive polymorphism in a monkey centromeric satellite, J. Mol. Biol., 179:629-649.

Moore, G. P., Costantini, F. D., Posakony, J. W., Davidson, E. H., and Britten, R. J., 1980, Evolutionary conservation of repetitive sequence expression in sea urchin egg RNA's, Science, 208:1046-1048.

Moore, G. P., Pearson, W. R., Davidson, E. H., and Britten, R. J., 1981, Long and short repeats of sea urchin DNA and their evolution, Chromosoma (Berl), 84:19-32.

Moore, G. P., Scheller, R. H., Davidson, E. H., and Britten, R. J., 1978, Evolutionary change in the repetition frequency of sea urchin DNA sequences, Cell, 15:649-660.

Moschonas, N., de Boer, E., and Flavell, R. A., 1982, The DNA sequence of the 5' flanking region of the human β-globin gene: Evolutionary conservation and polymorphic differences, Nucl. Acids Res., 10:2109-2120.

Nagl, W., and Capesius, I., 1976, Molecular and cytological characteristics of nuclear DNA and chromatin for angiosperm systematics: Basic data for Helianthus annuus (Asteraceae), Plant Syst. Evol., 126:221-237.

Nei, M., and Li, W.-H., 1979, Mathematical model for studying genetic variation in terms of restriction endonucleases, Proc. Natl. Acad. Sci. USA, 76:5269-5273.

Ohta, T., 1980, "Lecture notes in Biomathematics: Evolution and Variation of Multigene Families," Springer-Verlag, New York.

Okamuro, J. K., and Goldberg, R. B., 1985, Tobacco single-copy DNA is highly homologous to sequences present in the genomes of its diploid progenitors, Mol. Gen. Genet., 198:290-298

Posakony, J. W., Flytzanis, C. N., Britten, R.J., and Davidson, E. H., 1983, Interspersed sequence organization and developmental representation of cloned poly(A) RNAs from sea urchin eggs, J. Mol. Biol., 167:361-389.

Price, H. J., Bachmann, K., Chambers, K., and Riggs, J., 1980, Detection of intraspecific variation in nuclear DNA content in Microseris douglasii, Bot. Gaz., 141:195-198.

Price, H. J., Chambers, K. L., Bachmann, K., and Riggs, J., 1983, Inheritance of 2C nuclear DNA content variation in intraspecific and interspecific hybrids of Microseris (Asteraceae), Am. J. Bot., 70:1133-1138.

Raina, S. N., and Rees, H., 1983, DNA variation between and within chromosome complements of Vicia species, Heredity, 51:335-346.

Rivin, C. J., Zimmer, E. A., Cullis, C. A., Walbot, V., Huynh, T., and Davis, R. W., 1983, Evaluation of genomic variability at the nucleic acid level, Plant Mol. Biol. Reporter, 1:9-16.

Roberts, J. W., Johnson, S. A., Kier, P., Hall, T. J., Davidson, E. H., and Britten, R. J., 1985, Evolutionary conservation of DNA sequences expressed in sea urchin eggs and early embryos, J. Mol. Evol., submitted.

Rotwein, P., Chyn, R., Chirgwin, J., Cordell, B., Goodman, H. M., and Permutt, M. A., 1981, Polymorphism in the 5'-flanking region of the human insulin gene and its possible relation to type 2 diabetes, Science, 213:1117-1120.

Shepherd, N. D., Schwarz-Sommer, Zs., Blumberg vel Spalve, J. Gupta, M., Wienand, U., and Saedler, H., 1984, Similarity of the Cinl repetitive family of Zea mays to eukaryotic transposable elements, Nature, 307:185-187.

Slightom, J. L., Blechl, A. E., and Smithies, O., 1980, Human fetal $^{G}\gamma$- and $^{A}\gamma$-globin genes: Complete nucleotide sequences suggest that DNA can be exchanged between these duplicated genes, Cell, 21:627-638.

Smith, M. J., Nicholson, R., Stuerzl, M., and Lui, A, 1982, Single copy DNA homology in sea stars, J. Mol. Evol., 18:92-101.

Sober, H. A., 1968, "Handbook of Biochemistry," Chemical Rubber Co., Cleveland.

Strobel, E., Dunsmuir, P., and Rubin, G. M., 1979, Polymorphisms in the chromosomal locations of elements of tge 413, copia and 297 dispersed repeated gene families in Drosophila, Cell, 17:429-439.

van Delden, W. 1982, The alcohol dehydrogenase polymorphism in Drosophila melanogaster: selection at an enzyme locus, Evol. Biol., 15:187-222.

STRUCTURE AND FUNCTION OF THE HUMAN RETROVIRUSES

Lee Ratner and Robert C. Gallo

Laboratory of Tumor Cell Biology
National Cancer Institute
Bethesda, Md. 20205

INTRODUCTION

Retroviruses, though discovered and studied in a number of animal species since the beginning of this century (Rous, 1911; Gross, 1951), have only recently been demonstrated in humans (Gallo, 1984). The delay in their identification in humans is due to differences in the biological properties of the human retroviruses, designated human T-cell lymphotropic viruses (HTLV), and a few other animal retroviruses such as bovine leukemia virus (BLV) from the majority of studied species. First, HTLV and BLV are unlike most avian and murine retroviruses in that abundant virus replication and viremia do not accompany the development of a tumor (Weiss et al., 1982; Wong-Staal et al., 1984). Thus, in order to detect the virus, it was important to develop methods for _in vitro_ cultivation of cells derived from human tumors, _allowing for_ activation and amplification of these viruses. These efforts were first applied to T lymphocytic tumors and their success was dependent upon the identification and utilization of T-cell growth factor (TCGF), which is also known as interleukin 2 (IL 2) (Morgan et al., 1976; Mier et al., 1980; Poiesz et al., 1980a,b, 1981).

Second, the human retroviruses are distinct from most other retroviruses in the structure of their proteins and nucleic acids (Kalyanaraman et al.; 1981 Reitz et al.; 1981). Thus, their detection required the development of a new set of sensitive probes. The ability to assay RNA-dependent DNA polymerase (reverse transcriptase) provided such an assay and has been critical for the identification and isolation of these viruses (Gallo and Gallagher, 1974; Gallo and Reitz 1976).

With these extensive prelimary developments, it was then possible to isolate human retroviruses. There are now hundreds of isolates which generally fall into three classes (Gallo, 1984) though variants within the classes have been described (Hahn et al., 1985; Ratner et al., 1985c). The first class is designated HTLV-I, a virus closely associated with and likely the etiological cause of a distinct human malignancy, adult T-cell leukemia/ lymphoma (ATL) (Tajima et al., 1979; Takatsuki, et al., 1979; Uchiyama, et al., 1979; Catovsky et al., 1982; Blattner et al., 1983; Blayney et al., 1983a,b). Infection with the virus is endemic in Southern Japan, the Caribbean Islands and surrounding areas of Southeastern United States, Central America, and portions of South America, as well as parts of Africa (Blattner et al., 1982; Kalyanaraman et al., 1982a; Robert-Guroff et al., 1982, 1983; Schupbach et al., 1983; Blayney et al., 1984; Saxinger et al., 1984). Only 0.05-1.0% of infected individuals have developed ATL. Whether the development of ATL is a chance occurrence or other factors might be involved in initiation of transformation remains to be determined.

The conclusion that HTLV-I is etiologically related to ATL is based upon the following observations:

1) There is a close geographical association of the occurrence of ATL and HTLV-I infection (Hinuma et al., 1982; Robert-Guroff et al., 1983).

2) There is an increased incidence of HTLV-I infection in family members of patients with ATL (Robert-Guroff et al., 1982, 1984).

3) Antibodies are invariably present in the sera of ATL patients reactive with group-specific antigens (*gag*) i.e. structural proteins of HTLV-I, as well as envelope proteins (Hinuma et al., 1981; Robert-Guroff et al., 1981; Kalyanaraman et al., 1982a; Lee et al., 1984a; Schupbach et al., 1984b).

4) One is able to isolate and propagate HTLV-I from tumor tissues of virtually all ATL patients (Miyoshi et al., 1981a, Vyth-Drees and de Vries, 1982; Yoshida et al., 1982; Haynes et al., 1983; Popovic et al., 1983b; our unpublished data with M. Popovic, Z. Salahuddin, and P. Markham).

5) There are one or a few copies of the HTLV-I proviral genome monoclonally integrated in the tumor tissues of all patients (Wong-Staal et al., 1983; Yoshida et al., 1984).

6) HTLV-I is able to transform normal T lymphocytes *in vitro* resulting in cell lines with similar morphologic, antigenic, and biochemical characteristics to those of the primary tumors

or primary cell lines derived from ATL patients (Miyoshi et al., 1981a,b; Mann et al., 1983a,b; Markham et al., 1983, 1984; Popovic et al., 1983a,b).

HTLV-II has been identified on only three occasions and is unlike other members of the HTLV family in its lack of a clearly defined disease association (Kalyanaraman et al., 1982b; Chorba et al., 1984; Hahn et al., 1984a). Though its proteins and nucleic acids are less closely related structurally to those of HTLV-I isolates, than are those of HTLV-I isolates to each other, its *in vitro* transforming abilities and biochemical effects are remarkably similar to those of HTLV-I (Chen et al., 1983a,b; Gelmann et al., 1984; Haseltine et al., 1984b, Shaw et al., 1984a; Shimotohno et al., 1984; Sodroski et al., 1984a).

HTLV-III is a strongly cytopathic but non-transforming retrovirus unlike HTLV-I and -II which are weakly cytopathic but transforming (Popovic et al., 1983a,b, 1984a,b; Mitsuya et al., 1984; DeVecchis et al., 1985). HTLV-III is closely associated with the acquired immunodeficiency syndrome (AIDS) and related disease states (e.g. AIDS related complex or ARC, lymphadenopathy syndrome or LAS) and is likely the etiological agent for this immune disorder (Gallo et al., 1984; Sarngadharan et al., 1984; Schupbach et al., 1984a). This conclusion is based on a number of findings outlined below:

1) Antibodies to HTLV-III proteins (predominately gp160, the precursor of the envelope protein; gp120, the presumed exterior portion of the envelope; gp41, the transmembrane portion of the envelope; and p24, the major capsid *gag* protein) are found in 85-95% of patients with AIDS, ARC, or LAS, or mothers of juvenile AIDS patients, but only 20-50% of asymptomatic homosexuals and less than 1% of heterosexuals lacking risk factors for AIDS (Kitchen et al., 1984; Safai et al., 1984; Sarngadaharan et al., 1984; Schupbach et al., 1984a).

2) Antibodies to HTLV-III proteins were found in patients who acquired AIDS by transfusion (Jaffe et al., 1984). These individuals lacked risk factors for AIDS, and also were not generally infected with a number of other viruses, e.g. herpes simplex, Epstein-Barr virus (EBV), or cytomegalovirus (CMV), which had been previously implicated in the etiology of AIDS.

3) Antibodies to HTLV-III proteins were identified in at least one transfusion donor of each of the above noted transfusion recipients who developed AIDS (Jaffe et al., 1984). In each case, the donor identified by antibody assays (done in a blinded fashion) was noted to have at least one risk factor for the development of AIDS.

4) The prevalence of antibodies to HTLV-III in hemophiliacs and homosexuals has increased during the period 1980-1984 when the AIDS epidemic became manifest (Chamberland et al., 1984; Eyster et al., 1985).

5) HTLV-III can be isolated from 30-90% of patients with AIDS or related conditions but from less than 1% of controls (Gallo et al., 1984; Groopman et al., 1984; Ho et al., 1984b; Levy et al., 1984; Zagury et al., 1984; Salahuddin et al., 1985).

6) _In vitro_, HTLV-III binds to the T4 antigen and is cytopathic towards OKT4(+) lymphocytes, the same cell type which is depleted in patients with AIDS (Gottlieb et al., 1981; Dalgleish et al., 1984; Fauci et al., 1984; Klatzmann et al., 1984; Popovic et al., 1984b; Seligman et al., 1984).

The structure and function of HTLV-III differs from that of the transforming viruses in this family in a number of ways. However, a number of features of HTLV-III are similar to those of HTLV-I, -II, and BLV and the lentiviruses suggesting that they are related. This review will focus on these characteristics.

STRUCTURE OF HTLV-I

Cloning of the HTLV-I genome made possible detailed studies of the structure and function of this virus (Manzari et al., 1983). The first data with these probes together with results of biological experiments showed that HTLV-I is an exogenous retrovirus (Gallo et al., 1982) that fit neatly into neither of the categories of acute or chronic leukemia viruses as is summarized in Table 1.

Though HTLV-I integrates at the same locus (loci) in every tumor cell of an individual or cell line, no integration site is found in common among different individuals (Seiki et al., 1984; Siegel et al., 1984). This contrasts with the findings of a common integration site for many of the transforming chronic leukemia viruses which activate the transcription of an adjacent cellular gene which may mediate the transforming activity (Fung et al., 1981, 1982; Hayward et al., 1981; Neel et al., 1981; Noori-Dalooi et al., 1981; Nusse et al., 1982; Tsichlis et al., 1983; Van Ooyen and Nusse, 1984). This was the first clue that the mechanism of transformation by HTLV-I is distinct from that of other retroviruses.

Table 1. Properties of HTLV-I

Properties typical of acute leukemia viruses:	Properties typical of chronic leukemia viruses:
rapid *in vitro* transformation of lymphocytes[1]	non-defective retrovirus[2]
	lack of a gene homologous to a cellular sequence (i.e. an oncogene)[3]
	monoclonal integration of provirus in tumor cells[4]

[1]Miyoshi et al., 1981a,b; Markham et al., 1983; Popovic et al., 1983a,b; DeRossi et al., 1985
[2]Popovic et al., 1983a,b; Wong-Staal et al., 1983
[3]Seiki et al., 1983
[4]Wong-Staal et al., 1983

With the availability of cloned sequences for the entire HTLV-I genome, nucleotide sequence analysis was carried out and provided a detailed description of the virus (Seiki et al. 1982, 1983). This demonstrated that the integrated provirus was 9032 base pairs (bp) in length flanked by long terminal repeat sequences (LTR). The proviral structure is summarized in Fig. 1. The structure of the LTR clearly differs from that of most retroviruses in a number of aspects (Table 2). First the LTR is rather large, especially the R region. Second, the inverted repeat sequences found at the 5' boundary of U5 and 3' boundary of U3 were shorter than those of most other retroviruses. Third, the HTLV-I LTR lacks a typical polyadenylation addition signal (AATAAA) 25-30 bp upstream of the polyadenylation site (R-U5 boundary). However, such a signal is found within U3 276 bp upstream of the polyadenylation site. A relatively stable hairpin loop may be hypothesized which would juxtapose this signal with the polyadenylation site. Experimental proof for the existence of such a secondary structure is not yet available. Several of these properties are also found found for the LTRs of HTLV-II and BLV (Couez et al., 1984; Sagata et al., 1984; Sodroski et al., 1984c). The LTR of HTLV-I is similar to those of other retroviruses in the presence of a) direct repeat sequences in U3 which may have transcriptional enhancer activity and b) a $tRNA^{proline}$ primer binding sequence immediately 3' of the 5'LTR (Temin et al., 1982; Varmus et al., 1982).

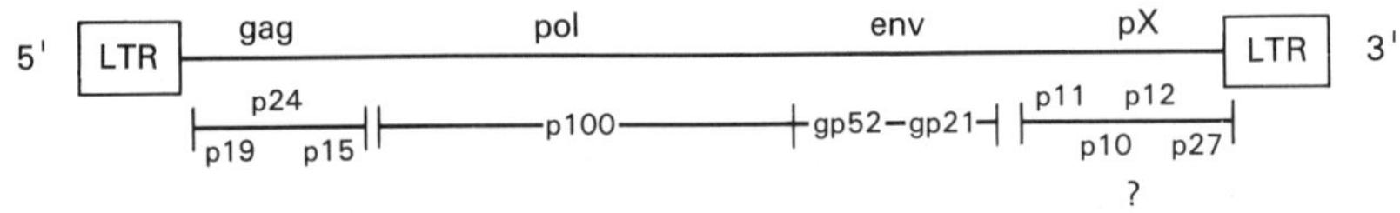

Fig. 1. Outline of the HTLV-I genome based on sequence data of Seiki et al. (1983). Reprinted with permission from Ratner et al., J. Soc. Prev. Oncol., in press.

The 5'LTR of HTLV-I is followed by a non-coding leader sequence which may contain a sequence analogous to that demonstrated for other retroviruses to be important for packaging of the RNA genome in the viral particle (Mann et al., 1983).

The leader sequence is followed by the *gag* gene which encodes a precursor protein that is processed into three *gag* proteins, a myristylated p19 or myrgagtin derived from the N-terminus, a major capsid protein, p24 or largagtin, and a nucleic acid binding protein, p15 or nugagtin derived from the C-terminus (Oroszolan et al., 1982; Copeland et al., 1983; Seiki et al., 1983; terminology for *gag* proteins reviewed in Oroszolan et al., 1985). A similar structure has recently been described for the *gag* proteins of BLV (Sagata et al., 1985); sequence information for the HTLV-II *gag* gene is not yet available. The HTLV-I and BLV *gag* proteins both differ from those of other retroviruses in a) the lack of a sequence encoding a small phosphoprotein interposed between the myrgagtin and largagtin sequences, and b) the size of the largagtin of 24 kilodaltons (kd) in comparison to p27 of Rous sarcoma virus (RSV) and p30 of Moloney murine leukemia virus (MoMuLV) (Shinnick et al., 1981; Oroszolan et al., 1982; Weiss et al., 1982; Schwartz et al., 1983; Seiki et al., 1983; Haseltine et al., 1984a).

Unlike most other retroviruses, a non-coding sequence of 408 bp is situated between the *gag* and *pol* genes (Seiki et al., 1983). A similar non-coding region between *gag* and *pol* is found in a variant of HTLV-I, HTLV-Ib (Ratner et al., 1985c) as well as in BLV (Sagata et al., 1985). Thus, this characteristic is likely a property of the fully competent HTLV-I virus, not a cloning artefact, nor a feature of a defective provirus. The *gag* and *pol* genes of HTLV-I are both found in reading frame 1 though ten termination codons are located between these two

Table 2. Comparison of retroviral LTRs.

Virus	Length (bp)					tRNA Primer
	LTR	Inverted Repeat	U3	R	U5	
HTLV-I	754	2	353	221	180	Pro
HTLV-II	763	2	314	247	203	Pro
HTLV-III	634	2	453	98	83	Lys
BLV	535	6	215	233	86	Pro
Avian myelocytomatosis virus	312	9	217	8	87	Trp
Rous sarcoma virus	334	15	233	22	79	Trp
Avian sarcoma virus	330	15	229	21	80	Trp
AKR	626	13	479	70	76	Pro
Moloney murine leukemia virus	594	13	449	68	77	Pro
Murine sarcoma virus	588	11	444	68	76	Pro
Feline leukemia virus B	539	11	394	68	77	Pro
Feline sarcoma virus	482	12	340	68	74	Pro
Spleen necrosis virus	570	3	396	79	95	Pro
Mouse mammary tumor virus	1328	6	1194	11	118	Lys
Simian sarcoma virus	504	7	360	70	76	Pro

Modified and reprinted with permission from Starcich et al. (1985).

genes. In MoMuLV, a single termination codon is present between *gag* and *pol* in the same reading frame (Shinnick et al., 1981). In cells infected by this virus, the *pol* gene product is derived at least in part from a *gag-pol* precursor protein synthesized by suppression of this termination codon (S. Oroszolan, personal communication). However, suppression of ten termination codons in HTLV-I would seem unlikely. Thus, *pol* is probably made as a *gag-pol* precursor like that of many other retroviruses (Philipson *et al.*, 1978; Weiss et al., 1978, 1982) with splicing around the non-coding region. However, such a mRNA must still be identified and accurately defined.

Of further interest is that the non-coding sequence between *gag* and *pol* represents a region in the viral genome which in other retrovirses likely encodes a protease (Haseltine et al, 1984a; Ratner et al., 1985b). In the case of MoMULV, the protease is derived from the N-terminus of the *pol* precursor (Shinnick et al., 1981; Levin et al., 1984). A deletion in this sequence results in defective proteolytic processing of the *gag* precursor, the *pol* precursor, and the *env* p15E transmembrane protein (Crawford and Goff, 1985). A homologous sequence is encoded by the 3' portion of the *gag* gene of RSV though it remains to be demonstrated that it encodes a protein that has proteolytic activity (Schwartz et al, 1983). It is interesting that nucleotide sequences within the presumed non-coding sequences between the *gag* and *pol* genes of HTLV-I predict peptide sequences within reading frames 2 and 3 which are homologous to those predicted for protease sequences of MoMULV and RSV (Table 3) (Shinnick et al., 1981, Schwartz et al., 1983; Seiki et al., 1983; Ratner et al., 1985b). It is unclear whether these sequences represent a vestige of a protease sequence which is no longer functional, or whether a protease might be encoded at least partly from these sequences by a complicated mRNA splicing mechanism. If in fact, BLV and HTLV-I do lack a viral protease and must use a cellular protease for viral protein processing, could this explain the low levels of replication of these viruses seen both *in vitro* and *in vivo*? If so, what biological advantage would this provide for a retrovirus to have evolved in this manner?

The *pol* gene of HTLV-I encodes the reverse transcriptase activity which differs from that of most mammalian retroviruses in its preference for magnesium rather than manganese and its large size (100 kd) (Rho et al., 1981). However, there are extensive sequence homologies throughout the sequence of the predicted HTLV-I *pol* product and those of other retroviruses, as well as the DNA viruses, type B hepatitis viurs and cauliflower mosaic virus (Gardner et al., 1981; Ono et al., 1983; Toh et al., 1983; Chiu et al., 1984; Patarca and Haseltine, 1984a). The C-terminus of this HTLV-I predicted *pol* product is also homologous to the C-terminus of the MoMULV predicted *pol* product which encodes an endonuclease that mediates chromosomal integration of the circular dsDNA viral intermediate (Levin et al., 1984; Panganiban and Temin, 1984; Schwartzberg et al., 1984). Such an endonuclease activity is likely to be repesent in HTLV-I infected cells as well.

Table 3. Amino acid homologies among proposed retroviral protease products.

Virus	Gene	Amino Acid Residue	Homologous Sequence
HTLV-III	Pol	80	T I K I G G Q L K E A L L D T G A D D T V L E E M S L P G R W
MoMuLV	Pol	14	T L K V G G Q P V T F L V D T G A Q H S V L T Q N P G P L S D
RSV	Gag	601	P V K Q R S V Y I T A L L D S G A D I T I I S E E D W P I D W
HTLV-I	Between Gag and Pol		E A L L D T G A D M T V L P I A L F S S N T
HTLV-III	Pol	117	G I G G F I K V R
MoMuLV	Pol	51	G A T G G K R Y R
RSV	Gag	643	G I G G G I P M R
HTLV-III	Pol	152	I I G R N L L T Q I G
MoMuLV	Pol	87	L L G R D L L T K L K
RSV	Gag	685	I L G R D C L Q G L G
HTLV-I	Between Gag and Pol		I I G R D A L Q Q C Q

Boxes identicate amino acid residue identitites. Modified and reprinted with permission from Ratner et al., 1985b.

The env gene of HTLV-I overlaps with the pol gene by 5 bp (Seiki et al., 1983). The first 20 predicted amino acids are hydrophobic and likely encode a signal peptide mediating insertion of the protein into the membrane (Lee et al., 1984a). The N-terminus of the env precursor encodes the exterior glycoprotein, which has four potential N-linked glycosylation sites (Haseltine et al., 1984a; Seiki et al., 1983). The C-terminus encodes a tranmembrane protein, p21, which anchors the exterior protein to the membrane through specific disulfide bonds. The sequence of this portion of the env protein is homologous to that of transmembrane proteins of HTLV-II and a number of other retroviruses (Ciancolo et al., 1984; Patarca and Haseltine, 1984b; Sodroski et al., 1984a).

Perhaps the most unusual aspect of the structure of the HTLV-I genome is the presence of a 1.5 kb sequence between the env gene and the 3'LTR, which has been called pX (Seiki et al., 1983). It includes 4 small open reading frames, each initiated with an ATG codon, designated pX-I, pX-II, pX-III, and pX-IV. The open reading frame for pX-IV actually starts 186 bp before the first methionine codon (Haseltine et al., 1984b). In fact this long open reading frame, now designated X-lor, likely encodes a protein as suggested by a number of observations:

1) A comparison of the pX region of HTLV-II with HTLV-I reveals that the 5'most 546 bp of this region is poorly conserved whereas the 3'most 1011 bp of HTLV-II is 70% homologous to the corresponding sequence of HTLV-I (Haseltine et al., 1984b; Shimotohno et al., 1984). The sequence homology begins rather abruptly with the beginning of the open reading frame, X-lor.

2) A splice acceptor site in the 5'portion of the X-lor open reading frame is utilized in both HTLV-I and -II (Wachsman et al., 1984; Okamoto et al., 1985).

3) Antibodies to synthetic peptides predicted by the X-lor open reading frame specifically immunoprecipitate 42 and 38 kd products from HTLV-I and -II infected cells, respectively, in a wide range of species (Miwa et al., 1984; Slamon et al., 1984; Shimotohno et al., 1985).

4) Partial amino acid sequences of a 42 kd product of HTLV-I infected cells are in agreement with those predicted by the DNA sequence of the X-lor gene (Lee et al., 1984b).

The initiator methionine codon and one additional amino acid for the X-lor product are likely donated by the first two codons of the env gene (Aldovini et al., 1985). The mRNA encoding X-lor thus arises as a result of a novel double splice mechanism, utilizing in at least some cases a donor splice sequence within R in the 5'LTR and an acceptor splice sequence in the 3' portion of the pol gene, and a donor splice site in the 5' portion of the env gene and the splice acceptor already described at the beginning of the X-lor gene (Fig. 2) (our unpublished data with G. Franchini and M. Feinberg; personnal communication with M. Yoshida). There is evidence for variability in the first splice acceptor site utilized in 3'pol which accounts for at least some of the heterogeneity in the size of the X-lor mRNA in different HTLV-I infected cells (Aldovini et al., 1985). Whether other splice sites for X-lor transcripts may vary remains to be shown.

As already mentioned, HTLV-II has an X-lor sequence highly homologous with that of HTLV-I. In addition, sequences from the BLV genome show a sequence interposed between env and the 3'LTR (Rice et al., 1985; Sagata et al., 1985). Though only distant DNA sequence homology with X-lor of HTLV-I and -II are present in this region, an open reading frame of 924 bp is found. While the predicted protein product of this open reading frame shows no close relationship of amino acid sequences to those of HTLV-I and -II, the general distribution of hydrophobic and hydrophilic residues is quite similar (Haseltine et al., 1984b).

TRANSCRIPTIONAL TRANS-ACTIVATION

The presence of an X-lor sequence is likely a property which is relatively specific to members of the HTLV-BLV family of retroviruses. The function of these sequences and their relationship to tumorigenesis remains to be defined. However, one property of HTLV infected cells, designated trans-acting transcriptional (TAT) activity, has been described which may

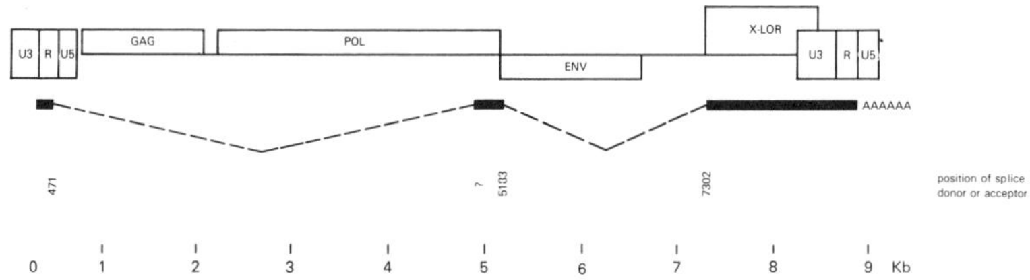

Fig. 2. Double-spliced X-lor mRNA of HTLV-I. A model for splicing is shown based on S1 nuclease mapping of the second splice acceptor and sequences determined for the second splice donor from a defective transforming HTLV-I virus. The position of the first splice donor is based on Northern blot hybridization data. The position of the first splice acceptor can not be located precisely but its approximate position is based on the size of the mRNA. The position of this latter site may be variable (Aldovini et al., 1985; our unpublished data with G. Franchini and M. Feinberg).

be directly or indirectly mediated by X-lor gene products (Sodroski et al., 1984b).

The TAT phenomenon was described utilizing a transient expression assay based on a plasmid construct with a transcriptional enhancer and promoter linked to the bacterial gene chloramphenicol acetyltransferase (CAT). CAT activity is easily measured by the conversion of ^{14}C-chloramphenicol to its acetylated forms which may be identified by thin layer chromatography and quantitated (Gorman et al., 1982a,b). CAT enzyme activity has been shown in a number of instances to be a direct reflection of CAT mRNA levels, thus, under conditions where CAT mRNA processing is not perturbed, it represents a measurement of transcriptional regulation. When the HTLV-I LTR is used as the transcriptional enhancer and promoter with the CAT gene, activity can be detected in uninfected human lymphocytes and fibroblasts (Sodroski et al., 1984b). However, when these same cells are infected by HTLV-I, the level of CAT activity is enhanced about 100-fold. This level of increase in CAT enzyme activity correlates with the levels of CAT mRNA, and thus, likely reflects transcriptional activity, though this point remains to be proven. No significant increase in the activity of the RSV LTR was noted in the HTLV-I infected versus uninfected cells, after correction for differing efficiencies of transfection. TAT activity is thus, not related to tissue tropism but rather relates to viral infection, occuring in both infected lymphocytes and fibroblasts.

The portion of the HTLV-I LTR responsible for TAT response localizes at least in part within U3 and is only partially dependent on its orientation relative to the promoter element (Rosen et al., 1985b). This portion of U3 can also be separated from the promoter by several kilobases and still manifest TAT activity. Whether other regions of the HTLV-I LTR participate in the TAT activity remains to be investigated.

The HTLV-II LTR also exhibits TAT activity in HTLV-II infected lymphocytes (Sodroski et al., 1984b). While the HTLV-I LTR demonstrates transactivation in HTLV-II infected cells, the converse is not true, i.e. the HTLV-II LTR is not transcriptionally activated in HTLV-I infected cells. Moreover, BLV, HTLV-III, and visna LTRs also exhibit TAT activity in homologous but not heterologous infected cell types (Derse et al., 1985; Hess et al., 1985; Rosen et al., 1985a; Sodroski et al., 1985).

Though TAT activity is described for a number of DNA tumor viruses, including SV40, adenovirus, polyoma, and herpes simplex (Cogen, 1978; Berk et al., 1979; Jones and Shenk, 1979; Nevins, 1981; Kamen et al., 1982; Fenton and Basilico, 1982; Green et al., 1983; Kao and Nevins, 1983; Brady et al., 1984; ElKareh et al., 1985), a similar activity has not been described in other RNA tumor viruses (Celander and Haseltine, 1984). Thus, it is likely that TAT is a characteristic feature of HTLV, BLV, and lentiviruses (Haseltine et al., 1985). However, a comprehensive examination of many other retroviruses and non-transforming DNA and RNA animal viruses for this type of activity remains to be carried out.

Several major questions concerning TAT activity remain:

1) Does the X-lor gene mediate trans-activation directly or indirectly? The presence of TAT activity in an HTLV-I infected non-producer cell line which makes X-lor mRNA as the predominant viral transcript is suggestive for for the involvement of X-lor (Sodroski et al., 1984b). Definitive experiments with subgenomic expression plasmids or purified proteins remain to be carried out.

2) Is a cellular gene involved in mediating the trans-activation phenomenon? In the case of DNA tumor viruses, there is some preliminary evidence in accord with this idea.

3) Does TAT activity enhance viral replication?

4) Does TAT activity mediate cellular transformation?

5) What are the molecular details of TAT activity? Does it involve protein or nucleic acid-DNA interactions? What are

the specific sequences in the LTR which bind the trans-acting factor? What other LTR sequences are involved in the mediation of this activity?

STRUCTURE AND FUNCTION OF HTLV-III

The identification of HTLV-III as the likely etiological agent in AIDS represented a major advance in the understanding and hopefully in the treatment of this human disorder. It also was an important step in the study of retroviruses in general. Unlike HTLV-I which can immortalize cells *in vivo* or *in vitro*, HTLV-III is cytopathic. Thus, a novel method of propagation of this virus *in vitro* was required. This was accomplished with cells with a surface phenotype similar to those which are infected *in vivo*, but which are already immortalized *in vitro*. Mature OKT4(+) lymphocytic cell lines were subcloned and examined for their ability to support high level replication of HTLV-III (Popovic et al., 1984b). A subclone of the Hut 78 cell line, designated H9, was chosen for this purpose.

The ready propagation of HTLV-III in H9 cells provided a good detection system for virus infection (Gallo et al., 1984). Furthermore, it provided material allowing the purification of viral proteins, and cloning the viral genome and virus-specific transcripts (Hahn et al., 1984b, Shaw et al., 1984b; Muesing et al., 1985). Using the cloned viral sequences, a number of aspects of the pathogenesis of AIDS were elucidated as outlined below:

1) Virus specific sequences were identified in the fresh lymph node, bone marrow, spleen, and liver of AIDS patients, reaffirming that the virus is not a laboratory contaminant but is rather closely associated with the immunological disorder (Shaw et al., 1984b).

2) Only a minority of cells at any given time have detectable viral sequences, even lymph nodes involved in LAS (Shaw et al., 1984b). This shows that the lymph node enlargement is not a direct result of infection and persistence of HTLV-III in most or all lymph node cells, but rather is an indirect result.

3) Similarly, HTLV-III sequences were not seen in Kaposi's sarcoma tissues (Shaw et al., 1984b), again arguing against direct infection and persistence in all or most endothelial cells which give rise to the tumor. Thus, lymphadenopathy and Kaposi's sarcoma (and perhaps B cell lymphomas) in association with AIDS or LAS are a result of "hit and run" by HTLV-

III or more likely due to opportunistic infections by EBV, CMV, and/or other viruses as a result of the primary immune deficiency.

4) HTLV-III sequences were identified in brains of encephalopathic AIDS patients (Shaw et al., 1985). This result did not appear to be a result of contamination of brain specimens with infected lymphocytes. This raises a number of additional questions. What cell type in brain is infected? Do these brain cells share a surface antigen with OKT4(+) lymphocytes which act as a receptor for virus adsorption and penetration? Are these proliferating cells or can this retrovirus replicate in non-proliferating cells? Is HTLV-III directly related to encephalopathy?

5) The restriction enzyme map of the HTLV-III genome in fresh tissues or cell lines from different individuals show a spectrum of differences from closely related to more distantly related sequences (about 1-15% bp differences) (Shaw et al., 1984b; Ratner et al., 1985a; our unpublished data with F. Wong-Staal, G. Shaw, B. Hahn, and B. Starcich). The viral restriction enzyme pattern within a given individual or a given cell line remains relatively stable with time though definitive analyses of these points remain to be carried out. Questions related to these findings include the following. Do particular structural variants of HTLV-III have different pathogenetic potentials? Are particular variants of HTLV-III temporaly or geographically related to one another? Do viruses or infected cells obtained from individuals infected with different variants of HTLV-III react with immune mediators (i.e. antibodies, cytotoxic T cells) of a particular individual in different ways?

The availability of cloned HTLV-III sequences has also made possible a detailed structural analysis by DNA sequencing (Muesing et al., 1985; Ratner et al., 1985b; Sachez-Pescador et al., 1985; Starcich et al., 1985; Wain-Hobson et al., 1985). This revealed that the integrated provirus is 9729-9749 bp in length with flanking LTRs of 634 bp (Table 2). The HTLV-III LTR includes only short tandem direct repeat sequences located 77 bp upstream of the RNA initiation site, and a core enhancer-like sequence located 225 bp upstream of the RNA CAP site. The 5'LTR is flanked on the 3'boundary by a sequence complementary to 18 bp of the 3'terminus of $tRNA^{lysine}$ unlike all other characterized retrovirus LTRs except that of mouse mammary tumor virus. The HTLV-III LTR is unlike those of HTLV-I, -II, and BLV in its shorter R sequence and presence of a polyadenylation addition signal 27 bp upstream of the R-U5 junction. However, sequences with homology to those surrounding the TATA box of HTLV-I and -II are noted in the U3 region of HTLV-III. Furthermore, sequence homologies around the

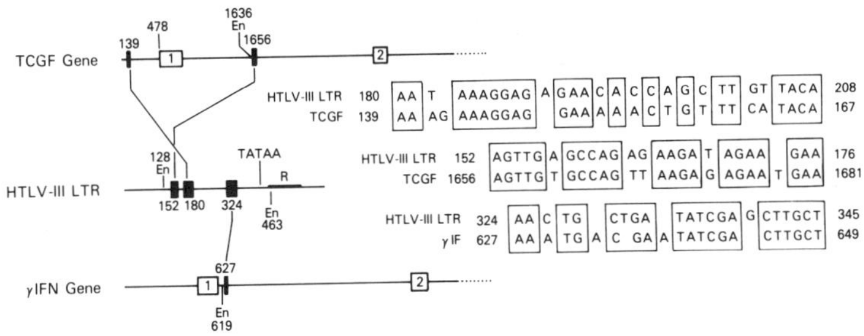

Fig. 3. Sequence homologies of HTLV-III LTR to genes expressed specifically in T lymphocytes. The HTLV-III LTR is shown in the center of the figure and the position of the R region is noted. Also shown schematically are exons 1 and 2 and surrounding regions of the TCGF and γ IFN genes. The numbers indicate the sequence positions; the RNA CAP sites are at position 478 in the TCGF gene, 454 in the HTLV-III LTR, and 1 in the γ IFN gene. Core enhancer-like sequences (En) are shown, as well as the position of the TATAA box in the HTLV-III LTR. Sequence alignments are shown on the right of the figure with identical nucleotides in boxes.

TATA box of HTLV-III are noted to the corresponding region of BLV.

Of interest is the presence of sequences within the HTLV-III U3 region 24 bp downstream of the core enhancer signal and 275 bp upstream of the RNA CAP site homologous to those in the TCGF gene 20 bp downstream of a core enhancer signal in the first intron and 293 bp upstream of the RNA CAP site, respectively (Fig. 3) (Starcich et al., 1985). A sequence in the HTLV-III U3 region is also noted to be homologous to one located immediately downstream of a core enhancer signal in the first intron of the γ-interferon (γ-IFN) gene. Homologous structures in the HTLV-III U3 region to corresponding regions of genes also expressed specifically in T lymphocytes suggest possible involvement of these elements in regulation of transcription and/or replication. These sequences may represent common binding sites for cellular or viral factors involved in inducing tissue-specific gene expression and/or TAT activity with viral infection (Fig. 4).

The 5'LTR of HTLV-III is followed by a 129 bp leader sequence which includes a predominant splice donor sequence for subgenomic

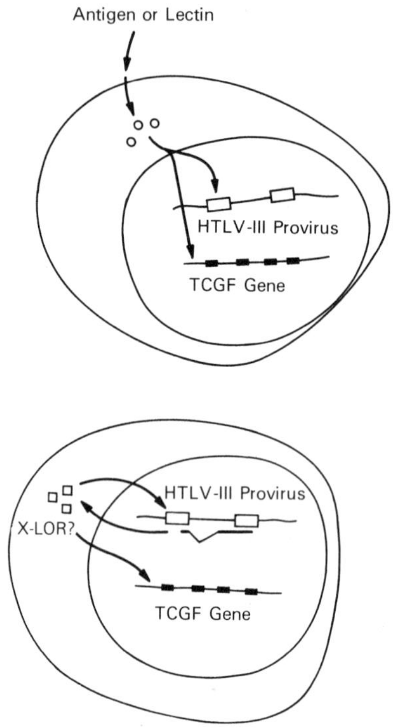

Fig 4. T lymphocyte specific gene and virus expression. Two models are presented explaining the possible significance of sequence homologies of the HTLV-III LTR to those within the TCGF gene. The top figures raises the possibility that these homologous sequences may mediate responsiveness to T lymphocyte stimulating factors. The bottom figure depicts a possible interaction of a viral product analogous to the X-lor protein with both the HTLV-III LTR and the TCGF gene.

mRNAs (Muesing et al., 1985; our unpublished observations with T. Okamoto, S. K. Arya, and S. F. Josephs) as well as a probable packaging signal.

The gag gene predicts a precursor of 500 or 512 amino acids depending on whether 1 or 2 copies of a tandemly repeated 12 amino acid sequence are present (Fig. 5) (Muesing et al., 1985; Ratner et al., 1985b; Sanchez-Pescador et al., 1985; Wain-Hobson et al., 1985). It encodes from N- to C-terminus a p17 myrgagtin, a p24 largagtin, and a p15 protein; the p15 protein is likely

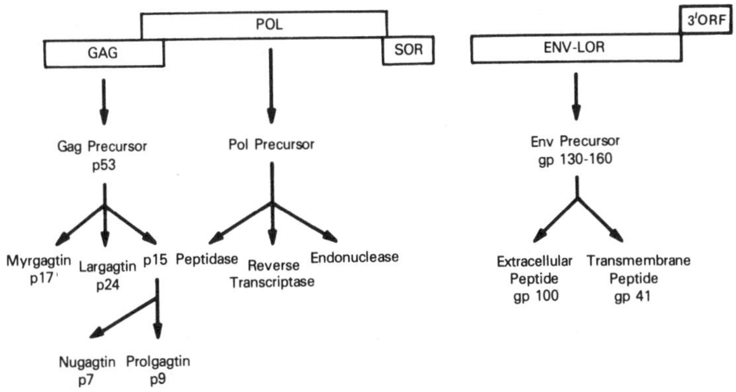

Fig. 5. Proposed scheme of HTLV-III viral protein processing. The nomenclature for *gag* proteins is described by Oroszolan et al. (1985).

processed to a p9 nugagtin and a p7 prolgagtin. This structure is unlike that of most other retroviruses but similar to that of HTLV-I and BLV in the size of largagtin and the lack of an intervening phosphoprotein between myrgagtin and largagtin sequences.

Unlike HTLV-I and BLV, the *pol* gene of HTLV-III overlaps the *gag* gene and is located in an alternate reading frame. There is no *pol*-specific mRNA (our unpublished data with S. K. Arya) and thus, a *gag-pol* precursor is likely to be encoded by a spliced mRNA. The N-terminus of the predicted *pol* gene product reveals homologies to MoMULV *pol* and RSV *gag* p15 proteases suggesting the presence of a similar activity in HTLV-III infected cells (Muesing et al., 1985; Ratner et al., 1985b; Sanchez-Pescador et al., 1985; Wain-Hobson et al., 1985). It is noteworthy that this portion of the predicted *pol* product shows homologies to the predicted amino acid sequences within short open reading frames located between *gag* and *pol* in HTLV-I (Table 3). Furthermore, the region of greatest DNA sequence homology between HTLV-I and HTLV-III is noted in this region (Fig. 6), and this represents the sequence similarty with the greatest homologies to those of any other retroviral sequence. The remainder of the predicted *pol* protein shows homologies to those of other retroviruses in both the reverse transcriptase and endonuclease regions.

The third open reading frame in HTLV-III is unusual in that it overlaps with the *pol* gene in a manner similar to that of many retroviral *env* genes but is large enough to encode a primary translation product of only 23 kd. This short open reading

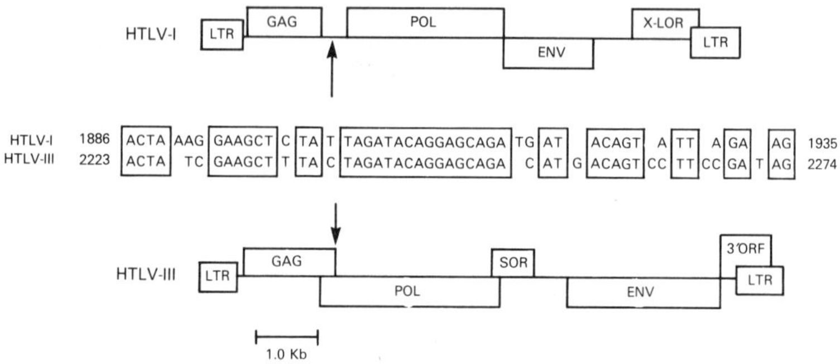

Fig. 6. Similarity of nucleotide sequence between the gag and pol genes of HTLV-I and those in the pol gene of HTLV-III. Sequences from each virus which show 76% nucleotide identity are shown and their positions in each viral genome are indicated by arrows.

frames, designated sor, is well conserved among different variants of HTLV-III (Ratner et al., 1985a) but whether or not it encodes a protein product, and if so the function of such a protein remains to be determined. Sor is followed by a non-coding region of 584 bp which includes a DNA sequence with homology to that of the HTLV-I env gene (Ratner et al., 1985b). These findings suggested that sor and the adjacent non-coding area may represent the vestige of an env gene, with no function or a new function in HTLV-III infected cells. This region of HTLV-III includes multiple splice acceptor and donor sites utilized in double or triple splicing to encode subgenomic mRNAs (Muesing et al., 1985; our unpublished data with T. Okamoto, S. K. Arya, and S. F. Josephs). Such complicated splice mechanisms have not been described for retroviruses other than HTLV-I, -II and -III.

The fourth open reading in HTLV-III encodes an 865 amino acid precursor to the envelope gp160, which is cleaved to a 481 amino acid exterior glycoprotein, gp120, and a 345 amino acid transmembrane protein, gp41. Comparison of different AIDS viral sequences has demonstrated that those encoding the exterior portion of the viral envelope are most variable and may determine different epitopes (Ratner et al., 1985a; our unpublished observations with B. Starcich). The large size of this open reading frame, most notably its 3'portion, is unusual for retroviral env genes, and raises the question as to whether a second function may be encoded by this portion of the gene.

Table 4. Structural and functional similarities of HTLV-III to HTLV-I, -II, BLV, and the Lentiviruses

1. Specific cell tropism i.e. OKT4(+) lymphocytes in the case of HTLV-I, -II, and -III

2. Ability to induce giant multinucleated cells *in vitro*

3. Trans-acting transcriptional activity

4. Juxtaposition of myrgagtin and largagtin due to the absence or an intervening phosphoprotein[a]

5. Similar size of largagtin[b]

6. Similar size reverse transcriptase which is magnesium dependent[b]

7. Immunological cross-reactivity in some cases of *gag* and *env* proteins[c]

8. Presence of potential coding sequences in addition to those for *gag*, *pol*, and *env*[b]

[a]no information available on this point for HTLV-II and lentiviruses
[b]no information available on this point for the lentiviruses
[c]Schupbach et al., 1984b; Mann et al., 1985

A fifth open reading frame of 402-648 bp, designated 3'*orf* is noted which overlaps the 3'LTR by 112-328 bp (Muesing et *al*., 1985, Ratner et al., 1985a,b,d; Sanchez-Pescador et al., 1985; Wain-Hobson et al., 1985). The function of a protein, if any, encoded by this open reading frame remains to be determined.

In summary, it is clear that the structure of the HTLV-III genome differs from that of HTLV-I, -II, and BLV, and these differences are likely to determine differences in the function of these viruses. Furthermore, the HTLV-III predicted viral proteins show no greater primary sequence homology to HTLV-I, -II, and BLV, than other retroviral proteins (Muesing et al., 1985; Ratner et al., 1985b; Sanchez-Pescador et al., 1985; Wain-Hobson et al., 1985). However, a number of structural and functional similarities of HTLV-III to HTLV-I, -II, and BLV, summarized in Table 4 suggest that it is reasonable to classify the AIDS virus as a member of this group of retroviruses. Furthermore, the recent demonstration of sequence homology of visna to HTLV-III (Gonda et al., 1985), and TAT activity in visna-infected cells for visna LTR-CAT constructs (Hess et al., 1985) suggests a relationship to the lentivirus family as well.

CONCLUSIONS

Detailed studies of two human diseases have revealed likely retroviral causes. Purification of viral proteins, cloning of viral genomes, and determination of their nucleotide sequence has revealed unusual structural features compared to those of most other retroviruses. Though HTLV-I, -II, and BLV differ from HTLV-III in some structural and functional features, a number of similarities suggest a relationship which may have arisen as a result of divergent evolution from a common ancestral virus or convergent evolution. The development of *in vitro* systems which reproduce certain aspects of each disease are permitting careful dissection of the transforming determinants of HTLV-I and the cytopathic determinants of HTLV-III. These studies should also provide clues to treatment or prevention of these retroviral induced diseases.

The human lymphotropic viruses infect predominantly OKT4(+) lymphocytes. Perturbations of the functions of this cell type, critical to immune regulation, may give rise to other diseases as a result of depression or alteration of these functions. Other cell types infectable by HTLV may account for other human disorders (Ho et al., 1984; Hoxie et al., 1984; Longo et al., 1984; Mann et al., 1984). Lastly the development of *in vitro* culture systems for other tissue types analogous to those described here for T-lymphocytes should allow the rational investigation of the etiology of a number of other pathological disorders in humans.

REFERENCES

Aldovini, A., DeRossi, A., Gallo, R. C., Wong-Staal, F., and

Franchini, G., 1985, Molecular analysis of an HTLV-I deletion mutant provirus isolated from a stably transformed T-cell line, Proc. Natl. Acad. Sci., U.S.A. submitted.

Berk, A. J., Lee, F., Harrison, T., Williams, J., and Sharp, P. A., 1979, Pre-early adenovirus 5 gene product regulates synthesis of early viral messenger RNAs, Cell, 17:935.

Blattner, W. A., Kalyanaraman, V. S., Robert-Guroff, M., Lister, T. A., Galton, D. A. G., Sarin, P. S., Crawford, M. H., Catovsky, D., Greaves, M., and Gallo, R. C., 1982, The human type C-retrovirus, HTLV, in blacks from the Caribbean region and relationship to adult T-cell leukemia/lymphoma, Int. J. Cancer, 30:257.

Blattner, W. A., Takatsuki, K., and Gallo, R. C., 1983, Human T-cell leukemia/lymphoma virus and adult T-cell leukemia, JAMA, 250:1074.

Blattner, W. A., Clark, J. W., Gibbs, W. N., Jaffe, E. S., Robert-Guroff, M., Saxinger, C., and Gallo, R. C., 1984, HTLV: epidemiology and relationship to human malignancy, in: "Cancer Cells: Human T-Cell Leukemia Viruses," R. C. Gallo and M. Essex, eds., Cold Spring Harbor Press, New York, vol. 3, pp.

Blayney, D. W., Jaffe, E. S., Blattner, W. A., Cossman, J., Robert-Guroff, M., Longo, D. L., Bunn, P. A., and Gallo, R. C., 1983a, The human T cell leukemia/lymphoma virus associated with American adult T cell leukemia/lymphoma, Blood, 62:401.

Blayney, D. W., Jaffe, E. S., Fisher, R. I., Schecter, G. P., Cossman, J., Robert-Guroff, M., Kalyanaraman, V. S., Blattner, W. A., and Gallo, R. C., 1983b, The human T-cell leukemia/lymphoma virus (HTLV), lymphoma, lytic bone lesions, and hypercalcemia, Ann. Intern. Med., 98:144.

Blayney, D. W., Blattner, W. A., Robert-Guroff, M., Jaffe, E. S., Fisher, R. I., Bunn P. A., Patton, M. G., Rorick, H. R., and Gallo, R. C., 1984, The human T-cell leukemia/lymphoma virus (HTLV) in the southeastern United States, JAMA, 250:1048.

Brady, J., Bolen, J. B., Radonovich, M., Salzman, N., and Khoury, G., 1984, Stimulation of simian virus 40 late gene expression by simian virus 40 tumor antigen, Proc. Natl. Acad. Sci., USA, 81:2040

Catovsky, D., Greaves, M. F., Rose, M., Galton, D. A. G,, Goolden, A. W. G., McCluskey, D. R., White, J. M., Lampert, I., Bourikas, G., Ireland, R., Brownell, A. I., Bridges, J. M., Blattner, W. A., and Gallo, R. C., 1982, Adult T-cell lymphoma-leukaemia in blacks from the West Indies, Lancet, i:639.

Celander, D. and Haseltine, W. A., 1984, Tissue-specific transcription preference as a determinant of cell tropism and leuaemogenic potential of murine retroviruses, Nature, 312:159.

Chamberland, M. E., Castro, K. G., Haverkos, H. W., Miller, B. I., Thomas, P. A., Reiss, R., Walker, J., Spira, T. J., Jaffe, H. W., and Curran, J. W., 1984, Acquired immunodeficiency syndrome in the United States: an analysis of cases outside high-incidence groups, Ann. Int. Med., 101:617.

Chen, I. S. Y., McLaughlin, J., Gasson, J. C., Clark, S. C., and Golde, D. W., 1983a, Molecular characterization of genome of a novel human T-cell leukaemia virus, Nature, 305:502

Chen, I. S. Y., Quan, S. G., and Golde, D. W., 1983b, Human T-cell leukemia virus type II transforms normal human lymphocytes, Proc. Natl. Acad. Sci., USA, 80:7006.

Chiu, I-M., Callahan, R., Tronick, S. R., Schlom, J., and Aaronson, S. A., 1984, Major pol gene progenitors in the evolution of oncoviruses, Science, 223:364.

Chorba, T., Brynes, R., Telfer, M., Kalyanaraman, V., Ramsey, R., Feorino, P., Preble, O., McDougal, J., Palmer, E., Narayanan, R., Cabradilla, C., Chen, A., Mawle, A., and Evatt, B., 1984, Characterization of T-lymphocytes infected by a novel isolate of human T leukemia virus type II (HTLV-II), Blood, (Suppl. 1):201a.

Cianciolo, G. J., Kipnis, R. J., and Snyderman, R., 1984, Similarity between p15E of murine and feline leukemia viruses and p21 of HTLV, Nature, 311, 515.

Cogen, B., 1978, Virus-specific early RNA in 3T6 cells infected by a ts mutant of polyoma virus, Virology, 85:222.

Copeland, T. D., Oroszlan, S., Kalyanaraman, V. S., Sarngadharan, M. G., and Gallo, R. C., 1983, Complete amino acid sequence of human T-cell leukemia virus structural protein p15, FEBS Lett., 162:390.

Couez, D., Deschamps, J., Kettmann, R., Stephens, R. M., Gilden R. V., and Burny, A., 1984, Nucleotide sequence analysis of the long terminal repeat of integrated bovine leukemia provirus DNA and of adjacent viral and host sequences, J. Virol., 49:615.

Crawford, S. and Goff, S. P., 1985, A deletion mutant in the 5' part of the pol gene of Moloney murine leukemia virus blocks proteolytic processing of the gag and pol polyproteins, J. Virol., 53:899.

Dalgleish, A. G., Beverley, P. C. L., Clapham, P. R., Crawford, D. H., Greaves, M. F., and Weiss, R. A., 1984, The CD4 (T4) antigen is an essential component of the receptor for the AIDS retrovirus, Nature, 312:763.

DeRossi, A., Aldovini, A., Franchini, G., Mann, D., Gallo, R. C., and Wong-Staal, F., Clonal selection of T lymphocytes infected by cell-free human T-cell leukemia/lymphoma virus type I: parameters of virus integration and expression, Virol. submitted.

Derse, D., Caradonna, S., and Casey, J. W., 1985, Bovine leukemia virus long terminal repeat: a cell type specific promoter, Science, 227:317.

DeVecchis, L., Graziani, G., Macchi, B., Grandori, C., Pastore, S., Popovic, M., Gallo, R. C., and Bonmassar, E., 1985, Decline of natural cytotoxicity of human lymphocytes following infection with human T-cell leukemia/lymphoma virus (HTLV), Leukemia Res., in press.

Dickson, C., Smith, R., Brookes, S., and Peters, G., 1984, Tumorigenesis by mouse mammary tumor virus: proviral activation of a cellular gene in the common integration region int-2, Cell, 37:529.

ElKareh, A., Murphy, A. J. M., Fichter, T., Efstratiadis, A., and Silverstein, S., 1985, "Transactivation" control signals in the promoter of the herpesvirus tymidine kinase gene, Proc. Natl. Acad. Sci., USA, 82:1002.

Eyster, M. E., Goedert, J. J., Sarngadharan, M. G., Weiss, S. H., Gallo, R. C., and Blattner, W.A., 1985, Development and early natural history of HTLV-III antibodies in persons with hemophilia, JAMA, in press.

Fauci, A. S., Macher, A. M., Longo, D. L., Lane, C. H., Rook, A. H., Masur, H., and Gelmann, E. P., 1984, Acquired immunodeficiency syndrome, epidemiologic, clinical, immunologic, and therapeutic considerations, Ann. Int. Med., 100:92.

Fenton, R. G. and Basilico, C., 1982, Regulation of polyoma virus early transcription in transformed cells by large T antigen, Virology, 121:384.

Fung, Y. K., Fadlay, A. N., Crittenden, L. B., and Kung, J. H., 1981, On the mechanism of retrovirus-induced avian lymphoid leukosis: deletion and integration of the provirus, Proc. Natl. Acad. Sci., 78:3418.

Fung, Y. K. T., Crittenden, L. B., and Kung, J. H., 1982, Orientation and position of avian leukosis virus DNA relative to the cellular oncogene c-myc in B-lymphoma tumors of highly susceptible $15I_2x7_2$ chickens, J. Virol., 44:742.

Gallo, R. C., 1984, Human T-cell leukaemia-lymphoma virus and T-cell malignancies in adults, Cancer Surv., 3:113.

Gallo, R. C. and Gallagher, R. E., 1974, Are viruses concerned in the etiology of human leukemia? in: "Series Hematologica, The Etiology of Leukemia," K. G. Jensen, S.A. Killman, and F. W. Gunz, eds., Munksgaard, Copenhagen, vol. 7, pp. 224-273.

Gallo, R. C., Mann, D., Broder, S., Ruscetti, F., Maeda, M., Kalyanaraman, V. S., Robert-Guroff, M., and Reitz, M. S., 1982, Human T-cell leukemia-lymphoma virus (HTLV) is in T- but not B-lymphocytes from a patient with cutaneous T cell lymphoma, Proc. Natl. Acad. Sci., USA, 79:4680.

Gallo, R. C. and Reitz, M. S., 1976, Molecular probes for tumor viruses in human cancer, Int. Rev. Exp. Pathol., 18:1.

Gallo, R. C., Salahuddin, S. Z., Popovic, M., Shearer, G. M., Kaplan, M., Haynes, B. F., Palker, T. J., Redfield, R., Oleske, J., Safai, B., White, G., Foster, P., and Markham, P. D., 1984, Frequent detection and isolation of cytopathic retroviruses (HTLV-III) from patients with AIDS and at risk for AIDS, Science, 224:500.

Gardner, R., Howarth, A. J., Hahn, P., Brown-Leudi, M., Shepherd, R. J., and Messing, J., 1981, The complete nucleotide sequence of an infectious clone of cauliflower mosaic virus by M13mp7 shotgun sequencing, Nucl. Acids Res., 9:2871.

Gelmann, E. P, Franchini, G., Manzari, V., Wong-Staal, F., and Gallo, R. C., 1984, Molecular cloning of a unique human T-cell leukemia virus (HTLV-II_{Mo}), Proc. Natl. Acad. Sci., USA, 81:993.

Gonda, M. A., Wong-Staal, F., Gallo, R. C., Clements, J. E., Narayan, O., and Gilden, R. V. 1985, Sequence homology and morphologic similarity of HTLV-III and visna virus, a pathogenic lentivirus, Science, 227:175.

Gorman, C. M., Merlino, G. T., Willingham, M. C., Pastan, I., and Howard, B., 1982a, The Rous sarcoma virus long terminal repeat is a strong promoter when introduced into a variety of eukaryotic cells by DNA mediated transfection, Proc. Natl. Acad. Sci., USA, 79:6777.

Gorman, C. M., Moffat, L. F., and Howard, B.H., 1982b, Recombinant genomes which express chloramphenicol acetyltransferase in mammalian genes, Mol. Cell. Biol., 2:1044.

Gottlieb, M. S., Schroff, R., Shanker, H., Weisman, J. D., Fan, P. T., Wolf, R. A., and Saxon, A., 1981, Pneumocystis carinii pneumonia and mucosal candidiasis in previously healthy homosexual men: evidence of a new acquired cellular immunodeficiency, New Engl. J. Med., 305:1425.

Green, M. R., Treisman, R., and Maniatis, T., 1983, Transcriptional activation of cloned human β-globin genes by viral immediate-early gene products, Cell, 35:137.

Groopman, J. E., Salahuddin, S. Z., Sarngadharan, M. G., Mullins, J. I., O'Hara, C. J., Cheeseman, S. H., Haverkos, H., Forgacs, P., Riedel, N., McLane, M. F., Essex, M., and Gallo, R.C., 1984, Virologic studies in a case of transfusion-associated AIDS, New Engl. J. Med., 311:1419.

Groopman, J. E., Mayer, K. H., Sarngadaharan, M. G., Ayotte, D., Devico, A. L., Finberg, A. L., Sliski, A. H., Allan, D., and Gallo, R. C., 1985, Seroepidemiology of human T lymphotropic virus type III among homosexual men with the acquired immunodeficiency syndrome or generalized lymphadenopathy and among asymptomatic controls in Boston, Ann. Int. Med., 102:334.

Gross, L., 1951, "Spontaenous" leukemia developing in C3H mice following inoculation in infancy with AK-leuemia extracts, or AK embryos, Proc. Soc. Exp. Biol. Med., 76:27.

Hahn, B. H., Popovic, M., Kalyanaraman, V. S., Shaw, G. M., Lomonico, A., Weiss, S. H., Wong-Staal, F., and Gallo, R. C., 1984a, Detection and characterization of an HTLV-II provirus in a patient with AIDS, in: "UCLA Symposia on Molecular and Cellular Biology, New Series, Acquired Immunodeficiency Syndrome," M. S. Gottlieb and J. E. Groopman, eds., Alan R. Liss, Inc., New York, pp. 73-81.

Hahn, B. H., Shaw, G. M., Arya, S. K., Popovic, M., Gallo, R. C., and Wong-Staal, F., 1984b, Molecular cloning and characterization of the HTLV-III virus associated with AIDS, Nature, 312:166.

Hahn, B. H., Shaw, G. M., Popovic, M., Lomonico, A., Gallo, R. C., and Wong-Staal, F., 1985, Molecular cloning and analysis of a new variant of human T-cell leukemia virus (HTLV-Ib) from an African patient with adult T-cell leukemia-lymphoma, Int. J. Cancer, in press.

Haseltine, W. A., Sodroski, J. G., and Patarca, R., 1984a, Structure and function of the genome of HTLV, in: "Current Topics in Microbiology: Human T-Cell Leukemia Viruses," P. K. Vogt, ed., Raven Press, New York, in press.

Haseltine, W. A., Sodroski, J., Patarca, R., Briggs, D., Perkins, D., and Wong-Staal, F., 1984b, Structure of the 3' terminal region of type II human T lymphotropic virus: evidence for a new coding region, Science, 225:419.

Haseltine, W. A., Gallo, R. C., Wong-Staal, F., Gonda, M., Sodroski, J., Clements, J., Rosen, C., Papas, T., Essex, M., Bolognesi, D., Yoshida, M., Burny, A., Jarrett, W., Gilden, R., Oroszlan, S., Vande Woude, G., Rossi, G., Sugimura, T., Toyashima, K., Vogt, P., and Weiss, R., 1985, A new type of retrovirus: the trans-acting, type T retroviruses, in preparation.

Haynes, B. F., Miller, S. E., Palker, T. J., Moore, J. O., Dunn, P. H., Bolognesi, D. P., and Metzgar, R. S., 1983, Identification of human T-cell leukemia virus in a Japanese patient with adult T cell leukemia and cutaneous lymphomatous vasculitis, Proc. Natl. Acad. Sci., USA, 80:2054.

Hayward, W. S., Neel, B. G., and Astrin, S. M., 1981, Induction of lymphoid leukosis by avian leukosis virus: activation of a cellular 'onc' gene by promoter insertion, Nature, 290:475.

Hess, J. L., Clements, J. E., and Narayan, O., 1985, Cis and trans-acting transcriptional regulation of visna virus, Science, submitted.

Hinuma, Y., Nagata, K., Misoka, M., Nakai, M., Matsumoto, T., Kinoshita, K. I., Shirakawa, S., and Miyoshi, I., 1981, Adult T-cell leukemia: antigen in an ATL cell line and detection of antibodies to the antigen in human sera, Proc. Natl. Acad. Sci., USA, 78:6476.

Hinuma, Y., Komoda, H., Chosa, T., Kondo, T., Kohakura, M., Takenaka, T., Kikuchi, M., Ichimura, M., Yunoki, K., Sato, I., Natsuo, R., Takiuchi, Y., Uchino, H., and Hanaoka, M., 1982, Antobodies to adult T-cell leukemia-virus associated antigen (ATLA) in sera from patients with ATL and controls in Japan: a nation-wide seroepidemiolgoic study, Int. J. Cancer, 29:631.

Ho, D., Rota, T. R., and Hirsch, M.S., 1984a, Infection of human endothelial cells by human T-lymphotropic virus type I, Proc. Natl. Acad. Sci., USA, 81:7588.

Ho, D. D., Schooley, R. T., Rota, T. R., Kaplan, J. C., Flynn, T., Salahuddin, S. Z., Gonda, M. A, and Hirsch, M. S., 1984b, HTLV-III in the semen and blood of a healthy homosexual man, Science, 226:451.

Hoxie, J. A., Matthews, D. M., and Cines, D. B., 1984, Infection of human endothelial cells by human T-cell leukemia virus type I, Proc. Natl. Acad. Sci., USA, 81:7591.

Jaffe, H. W., Sarngadaharan, M. G., Devico, A., Bruch, L., Getchell, J. P., Kalyanaraman, V. S., Kilbourne, B. W., Peterman, T. A., Haverkos, H. W., McDougal, J. S., Stoneburner, R. L., Gallo, R. C., Francis, D. P., and Curran, J. W., 1985, Serologic evidence of an etiologic role for a human T-lymphotropic retrovirus in transfusion-associated AIDS, Lancet, submitted.

Jones N. C. and Shenk, T., 1979, An adenovrius 5 early gene product function regulates expression of other early viral genes, Proc. Natl. Acad. Sci., USA, 76:3665.

Kalyanaraman, V. S., Sarngadharan, M. G., Poiesz, B. J., Ruscetti, F. W., and Gallo, R. C., 1981, Immunological properties of a type C retrovirus isolated from cultured human T-lymphoma cells and comparison to other mammalian retroviruses, J. Virol., 38:906.

Kalyanaraman, V. S., Sarngadharan, M. G., Nakao, Y., Ito, Y., Aoki, T., and Gallo, R. C., 1982a, Natural antibodies to the structural core protein (p24) of the human T cell leukemia (lymphoma) retrovirus found in sera of leukemia patients in Japan, Proc. Natl. Acad. Sci., USA, 79:1653.

Kalyanaraman, V. S., Sarngadharan, M. G., Robert-Guroff, M., Miyoshi, I., Blayney, D., Golde, D, and Gallo, R. C., 1982b, A new subtype of human T-cell leukemia virus (HTLV-II) associated with a T-cell variant of hairy cell leukemia, Science, 218:571.

Kamen, R., Jat, P., Treisman, R., Favarolo, J., and Folk, W. R., 1982, 5'Termini of polyoma virus early region transcripts synthesized in vivo by wild-type virus and viable deletion mutants, J. Mol. Biol., 159:189.

Kao, H.-T. and Nevins, J. R., 1983, Transcriptional activation and subsequent control of the human heat shock gene during adenovirus infection, Mol. Cell. Biol., 3:2058.

Kitchen, L. W., Barin, F., Sullivan, J. L., McLane, M. F., Brettler, D. B., Levine, P. H., and Essex, M., 1984, Aetiology of AIDS - antibodies to human T-cell leukaemia virus (type III) in haemophiliacs, Nature, 312:367.

Klatzmann, D., Barre-Sinoussi, F., Nugeyre, M. T., Dauguet, C., Vilmer, E., Griscelli, C., Brun-Vezinet, F., Rouzioux, C., Gluckman, J. C., Chermann, J.-C., and Montagnier, L., 1984, Selective tropism of lymphadenopathy associated virus (LAV) for helper-inducer T lymphocytes, Science, 225:59.

Lee, T. J., Coligan, J. E., Homma, T., McLane, M. F., Tachibana, N., and Essex, M., 1984a, Human T-cell leukemia virus associated membrane antigens: identity of the membrane antigens recognized after virus infection, Proc. Natl. Acad. Sci., USA, 81:3856.

Lee, T. H., Coligan, J. E., Sodroski, J. G., Haseltine, W. A., Salahuddin, S. Z., Wong-Staal, F., Gallo, R. C., and Essex, M., 1984b, Antigens encoded by the 3'-terminal region of human T-cell leukemia virus: evidence for a functional gene, Science, 226:57.

Levin, J. G., Hu, S. C., Rein, A., Messer, L. I., and Gerwin, B. I., 1984, Murine leukemia virus mutant with a frameshift in the reverse transcriptase coding region: implications for pol gene structure, J. Virol., 51:470.

Levy, J. A., Hoffman, A. D., Kramer, S. M., Landis, J. A., Shimabukuro, J. M., and Oshiro, L. S., 1984, Isolation of lymphocytopathic retroviruses from San Francisco patient with AIDS, Science, 225:840.

Longo, D. L., Gelmann, E. P., Cossman, J., Young, R. A., Gallo, R. C., O'Brien, S. J., and Matis, L. A., 1984, Isolation of HTLV-transformed B-lymphocyte clone from a patient with HTLV-associated adult T-cell leukaemia, Nature, 310:505.

Mann, D. L., Popovic, M., Murray, C., Neuland, C., Strong, D. M., Sarin, P., Gallo, R. C., and Blattner, W. A., 1983a, Cell surface antigen expression in newborn cord blood lymphocytes infected with HTLV, J. Immunol., 131:2021.

Mann, D. L., Popovic, M., Sarin, P. S., Murray, C., Reitz, M. S., Strong, D. M., Haynes, B. F., Gallo, R. C., and Blattner, W. A., 1983b, Cell lines producing human T-cell lymphoma virus show altered HLA expression. Nature, 305:58.

Mann, D. L., Clark, J., Clarke, M., Reitz, M., Popovic, M., Franchini, G., Trainor, C. D., Strong, D. M., Blattner, W. A., and Gallo, R. C., 1984, Identification of the human T cell lymphoma virus in B cell lines established from patients with adult T cell leukemia, J. Clin. Invest., 74:56.

Mann, D. L., DeSanctis, P., Newman, M., Gibbs, N., Popovic, M., Sarngadaharan, M. G., Gallo, R., Clark, J., and Blattner, W., 1985, Immunoglobulin "captured" from chronic lympocytic leukemia cells has antibody activity to HTLV-I, -II, -III., Science, submitted.

Mann, R., Mulligan, R. L., and Baltimore, D., 1983, Construction of a retrovirus packaging mutant and its use to produce helper-free defective retroviruses, Cell, 33:153.

Manzari, V., Wong-Staal, F., Franchini, G., Colombini, S., Gelmann, E. P., Oroszlan, S., Staal, S., and Gallo, R. C., 1983, Human T-cell leukemia-lymphoma virus (HTLV): cloning of an integrated defective provirus and flanking cellular sequences, Proc. Natl. Acad. Sci., USA, 80:1574.

Markham, P. D., Salahuddin, S. Z., Kalyanaraman, V. S., Popovic, M., Sarin, P. S., and Gallo, R. C., 1983, Infection and transformation of fresh human umbilical cord cells by multiple sources of human T-cell leukemia-lymphoma virus (HTLV), Int. J. Cancer, 31:413.

Markham, P. D., Salahuddin, S. Z., Macchi, B., Robert-Guroff, M., and Gallo, R. C., 1984, Transformation of different phenotypic types of human bone marrow T-lymphocytes by HTLV-I, Int. J. Cancer, 33:13.

Mier, J. W. and Gallo, R. C., 1980, Purification and some characteristics of human T-cell growth factor from phytohemaglutinin-stimulated lymphocyte conditioned media, Proc. Natl. Acad. Sci., USA, 77:6134.

Mitsuya, H., Guo, H.-G., Cossman, J., Megson, M., Reitz, M. S., Jr., and Broder, S., 1984, Functional properties of antigen-specific T cells infected by human T-cell leukemia-lymphoma virus (HTLV-I), Science, 225:1484.

Miwa, M., Shimotohno, K., Hoshino, H., Fujino, M., and Sugimura, T., 1984, Detection of pX proteins in human T-cell leukemia virus (HTLV)-infected cells by using antibody against peptide deduced from sequences of X-IV DNA of HTLV-I and Xc DNA of HTLV-II proviruses, Gann, 75:752.

Miyoshi, I., Kubonishi, I., Yoshimoto, S., Akagi, T., Ohtuski, Y., Shiraishi, Y., Nagato, K., and Hinuma, Y., 1981a, Type c virus particles in a cord T-cell line derived by co-cultivating normal human cord leukocytes and human leukemic T-cells, Nature, 294:770.

Miyoshi, I., Yoshimoto, S., Kubonishi, I., Taguchi, H., Shiraishi, Y., Ohtsuki, Y., and Akagi, T., 1981b, Transformation of normal human cord lymphocytes by co-cultivation with a lethally irradiated human T-cell line carrying type C virus particles, Gann, 72:997.

Morgan, D. A., Ruscetti, F. W., and Gallo, R. C., 1976, Selective in vitro growth of T-lymphocytes from normal human bone marrow, Science 193:1007.

Muesing, M. A., Smith, D. H., Cabradilla, C. D., Benton, C. V., Lasky, L. A., and Capon, D. J., 1985, Nucleic acid structure and expression of the human AIDS/lymphadenopathy retrovirus, Nature, 313:450.

Neel, B. G., Hayward, W. G., Robinson, M. L., Fang, J., and Astrin, S. M., 1981, Avian leukosis virus-induced tumors have common proviral integration sites and synthesize discrete new mRNAs: oncogenesis by promoter insertion, Cell, 23: 323.

Nevins, J. R., 1981, Mechanism of activation of early viral transcripts by the adenovirus E1A gene product, Cell, 26:213.

Noorii-Daloii, M., Swift, R. A., Kung, J. H., Crittenden, L. B., and Witter, R. L., 1981, Specific integration of REV provirus in avian bursal lymphomas, Nature, 294:574.

Nusse, R. and Varmus, H. E., 1982, Many tumors induced by the mouse mammary tumor virus contain a provirus integrated in the same region of the host genome, Cell, 31:93.

O'Hare, P. and Hayward, G. S., 1985, Evidence for a direct role for both the 175,000 and 110,000-molecular weight immediate-early proteins of herpes simplex virus in the transactivation of delayed-early promoters, J. Virol., 53:751.

Okamoto, T., Josephs, S. F., Kawanishi, M., Gallo, R. C., and Wong-Staal, F., 1985, Determination of a splice acceptor site of pX gene in HTLV-I infected cells, Virology, in press.

Ono, Y., Onda, H., Sasada, R., Igaroshi, K., and Sugino, Y., 1983, The complete nucleotide sequences of the cloned hepatitis B virus DNA; subtype adr and ayw, Nucl. Acids Res., 11:1747.

Oroszlan, S., Sarngadharan, M. G., Copeland, T. D., Kalyanaraman, V. S., Gilden, R. V., and Gallo, R. C., 1982, Primary structure analysis of the major internal protein p24 of human type C T-cell leukemia virus, Proc. Natl. Acad. Sci., USA, 79:1291.

Oroszolan, S., Copeland, T. D., Rice, N. R., Smythers, G. W., Tsai, W. P., Yoshinaka, Y., and Shimotohno, K., 1985, Structural and antigenic characterization of the proteins of human T-cell leukemia virus and their relationship to the gene products of other retroviruses, in: "Retroviruses in Human Lymphoma/Leukemia," M. Miwa, ed., Japan Sci. Soc. Press, Tokyo, in press.

Panganiban, A. T. and Temin, H. M., 1984, The retrovirus pol gene encodes a product required for DNA integration: identification of a retrovirus int locus, Proc. Natl. Acad. Sci., USA, 81:7885.

Patarca, R. and Haseltine, W. A., 1984a, Sequence similarity among retroviruses Nature, 309:728.

Patarca, R. and Haseltine, W. A., 1984b, Similarities among retrovirus proteins, Nature, 312:496.

Philipson, L., Anderson, P., Olshevsky, U., Weinberg, R., and Baltimore, D., 1978, Translation of MuLV and MSV RNAs in nuclease-treated reticulocyte extracts: enhancement of the gag-pol polypeptide with yeast suppressor tRNA, Cell, 13:189.

Poiesz, B. J., Ruscetti, F. W., Gazdar, A. F., Bunn, P. A., Minna, J. D., and Gallo, R. C., 1980a, Detection and isolation of type C retrovirus particles from fresh and cultured lymphocytes of a patient with cutaneous T-cell lymphoma, Proc. Natl. Acad. Sci., USA, 77:7415.

Poiesz, B. J., Ruscetti, F. W., Mier, J. W., Woods, A. M., and Gallo, R. C., 1980b, T-cell lines established from human T-lymphocytic neoplasias by direct response to T-cell growth factor, Proc. Natl. Acad. Sci., USA, 77:6815.

Poiesz, B. J., Ruscetti, F. W., Reitz, M. S., Kalyanaraman, V. S., and Gallo, R. C., 1981, Isolation of a new type-C retrovirus (HTLV) in primary uncultured cells of a patient with Sezary T-cell leukemia, Nature, 294:268.

Popovic, M., Lange-Watzin, G., Sarin, P. S., Mann, D., and Gallo, R. C., 1983a, Transformation of human umbilical cord blood T cells by human T-cell leukemia/lymphoma virus, Proc. Natl. Acad. Sci., USA, 80:5402.

Popovic, M., Sarin, P. S., Robert-Guroff, M., Kalyanaraman, V. S., Mann, D., Minowada, J., and Gallo, R. C., 1983b, Isolation and transmission of human retrovirus (human T-cell leukemia virus), Science, 219:856.

Popovic, M., Flomenberg, N., Volkman, D. J., Mann, D., Fauci, A. S., DuPont, B., and Gallo, R. C., 1984a, Alteration of T cell functions by infection with HTLV-I or HTLV-II, Science, 226:459.

Popovic, M., Sarngadharan, M. G., Read, E., and Gallo, R. C., 1984b, Detection, isolation, and continuous production of cytopathic retroviruses (HTLV-III) from patients with AIDS and pre-AIDS, Science, 224:497.

Ratner, L., Gallo, R. C., and Wong-Staal, F., 1985a, HTLV-III, LAV, ARV are variants of the same AIDS virus, Nature, 313:636.

Ratner, L., Haseltine, W., Patarca, R., Livak, K. J., Starcich, B., Josephs, S. F., Doran, E. R., Rafalski, J. A., Whitehorn, E. A., Baumeister, K., Ivanoff, L., Petteway, S. R., Jr., Pearson, M. L., Lautenberger, J. A., Papas, T. S., Ghrayeb, J., Chang, N. T., Gallo, R. C., and Wong-Staal, F., 1985b, Complete nucleotide sequence of the AIDS virus, HTLV-III, Nature, 313:277.

Ratner, L., Josephs, S. F., Starcich, B., Hahn, B., Shaw, G. M., Gallo, R. C., and Wong-Staal, F., 1985c, Nucleotide sequence analysis of a variant human T-cell leukemia virus (HTLV-Ib) provirus with a deletion in pX-I, J. Virol., in press.

Ratner, L., Starcich, B., Josephs, S. F., Hahn, B. H., Reddy, E. P., Livak, K. J., Petteway, S. R., Jr., Pearson, M. L., Haseltine, W. A., Arya, S. K., Gallo, R. C., and Wong-Staal, F., 1985d, Polymorphism of the long terminal repeat sequence of human T lymphotropic virus type III, Virol, submitted.

Reitz, M. S., Poiesz, B. J., Ruscetti, F. W., and Gallo, R. C., 1981, Characterization and distribution of a novel type C retrovirus isolated from neoplastic human T lymphocytes, Proc. Natl. Acad. Sci., USA, 78:1887.

Rho, H. M., Poiesz, B. J., Ruscetti, F. W., and Gallo, R. C., 1981, Characterization of the reverse transcriptase from a new retrovirus (HTLV) produced by a human cutaneous T-cell lymphoma cell line, Virology, 126:688.

Rice, N., Stephens, R. M., Couez, D., Descharge, J., Kettmann, R., Burny, A., and Gilden, R., 1985, The nucleotide sequence of the env gene and post-env region of bovine leukemia virus, Virology, 138:82.

Robert-Guroff, M., Ruscetti, F. W., Posner, L. E., Poiesz, B. J., and Gallo, R. C., 1981, Detection of the human T-cell lymphoma virus p19 in cells of some patients with cutaneous T-cell lymphoma and leukemia using a monoclonal antibody, J. Exp. Med., 154:1957.

Robert-Guroff, M., Nakao, Y., Notake, K., Ito, Y., Sliski, A., and Gallo, R. C., 1982, Natural antibodies to human retrovirus HTLV in a cluster of Japanese patients with adult T cell leukemia, Science, 215:975.

Robert-Guroff, M., Kalyanaraman, V. S., Blattner, W. A., Popovic, M., Sarngadharan, M. G., Maeda, M., Blayney, D., Catovsky, D., Bunn, P. A., Shibata, A., Nakao, Y., Ito, Y., Aoki, T., and Gallo, R. C., 1983, Evidence for human T cell lymphoma-leukemia virus infection of family members of human T cell lymphoma-leukemia virus positive T cell leukemia-lymphoma patients, J. Exp. Med., 157:248.

Robert-Guroff, M., Schupbach, J., Blayney, D. W., Kalyanaraman, V. S., Merino, F., Lanier, A., Sarngadharan, M. G., Clark, J., Saxinger, W. C., Blattner, W. A., and Gallo, R. C., 1984, Sero-epidemiologic studies on HTLV-I, in: "Human T-Cell Leukemia Viruses," Cold Spring Harbor Laboratory, New York, pp. 285.

Rosen, C. A., Sodroski, J. G., Kettman, R., Burny, A., and Haseltine, W. A., 1985a, Trans-activation of the bovine leukemia virus long terminal repeat in BLV-infected cells, Science, 227:320.

Rosen, C. A., Sodroski, J. G., and Haseltine, W. A., 1985b, Functional analysis of the human T cell leukemia virus long terminal repeat, Nature, in press.

Rous, P., 1911, A sarcoma of the fowl transmissable by an agent from the tumor cells, J. Exp. Med., 13:397.

Safai, B., Groopman, J. E., Popovic, M., Schupbach, J., Sarngadharan, M. G., Arnett, K., Sliski, A., and Gallo, R. C., 1984, Seroepidemiological studies of human T-lymphotropic retrovirus type III in acquired immunodeficiency syndrome, Lancet, i:1438.

Sagata, N., Yasunaga, T., Ogawa, Y., Tsuzuku-Kawamura, J., and Ikawa, Y., 1984, Bovine leukemia virus: unique structural features of its long terminal repeats and its evolutionary relationship to human T-cell leukemia virus, Proc. Natl. Acad. Sci., USA, 81:4741.

Sagata, N., Yasunaga, T., Tsuzuku-Kawamura, J., Ohishi, K., Ogawa, Y., and Ikawa, Y., 1985, Complete nucleotide sequence of the genome of bovine leukemia virus: its evolutionary relationship to other retroviruses, Proc. Natl. Acad. Sci., USA, 82:677.

Salahuddin, S. Z., Markham, P. D., Popovic, M., Sarngadharan, M. G., Orndorff, S., Fladagar, A., Patel, A., Gold, J., and Gallo, R. C., 1985, One hundred-one infectious HTLV-III isolates from AIDS and ARC patients, and clinically normal donors at risk for AIDS, Proc. Natl. Acad. Sci., USA, in press.

Sanchez-Pescador, R., Poner, M. D., Barr, P. J., Stempier, K. S., Brown-Shimer, S. L., Gee, W. W., Renard, A., Randolph, A., Levy, J. A., Dina, D., and Luciw, P. A., 1985, Sequence and expression of an AIDS-associated retrovirus, Science, 227:484.

Sarngadharan, M. G., Popovic, M., Bruch, L., Schupbach, J., and Gallo, R. C., 1984, Antibodies reactive with human T-lymphotropic retroviruses (HTLV-III) in the serum of patients with AIDS, Science, 224:506.

Saxinger, W., Blattner, W. A., Levine, P. H., Clark, J., Biggar, R., Hoh, M., Moghissi, J., Jacobs, P., Wilson, L, Jacobson,

R., Crookes, R., Strong, M., Ansari, A. A., Dean, A. G., Nkrumah, F. K., Mourali, N., and Gallo, R. C., 1984, Human T-cell leukemia virus (HTLV-I) antibodies in Africa, Science, 225:1473.

Schupbach, J., Kalyanaraman, V. S., Sarngadharan, M. G., Blattner, W. A., and Gallo, R. C., 1983, Antibodies against three purified proteins of the human type C retrovirus, human T-cell leukemia-lymphoma virus, in adult T-cell leukemia-lymphoma patients and helathy blacks from the Caribbean, Cancer Res., 43:886.

Schupbach, J., Popovic, M., Gilden, R. V., Gonda, M. A., Sarngadharan, M. G., and Gallo, R. C., 1984a, Serological analysis of a subgroup of human T-lymphotropic retroviruses (HTLV-III) associated with AIDS, Science, 226:503.

Schupbach, J., Sarngadharan, M. G., and Gallo, R. C., 1984b, Antigens on HTLV-infected cells recognized by leukemia and AIDS sera are related to HTLV viral glycoprotein, Science, 224:607.

Schwartz, D. E., Tizard, R., and Gilbert, W., 1983, Nucleotide sequence or Rous sarcoma virus, Cell, 32:853.

Schwartzberg, P., Colicelli, J., and Goff, S. P., 1984, Construction and analysis of deletion mutations in the pol gene of Moloney murine leukemia virus: a new viral function requried for productive infection, Cell, 37:1043.

Seiki, M., Hattori, S., and Yoshida, M., 1982, Human adult T-cell leukemia virus: molecular cloning of the provirus DNA and the unique terminal structure, Proc. Natl. Acad. Sci., USA, 79:2031.

Seiki, M., Hattori, S., Hirayama, Y., and Yoshida, M., 1983, Human adult T-cell leukemia virus: complete nucleotide sequence of the provirus genome integrated in leukemia cell DNA, Proc. Natl. Acad. Sci., USA, 80:3618.

Seiki, M., Eddy, R., Shows, T. B., and Yoshida, M., 1984, Non-specific integration of the HTLV provirus genome into adult T-cell leukaemia cells, Nature, 309:640.

Seligman, M., Chess, L., Fahey, J. L., Fauci, A. S., Lachmann, P. J., L'Age-Stehr, J., Ngu, J., Pinching, A. J., Rosen, F. S., Spira, T. J., and Wybran, J., 1984, AIDS - an immunological reevaluation, New Engl. J. Med., 311:1286.

Shaw, G. M., Gonda, M. A., Flickinger, G. H., Hahn, B. H., Gallo, R. C., and Wong-Staal, F., 1984a, The genomes of evolutionary divergent members of human T-cell leukemia virus family (HTLV-I and HTLV-II) are highly conserved especially in pX, Proc. Natl. Acad. Sci., USA, 225:419.

Shaw, G. M., Hahn, B. H., Arya, S. K., Groopman, J. E., Gallo, R. C., and Wong-Staal, F., 1984b, Molecular characterization of human T-cell leukemia (lymphotropic) virus type III in the acquired immune defiency syndrome, Science, 226:1165.

Shaw, G. M., Harper, M. E., Hahn, B. H., Epstein, L. G., Gajdusek, D. C., Price, R. W., Navia, B. A., Petito, C. K., O'Hara, C. J., Groopman, J. E., Cho, E.-S., Oleske, J. M., Wong-Staal, F., and Gallo, R. C., 1985, HTLV-III infection in brains of children and adults with AIDS encephalopathy, Science, 227:177.

Shimotohno, K., Wachsman, W., Takahashi, Y., Golde, D. W., Miwa, M., Sugimura, T., and Chen, I. S. Y., 1984, Nucleotide sequence of the 3' region of an infectious human T cell leukemia virus type II genome, Proc. Natl. Acad. Sci., USA, 80:6657.

Shimotohno, K., Miwa, M., Slamon, D. J., Chen, I. S. Y., Hoshino, H.-O., Takano, M., Fujino, M., and Sugimura, T., 1985, Identification of new gene products encoded from X regions of human T-cell leukemia viruses, Proc. Natl. Acad. Sci., USA, 82:302.

Shinnick, T. R., Lerner, R. A., and Stucliffe, J. G., 1981, Nucleotide sequence of moloney leukemia virus, Nature, 293:543.

Siegel, L. J., Nash, W. G., Manzari, V., Wong-Staal, F., Gallo, R. C., and O'Brien, J., 1984, A genetic analysis of human T-cell leukemia virus integration in Hut 102 and the localization of the structural gene for T-cell growth factor to chromosome 4, in: "Human T-Cell Leukemia/Lymphoma Virus," R. C. Gallo, M. E. Essex, and L. Gross, eds., Cold Spring Harbor Press, New York, pp. 167.

Slamon, D., Shimotohno, K., Cline, M. J., Golde, D. W., and Chen, I. S. Y., 1984, Identification of the putative transforming protein of the human T-cell leukemia virus HTLV-I and HTLV-II, Science, 226:61.

Sodroski, J. R., Patarca, R., Perkins, D., Briggs, D., Lee, T.-H., Essex, M., Coligan, J., Wong-Staal, F., Gallo, R. C., and Haseltine, W. A., 1984a, Structure of the envelope gene of type II human T lymphotropic virus, Science, 224:421.

Sodroski, J. G., Rosen, C. A., and Haseltine, W. A., 1984b, Trans-acting transcriptional activation of the long terminal repeat of human T lymphotropic viruses in infected cells, Science, 225:381.

Sodroski, J., Trus, M., Perkins, D., Patarca, R., Wong-Staal, F., Gelmann, E., Gallo, R. C., and Haseltine, W. A., 1984c, Repetitive structure in the long-terminal repeat element of a type II human T-cell leukemia virus, Proc. Natl. Acad. Sci., USA, 81:4617.

Sodroski, J. G., Rosen, C. A., and Haseltine, W. A. 1985, Trans-acting transcriptional regulation of human T-cell leukemia virus type III long terminal repeat, Science, 227:171.

Starcich, B., Ratner, L., Josephs, S. F., Okamoto, T., Gallo, R. C., and Wong-Staal, F., 1985, Characterization of the long terminal repeat sequences of the AIDS virus human T-cell leukemia (lymphotropic) virus type III (HTLV-III), Science, 227:538.

Tajima, K., Tominaga, S., Kuroishi, T., Shimizu, H., and Suchi, T., 1979, Geographical features and epidemiological approach to endemic T-cell leukemia/lymphoma in Japan, Jap. J. Clin. Oncol. 9(Suppl):495.

Takatsuki, A., Uchiyama, T., Ueshima, Y., and Hattori, T., 1979, Adult T-cell leukemia: further clinical observations and cytogenetic and functional studies of leukemic cells, Jap. J. Clin. Oncol., 9(Suppl):317.

Temin, H., 1982, Function of the retrovirus long terminal repeat, Cell, 28:3.

Tsichlis, P. N., Strauss, P. G., and Hu, L. F., 1983, A common region for proviral DNA integration in MoMuLV induced rat thymic lymphomas, Nature, 302:445.

Toh, H., Hayashida, H., and Miyata, T., 1983, Sequence homology between retroviral reverse transcriptase and putative polymerases of hepatitis B virus and cauliflower mosaic virus, Nature, 305:827.

Uchiyama, T., Yodoi, J., Sagawa, K., Takatsuki, K., and Uchino, H., 1977, Adult T-cell leukemia: clinical and hematologic features of 16 cases, Blood, 50:481.

Van Ooyen, A. and Nusse, R., 1984, Structure and nucleotide sequence of the putative mammary oncogene int-1; proviral insertions leave the protein-encoding domain intact, Cell, 39:233.

Varmus, H. E., 1982, Form and function of retroviral proviruses, Science, 216:812.

Vyth-Drees, F. A. and de Vries, J. E., 1982, Human T-cell leukemia virus in lymphocytes from a T-cell leukaemia patient originating from Surinam, Lancet, ii:993.

Wachsman, W., Shimotohno, K., Clark, S. C., Golde, D. W., and Chen, I. S. Y., 1984, Expression of the 3'terminal region of human T-cell leukemia viruses, Science, 226:177.

Wain-Hobson, S., Sonigo, P., Danos, O., Cole, S., and Alizon, M., 1985, Nucleotide sequence of the AIDS virus, LAV, Cell, 40:9.

Weiss, S. R., Hackett, P. B., Oppermann, H., Ullrich, A., Lewintow, L., and Bishop, J. M., 1978, Cell free translation of avian sarcoma virus RNA: suppression of the gag terminator codon does not augment synthesis of the joint gag-pol product, Cell, 15:607.

Weiss, R., Teich, N., Varmus, H., and Coffin, J., eds., 1982, "RNA Tumor Viruses," Cold Spring Harbor Press, New York.

Wong-Staal, F., Hahn, B., Manzari, V., Colombini, S., Franchini, G., Gelmann, E. P., and Gallo, R. C., 1983, A survey of human leukaemias for sequences of a human retrovirus, Nature, 302:626.

Wong-Staal, F., F., Westin, E., and Gallo, R. C., 1984, HTLV, a human leukemia retrovirus, in: "Contemporary Hematology/ Oncology," R. Silver, A. S. Gordon, and J. LoBue, eds., Plenum Press, New York, vol. 3, pp. 155.

Yoshida, M., Miyoshi, I., and Hinuma, Y., 1982, Isolation and characterization of retrovirus from cell lines of human adult T-cell leukemia and its implication in the disease, Proc. Natl. Acad. Sci., USA, 79:2031.

Yoshida, M., Seiki, M., Yamaguchi, K., and Takatsuki, K., 1984, Monoclonal integration of human T-cell leukemia provirus in all primary tumors of adult T-cell leukemia suggests causative role of human T-cell leukemia virus in the disease, Proc. Natl. Acad. Sci., USA, 81:2534.

Zagury, D., Bernard, J., Leibowitch, J., Safai, B., Groopman, J. E., Feldman, M., Sarngadharan, M. G., and Gallo, R. C., 1984, HTLV-III in cells cultured from semen of two patients with AIDS, Science, 226:447.

THE T-DNA GENES OF *AGROBACTERIUM* PLASMIDS APPEAR TO BE OF A COMPLEX EVOLUTIONARY ORIGIN

Jeff Schell

Max-Planck-Institut für Züchtungsforschung
Köln (FRG)

Rijksuniversiteit Gent, Laboratorium voor Genetika
Gent (Belgium)

INTRODUCTION

Crown galls are typical neoplasmic tissues of plants. These plant tumours are the result of a complex interaction between susceptible plants and gram-negative bacteria (*Agrobacterium tumefaciens*). During this interaction a defined segment (T-DNA) of a large extrachromosomal plasmid harbored by *Agrobacteria* (Ti-plasmid) is transferred from the bacteria to the nucleus of infected plant cells. The transferred DNA segment is stably integrated in one of many possible loci on different chromosomes (for recent reviews see Caplan et al., 1983; Gheysen et al., 1985; Nester et al., 1984; Helle et al., 1984).

Plant cells transformed by *Agrobacterium* Ti-plasmids and harboring the T-DNA segment in their genome, do not require supplementation with cellular growth-factors (or growth hormones) such as auxins and cytokinins for continued growth and division in tissue culture. It was clearly established that this hormone independent neoplastic type of growth is the direct consequence of the activity of a number of genes carried by the T-DNA segment. A nine kilobase region of this T-DNA is highly conserved between different types of Ti plasmids (Engler et al., 1981) and is therefore referred to as "common T-DNA". The genes 1, 2 and 4 (Willmitzer et al., 1982, 1983) play the major role in the control of neoplastic growth (Garfinkel et al., 1981, Ooms et al., 1981; Leemans et al., 1982; Joos et al., 1983; Akiyoshi et al., 1983).

Mutant T-DNA's in which gene 4 (called previously either *tms* or roi) is inactive, produce transformed plant cells that sprout roots. The product of the wild-type gene 4 was therefore thought to inhibit specifically root-formation (roi for root inhibiting). In analogy to the known effect of growth-factors on normal plant tissues, the effect of the active gene 4 could therefore be interpreted as "cytokinin-like". Similarly T-DNA's mutated for either gene 1 or 2 or for both of these genes, produce tissues that sprout shoots but no roots. These genes therefore must determine products that prevent shoot formation. This is a known property of auxins and hence the combined effect of genes 1 and 2 could be described as "auxin-like". Plant cells containing no other T-DNA genes than 1 and 2 and 4 will produce undifferentiated tumors, practically undistinguishable from wild-type crown-gall tumors. Plant cells containing active T-DNA genes 1 and 2 will produce root sprouting tissues, whereas plant cells containing gene 4 as the only T-DNA gene will grow as so-called shoot teratomas (Inzé et al., 1984; Follin et al., 1985).

On the basis of these observations it became apparent that the T-DNA genes involved in controlling oncogenicity in plants were probably somehow involved either in the synthesis of plant growth factors or of hypothetical growth-factor receptors.

The T-DNA genes 1 and 2 code for enzymes that establish a new pathway for the synthesis of indole-3-acetic acid (IAA; auxin) in transformed plant cells

The direct inspiration for the experiments that established that the T-DNA genes 1 and 2 are coding for enzymes involved in auxin biosynthesis came from a comparison with another gram-negative soil plant pathogen: *Pseudomonas savastanoi*. This organism produces pathogenic outgrowths on olive and oleander plants. In contrast to tumors induced by *Agrobacterium*, the *P. savastanoi* induced tissue proliferations depend on the continuous production of IAA by the infecting bacteria (Smidt and Kosuge, 1978). *P. savastanoi* was shown to produce IAA in two steps from tryptophan. A tryptophan 2-mono-oxygenase produces indole-3-acetamide which is converted by an hydrolase to indole-3-acetic acid (Comai and Kosuge, 1980).

The following set of experiments demonstrated that the T-DNA genes 1 and 2 establish within the transformed plant cells a new auxin biosynthetic pathway which is in fact identical to the auxin pathway active in *P. savastanoi*. Schröder et al. (1983) established that T-DNA gene 2 produces a defined protein of M 48,000 when expressed in *E. coli*. Later it was shown (Schröder et al., 1984) that only extracts from *E. coli* expressing gene 2 were able to hydrolyze indole-3-acetamide into a substance that was unambiguously identified as indole-3-acetic acid. The same reaction was also found in *Agrobacteria* containing gene 2, but not in isogenic strains lacking this gene. Finally it was shown that extracts from tobacco crown-gall cells in contrast to isogenic untransformed tobacco cultures, exhibited the same enzyme activity. Independent work by Thomashow et al. (1984) lead to the same conclusions.

That gene 1 is directly or indirectly involved in the production of indole-3-acetamide, the substrate for the hydrolase coded for by gene 2, was first shown by genetic studies (Inzé et al., 1984). These experiments were designed to study separately the contribution of each T-DNA gene to the development of abnormal growth. Therefore cloned tobacco tissue containing either gene 1 alone or gene 2 alone were obtained using Ti plasmid derived vectors which only contained one defined T-DNA gene.

Tobacco cells containing either gene 1 or gene 2 alone do not proliferate in the absence of externally supplied auxins and cytokinins and differentiate into normal plants. Whereas the external addition of alpha-naphthalene acetic acid (an analogue of indole-3-acetic acid and an active auxin) would induce a tumorous growth in untransformed cells as well as in cells containing either gene 1 or gene 2, the addition of alpha-naphthalene acetamide (which is not active by itself as an auxin) would only induce tumorous growth in cells containing gene 1. In subsequent studies it was shown (Van Onckelen et al., 1985) that tobacco plants harbouring gene 1 contained 500 to a 1000 fold more IAM (indole-3-acetamide) than control untransformed tobacco or than tobacco plants containing the T-DNA genes 6a or 6b. However, no significant difference was observed in the level of indole-3-acetamide (IAA) in plants containing gene 1 as compared to controls, indicating that tobacco does not normally have the capacity to convert IAM into the active auxin IAA. Kemper et al. (1985) studied in more detail the amidohydrolase coded for by gene 2 (<u>IaaH</u> gene) and found that it requires no cofactors and can hydrolyze not only indole-3-acetamide but also indole-3-acetonitrile, conjugates of indole-3-acetic acid with glucose and myo-inositol, alpha-naphthaleneacetamide and phenylacetamide. Because the <u>IaaH</u> encoded hydrolase can convert some auxin-derivates (conjugates) with low growth-factor activity into more active auxins, it is conceivable that cells harbouring the combination of genes 1 and 2 not only continuously produce active auxins but also cannot control the auxin level (as normal plant cells might do) by converting active auxins into less active conjugates, because these conjugates are converted back to auxin by the <u>IaaH</u> encoded hydrolase.

The observation (Inzé et al., 1984) that plants containing the <u>IaaH</u> gene alone are apparently normal, would indicate that significant reservoirs of inactive auxin conjugates that could be activated by the <u>IaaH</u> coded hydrolase either do not exist in these plants or are not accessible to the hydrolase.

T-DNA gene 4 codes for an enzyme of cytokinin biosynthesis (Isopentyl-transferase)

Cytokinins are involved in the regulation of cell division in plants and play a regulatory role in several other processes such as stimulation of shoot differentiation and inhibition of root formation. One of the enzymes involved in the synthesis of cytokinins in plants is a phenyl-transferase catalyzing the formation of N6-(Δ2-isopentyl)-5'-AMP or isopentenyl-adenine (iPeA) from 5'-AMP and Δ2-isopentenylpyrophosphate (Taya et al., 1978). Such an enzyme activity was indeed detected in grown gall cells (Morris et al., 1982). Several experiments based on the expression of gene 4 in *E. coli* have demonstrated that this T-DNA gene indeed codes for an isopentenyl transferase. Akiyoshi et al. (1984) identified the isopentenyladenine as the product of a weak enzyme reaction by h.p.l.c. chromatography and further identified the product by binding studies with antibodies specific for isopentenyl-AMP.

Barry et al. (1984) used expression vectors to study the product of gene 4 in *E. coli* and by supplying 5'AMP as radioactive substrate they found that a radioactive product co-migrated with authentic isopentenyl-AMP in h.p.l.c. chromatography. More recently Buchmann et al. (1985) confirmed the earlier observations and firmly established that the product of gene 4 is an isopentenyl transferase with a mol. wt. of 27,000. These authors also expressed gene 4 in *E. coli* and identified the gene product by reaction with antiserum raised against a decapeptide the amino acid sequence of which had been derived from the known DNA sequence of gene 4. Extracts from *E. coli* cells expressing the gene contained high isopentenyl-transferase activity and the iPeA product was identified by h.p.l.c. and mass spectrometry.

Interestingly it was also demonstrated that tobacco crown gall tissues contained a protein of mol. wt. of 27,000 binding to the gene 4 specific decapeptide antiserum and that the isopentenyl-transferase activity of these tobacco cells was specifically inhibited by anti-gene-4-serum. Tissues of tobacco cells which do not contain any T-DNA genes, but are epigenetically cytokinin-autotrophic do not contain a protein which can cross-react with the anti-gene-4-serum.

These results therefore firmly established the T-DNA gene 4 is responsible for the production in transformed plant cells of a *de novo* pathway leading to the synthesis of cytokinin growth hormone.

The T-DNA genes have evolved from similar procaryotic genes primarily by modification of the 5' flanking sequences into plant promotor sequences

T-DNA genes, although carried by bacterial Ti-plasmids, are flanked by 5' expression signals that allow these genes to be expressed in plant cells (Willmitzer et al., 1981, 1982, 1983; Depicker et al., 1982; De Greve et al., 1982; Dhaese et al., 1983; 1984). One can therefore imagine two fundamentally different ways that could have led to the evolvement of such genes. Either these genes would be similar to some animal onc genes in that they would be derived from normal plant genes involved in the regulated synthesis of cellular growth-factors. One would then have to accept the notion that bacterial plasmids somehow acquired these plant genes. Alternatively the T-DNA genes are of procaryotic origin. One must then assume that their 5' flanking expression signals evolved to be recognized by the plant DNA transcriptional system.

The latter but not the former hypothesis is in fact supported by the following observations. Several soil bacteria that live in close association with plants either as symbionts or as pathogens, do produce a number of phytohormones such as auxins and cytokinins (Kosuge et al., 1966; Surico et al., 1975; Cherayil and Lipsett, 1977; Murai et al., 1980). It is therefore conceivable that by association with a natural plant gene vector system such as the Ti-plasmid of Agrobacteria, bacterial phytohormone genes got to be modified such as to become functional in a eucaryotic (plant) nuclear and cellular environment. This is most probably the case for the T-DNA genes IaaM (gene 1) and IaaH (gene 2). Indeed as described above these genes were shown to code for an auxin biosynthesis pathway which is identical to a pathway used by Pseudomonas savastanoi.

As a personal communication Prof. T. Kosuge and Dr. Yamada (Department of Plant Physiology, University of California, Davis) provided data showing that the cloned iaaM and iaaH genes from P. savastanoi are indeed related to the equivalent T-DNA genes from Agrobacterium tumefaciens. The open reading frame of the P. savastanoi iaaM gene encodes a protein of 557 amino acids with a molecular weight of 61.783. The sequence of the equivalent T-DNA iaaM gene 1 encodes a protein of 755 amino acids with a mol. wt. of 83.769. The deduced amino acid sequences for both iaaM genes show strong homology throughout the entire length of the coding region, provided a Tyr at position 90 and the 3 consecutive amino acids in the Pseudomonas iaam gene are skipped. Overall, the percentage is 50% for perfect matches of the deduced amino acid sequences. Homology is even higher if matches are based on amino acid groups or families. The homology between the Pseudomonas and Agrobacterium iaaH genes is less clear. The deduced amino acid sequences for these proteins of similar mol. wt. is 27% for perfect matches if Thr at position 251 in the T-DNA gene 2 is skipped. Recently (Follin et al., 1985) genetic evidence was obtained showing that the tryptophan 2-monooxygenase gene of *P. savastanoi* is in fact functionally equivalent to the *iaaM* gene 1 of the T-DNA. Tobacco plants containing either the T-DNA *iaaM* gene or the T-DNA *iaaH* gene 2 alone were shown to grow and develop in a normal fashion. However, when tobacco plants containing the *iaaH* T-DNA gene 2 were infected with either *E. coli* or *A. tumefaciens* carrying the *Pseudomonas savastanoi iaaM* gene, important root and callus formation occured at the site of infection.

Taken together all these data clearly support the hypothesis that the T-DNA genes *iaaM* and *iaaH* are of bacterial origin. A similar albeit less documented argument can be made with regard to the origin of gene 4 (*ipt*). Cultures of *Agrobacterium* strains were shown to contain auxins (Kao et al., 1982) and cytokinins (Chapmann and Morris, 1977; Claeys et al., 1978; Weiler and Spanier, 1981). Isopentenyladenine (iPeA) and *trans*-zeatin is located in the so-called virulence region of the nopaline Ti-plasmids (Regier and Morris, 1980). This *tzs* locus was recently cloned by Akiyoshi et al. (1985) and its nucleotide sequence was determined. The sequence analysis showed an open reading frame of 729 nucleotides encoding a protein of 27.545 dalton. When the deduced amino acid sequence of this *tzs* gene was compared with that of the *ipt* gene 4 of the T-DNA, a significant amount of homology was readily detected. I. Buchmann, W. Heinemeyer, D. Tonge and J. Schroder (pers. comm.) have expressed vir-region of the pTi-C58 plasmid in *E. coli* and demonstrated that it contains a cytokinin gene which is functional in *E. coli* and causes *trans*-zeatin formation in these bacteria. The relevant gene was shown to encode a protein of 27 kd which has isopentenyl-transferase activity. The procaryotic enzyme encoded by the vir-region of the Ti-plasmid was compared to the equivalent enzyme derived from the eucaryotically active T-DNA *ipt* gene 4 and expressed in *E. coli* by means of a high expression vector. Both enzymes have an isopentenyl-transferase activity and their kinetic properties were found to be identical for almost all parameters tested!

However, whereas the vir-gene is clearly procaryotic and quite active in *Agrobacterium* as well as in *E. coli* cultures, the T-DNA gene is only weakly if at all active in a procaryotic cellular environment but is clearly active in plant cells after transfer via the Ti-plasmid. These data are most easily interpreted by assuming that the procaryotic *tzs* gene of the vir-region of Ti-plasmids got duplicated during evolution and associated with the transferrable DNA region (T-DNA region) of the Ti-plasmid. Here the 5' upstream flanking prometer sequences of this gene apparently evolved to become functional after transfer in a eucaryotic cellular environment!

General conclusions

The data discussed in this paper illustrate and to some extent explain an unexpected situation in evolution. Bacteria such as *Agrobacterium tumefaciens* are shown to contain a number of oncogenicity genes which are transferred to eucaryotic cells (plants). As a direct consequence of the presence of these oncogenicity genes in the plant cells, these undergo a neoplasmic transformation. The bacterially derived oncogenicity genes indeed code for enzymes that will continuously supply the transformed plant cells with cellular growth factors. These bacterially derived plant oncogenicity genes have most likely evolved from similar genes present in a number of bacterial plant symbionts and pathogens by modifications of 5' upstream promoter sequences. We must therefore assume that exchange of DNA between procaryotes and eucaryotes such as plants must have been a fairly common event during evolution

ACKNOWLEDGEMENT

The author wishes to acknowledge and thank the following colleagues and collaborators who contributed much of the data and conclusions discussed in this review: M. Van Montagu, D. Inzé, A. Janssens, A. Follin of the University of Gent, Belgium; J. Schröder, G. Schröder, I. Buchmann, MPI Köln and presently University of Freiburg (FRG), J. P. Hernalsteens and R. Deblaere, University of Brussels (Belgium). Prof. T. Kosuge is thanked for the inspriation his work has provided and for making data available prior to publication.

REFERENCES

Akiyoshi, D. E., Regier, D. A., Jen, G., and Gordon, M. P., 1985, cloning and nucleotide sequence of the *tzs* gene from *Agrobacterium tumefaciens* strain T37, *Nucleic Acids Res.*, 13:2773.

Akiyoshi, D. E., Klee, H., Amasino, R. M., Nester, E. W., and Gordon, M. P., 1984, T-DNA of *Agrobacterium tumefaciens* encodes an enzyme of cytokinin biosynthesis, *Proc. Natl. Acad. Sci. USA*, 81:5994, 13:2773.

Akiyoshi, D. E., Morris, R. O., Hinz, R., Mischke, B. S., Kosuge, T., Garfinkel, D. J., Gordon, M. P., and Nester, W., 1983, Cytokinin/auxin balance in crown gall tumors is regulated by specific loci in the T-DNA, *Proc. Natl. Acad. Sci. USA*, 80:407.

Barry, G. F., Rogers, S. G., Fraley, R. T., and Brand, L., 1984, Identification of a cloned cytokinin biosynthetic gene, Proc. Natl. Acad. USA, 81:4776.

Buchmann, I., Marner, F.-J., Schröder, G., Waffenschmidt, S., and Schröder, J., 1985, Tumor genes in plants: T-DNA encoded cytokinin biosynthesis, EMBO J., 4:853.

Caplan, A., Herrera, Estrella, L., Inzé, D., Van Haute, E., Van Montagu, M., Schell, J., and Zambryski, P., 1983, Introduction of genetic material into plant cells. Science, 222:815.

Chapman, R. W., and Morris, R. O., 1977, Trans-zeatin in culture filtrates of Agrobacterium tumefaciens, Biochem. Biophys. Res. Commun., 76:453-459.

Cherayil, J. D., and Lipsett, M. N., 1977, Zeatin ribonucleotides in the transfer ribonucleic acid of Rhizobium leguminosarum, Agrobacterium tumefaciens, Corynebacterium fascians and Erwinai amylovora, J. Bacteriol., 131:741.

Claeys, M., Messens, E., Van Montagu, M., and Schell, J., 1978, Quantitative determination of cytokinins in culture media of Agrobacterium tumefaciens, Quant. Mass Spec. Lift Sci., 2:409.

Comai, L., and Kosuge, T., 1980, Involvement of plasmid deoxyribonucleic acid in indeoleacetic acid synthesis in Pseudomonas savastanoi, J. Bacteriol., 143:950.

De Greve, H., Dhaese, P., Seurinck, J., Lemmers, M., Van Montagu, M., and Schell, J., 1982, Nucleotide sequence and transcript map of the Agrobacterium tumefaciens Ti plasmid-encoded octopine synthase gene, J. Mol. Appl. Genet., 1:499.

Depicker, A., Stachel, S., Dhaese, P., Zambryski, P., and Goodman, H. M., 1982, Nopaline synthase: transcript mapping and DNA sequence. J. Mol. Appl. Genet. 1:561

Dhaese, P., De Greve, H., Gielen, J., Seurinck, J., Van Montagu, M., and Schell, J., 1983, Identification of sequences involved in the polyadenylation of higher plant nuclear transcripts using Agrobacterium T-DNA genes as models, EMBO J., 2:419.

Engler, G., Depicker, A., Maenhaut, R., Villarroel-Mandiola, R., Van Montagu, M. Schell, J. 1981, Physical mapping of DNA base sequence homologies between an octopine and a nopaline Ti-plasmid of Agrobacterium tumefaciens, J. Mol. Biol., 152:183.

Follin, A., Inzé, D., Budar, F., Genetello, C., Van Montagu, M., and Schell, J., 1985, Genetic evidence that the tryptophan 2-mono-oxygenase gene of Pseudomonas savastanoi is functionally equivalent to one of the T-DNA genes involved in plant tumour formation by Agrobacterium tumefaciens, Mol. Gen. Genet., in press.

Garfinkel, D. J., Simpson, R. B., Ream, L. W., White, F. F., Gordon, M. P., and Nester, E. W., 1981, Genetic analysis

Gordon, M. P., and Nester, E. W., 1981, Genetic analysis of crown gall: fine structure map of the T-DNA by site-directed mutagenesis, Cell, 27:143.

Gheysen, G., Dhaese, P., Van Montagu, M., and Schell, J., 1985, DNA flux across genetic barriers: the crown gall phenomenon, in: "Genetic Flux in Plants" (Advances in Plant Gene Research, vol. 2), B. Hohn, and E. S. Dennis, eds., Springer Verlag, Wien, pp. 11-47.

Gielen, J., De Beuckeleer, M., Seurinck, J., Deboeck, F., De Greve, H., Lemmers, M., Van Montagu, M., and Schell, J., 1984, The complete nucleotide sequence of the TL-DNA of the Agrobacterium tumefaciens plasmid pTiAch5, EMBO J., 3:835.

Goldberg, S. B., Flick, J. S., and Rogers, S. G., 1984, Nucleotide sequence of the tmr locus of Agrobacterium tumefaciens pTiT37 T-DNA, Nucl. Acids Res., 12:4665.

Heidekamp, F., Dirkse, W. G., Hille, J., and Van Ormondt, H., 1983, Nucleotide sequence of the Agrobacterium tumefaciens octopine Ti-plasmid-encoded tmr gene, Nucl. Acids Res., 11:6211.

Hille, J., Hoekema, A., Hooykaas, P., and Schilperoort, R. A., 1984, Gene organisation of the Ti-plasmid, in: "Genes Involved in Microbe-Plant Interactions" (Advances in Plant Gene Research, Vol. 1), D.P.S. Verma, and Th. Hohn, eds., Springer, Wien, pp. 287-309.

Inzé, D., Follin, A., Van Lijsebettens, M., Simoens, C., Genetello, C., Van Montagu, M., and Schell, J., 1984, Genetic analysis of the individual T-DNA genes of Agrobacterium tumefaciens; further evidence that two genes are involved in indole-3-acetic acid synthesis, Mol. Gen. Genet., 194:265.

Joos, H., Inzé, D., Caplan, A., Sormann, M., Van Montagu, M., and Schell, J., 1983, Genetic analysis of T-DNA transcripts in nopaline crown galls, Cell, 32:1057

Kao, J. C., Perry, K. L., and Kado, C. I., 1982, Indoleacetic acid complementation and its relation to host range specifying genes on the Ti plasmid of Agrobacterium tumefaciens, Mol. Gen. Genet., 188:425.

Kemper, E., Waffenschmidt, S., Weiler, E. W., Rausch, T., and Schröder, J., 1985, T-DNA-encoded auxin formation in crown-gall cells, Planta, 163:257.

Kosuge, T., Heskett, M. G., and Wilson, E. E., 1966, Microbial synthesis and degradation of indole-3-acetic acid. The conversion of L-tryptophan to indole-3-acetamide by an enzyme system for Pseudomonas savastanoi, J. Biol. Chem., 241:3738.

Leemans, J., Deblaere, R., Willmitzer, L., De Greve, H., Hernalsteens, J.-P., Van Montagu, M., and Schell, J., 1982, Genetic identification of functions of TL-DNA transcripts in octopine crown galls, EMBO J., 1:147.

Morris, R. O., Akiyoshi, D. E., MacDonald, E.M.S., Morris, J. W., Regier, D. A., and Zaerr, J. B., 1982, Cytokinin metabolism in relation to tumor induction by Agrobacterium tumefaciens, in: "Plant Growth Substances 1982", P. F. Wareing, ed., Academic Press, New York, pp. 175-183.

Murai, N., Skoog, F., Doyle, M. E., and Hanson, R. S., 1980, Relationship between cytokinin production, presence of plasmids and fasciation caused by strains of C. fascians, Proc. Natl. Acad. Sci. USA, 77:619.

Nester, E. W., Gordon, M. P., Amasino, R. M., and Yanofsky, M. F., 1984, Crown gall: a molecular and physiological analysis, Ann. Rev. Plant Physiol., 35:387.

Ooms, G., Hooykaas, P. J., Moleman, G., and Schilperoort, R. A., 1981, Crown gall plant tumors of abnormal morphology, induced by Agrobacterium tumefaciens carrying mutated octopine Ti plasmids; analysis of T-DNA functions, Gene, 14:33.

Regier, D. A., and Morris, R. O., 1980, Secretion of trans-zeatin by Agrobacterium tumefaciens: a function determined by the nopaline Ti plasmid, Biochem. Biophys. Res. Commun., 104:1560.

Schröder, G., Klipp, W., Hillebrand, A., Ehring, R., Koncz, C., and Schröder, J., 1983, The conserved part of the T-region in Ti-plasmids expresses four proteins in bacteria, EMBO J., 2:403.

Schröder, G., Waffenschmidt, S., Weiler, E. W., and Schröder, J., 1984, The T-region of Ti plasmids codes for an enzyme synthesizing indole-3-acetic acid. Eur. J. Biochem., 138:387.

Smidt, M., and Kosuge, T., 1978, The role of indole-3-acetic acid accumulation by alpha methyl tryptophan-resistant mutants of Pseudomonas savastanoi in gall formation on oleanders Physiol. Plant Pathol., 13:203.

Surico, G., Comai, L., and Kosuge, T., 1975, Pathogenicity of strains of Pseudomonas syringae pv. savastanoi and their indoleacetic acid deficient mutants on olive and oleander, Phytopathology, 74:490.

Taya, Y., Tanaka, Y., and Nishimura, S., 1978, 5'-AMP is a direct precursor of cytokinin in Dictyostelium discoideum, Nature, 271:545.

Thomashow, L. S., Reeves, S., and Thomashow, M. F., 1984, Crown gall oncogenesis: evidence that a T-DNA gene from the Agrobacterium Ti plasmid pTiA6 encodes an enzyme that catalyzes synthesis of indoleacetic acid, Proc. Natl. Acad. Sci. USA, 81:5071.

Van Onckelen, H., Rüdelscheim, P., Inzé, D., Follin, A., Messens, E., Horemans, S., Schell, J., Van Montagu, M., and De Greef, J., 1985, Tobacco plants transformed with the Agrobacterium T-DNA gene 1 contain high amounts of indole-3-acetamide, FEBS Letters, 181:373.

Weiler, E. W., and Spanier, K., 1981, Phytohormones in the formation of crown gall tumors, Planta, 153:326.

Willmitzer, L., Dhaese, P., Schreier, P. H., Schmalenbach, W.,

Van Montagu, M., and Schell, J., 1983, Size, localisation, and polarity of T-DNA-encoded transcripts in nopaline crown gall tumors; evidence for common transcripts present in both octopine and nopaline tumors, Cell, 32:1045.

Willmitzer, L., Simons, G., and Schell, J., 1982, The TL-DNA in octopine crown gall tumours codes for seven well-defined polyadenylated transcripts, EMBO J., 1:139.

Willmitzer, L., Schmalenbach, W., and Schell, J., 1981, Transcription of T-DNA in octopine and nopaline crown gall tumours is inhibited by low concentrations of alpha-amanitin, Nucl. Acids Res., 9:4801.

INDEX